2.2 Measures of Central Tendency

the same amount and they were placed along a weightless number line, the mean would be that point at which the number line would be balanced. It will not, in general, be equal to any particular data point.

Finding the mean of a group of data is straightforward; you have probably done it before. All one has to do is sum the data points and divide by the number of data points. We can represent this by:

$$\overline{X} = \frac{\sum_{i=1}^{n} x_i}{n}$$

where,

x_i = value for data point i
n = number of data points
$\overline{X}$ = mean or average (sometimes called X-bar)

For a small amount of data, this can easily be accomplished by hand or performed on a calculator.

EXAMPLE 1. Find the mean of the following group of numbers:

35, 21, 42, 16, 71, 55

SOLUTION.

$$\overline{X} = \frac{35+21+42+16+71+55}{6} = \frac{240}{6} = 40$$

For larger groups of data this becomes tedious; obviously it is a job for the computer. The only tasks required of a computer program are to read in the data and make the calculation to find the mean. We do not present a subroutine here to find the mean. This is presented, with the other measures discussed, in a complete data anlaysis subroutine later in this chapter.

Median

The median is another measure of central tendency. It differs from the mean in that it is not the "center of gravity" of the data. Instead, it is a point on the line which "splits" the data. That is, the same number of data points are on either side of the median.

The median of a group of data is easily found. If we rank the data points in ascending order, we can count from either end until we have counted half the data points. At this point we have determined the median. There are two conditions we may face. If there is an odd number of data points then one of the data points will be the median. There will be one data point that has the same number of data points on either side of it. In this case the median is unique. The other condition is an even number of data points. In this case none of the data points may be the median. If there are n data points then there will be $n/2$ on either side of the median. Thus, if the data are ranked, the median will lie between the $n/2$ and $(n/2)+1$ data points. Theoretically, any point between these can be classified as the median. For our purposes, we will choose the point equidistant between the $n/2$ and $(n/2)+1$ points. This is the average, or mean, of these points.

EXAMPLE 2. Find the median of the following set of data.

3, 9, 2, 15, 22, 11, 7, 35, 4, 16

Compare this to the value of the mean.

SOLUTION. First, rank the data in ascending order.

Rank	1	2	3	4	5	6	7	8	9	10
Data	2,	3,	4,	7,	9,	11,	15,	16,	22,	35

There are $n=10$ data points, so the median will be the average of the $n/2(=5)$ and $(n/2)+1(=6)$ points. Thus,

$$\text{median} = X_m = \frac{9+11}{2} = 10.$$

We can calculate the mean as

$$\bar{X} = \frac{2+3+4+7+9+11+15+16+22+35}{10} = 12.4$$

Of course there is a difference between the median and the mean, although they are close. You can probably see why the median is sometimes a more useful measure than the mean. The mean is affected by the size of the data and the possibility that large numbers do not reflect the rest of the data. In this case, most of the data is less than 20, but the point 35 has a large impact on the mean, whereas its impact on the median is not nearly as large. Thus, either may be more appropriate in particular situations.

Mode

The least-used measure of central tendency is the mode. The mode is the data point which occurs most often. In the first two examples of this chapter, the mode is each data point, since no points are repeated. There may be only one mode, or many modes as in the examples seen thus far.

One would not want to use the mode as the sole measure of central tendency. It should be used with the mean or the median, or both.

EXAMPLE 3. Find the mode of the following set of data.

4, 2, 8, 10, 14, 8, 6, 14, 20, 6, 3, 5, 8, 6

Also find the mean and the median.

SOLUTION. To find the mode, like the median, it is helpful to rank the data in ascending order. Doing this we find:

2, 3, 4, 5, 6, 6, 6, 8, 8, 8, 10, 14, 14, 20

The values 6 and 8 each appear three times, more than any other values. These would both represent the mode of the data.

For the median, notice that there are 14 data points; n=14. The median is the average of the seventh and eighth ranked points. So:

$$X_m = \frac{6+8}{2} = 7$$

We calculate the mean as:

$$\overline{X} = \frac{\sum_{i=1}^{14} x_i}{14} = \frac{114}{14} = 8.1429$$

It will not always be the case that the mean, median, and mode are in such close agreement as in the above example. This is why it is often beneficial to examine all three measures. For instance, look at the difference in measures for the following set of data.

Data: 3, 4, 4, 5, 2, 4, 6, 5, 3, 20, 22, 20, 24, 26, 20, 25
Mean: $\overline{X} = 12$
Median: $X_m = 5$
Mode: 4 and 20

You can see that each by itself does not convey the structure of the data. By examining all measures you get a pretty good feel for the content of the data.

2.3 Measures of Dispersion

The measures of central tendency just discussed describe the way in which data are clustered around some central value. They do not tell us how close or how far apart the data values are; that is, how they are scattered. A statistic measuring the scatter or spread of a group of data is called a measure of dispersion. Two different groups of data may have identical means, medians, and modes, yet be vastly different because of differences in dispersion.

In this section we will discuss three measures of dispersion: range, variance, and standard deviation. The latter two measures are closely related and we will present them together.

Range

The range of a group of data is a simple measure both to understand and calculate. It is defined as the difference between the largest and smallest values in the data. You can see that this gives us a number which specifies the "width" of the data points. It does not contain any additional information that may be desirable, such as where the data points are concentrated.

EXAMPLE 4. Find the range for the following group of data:

3, 6, 2, 9, 5, 3, 16, 4, 15

SOLUTION. If we rank the data in ascending order, finding the range becomes a simple matter of finding the difference between the end points.

2, 3, 3, 4, 5, 6, 9, 15, 16
Range $X_R = 16-2 = 14$

Although the range is quite easy to calculate, in most cases it does not provide sufficient information about the variation of the data. Because only two data points are used, the range says nothing about the location and concentration of other data points except the distance between them is less than or equal to X_R. You have also noticed that the range does not depend on the number of data points. For these reasons, we now present two more measures of dispersion. These will provide more useful information than the range.

Variance and Standard Deviation

The variance and the standard deviation are quite different in concept and calculation from the range. They are based upon the distance of each individual point from the mean. Because of this, their calculation is significantly more involved than that for the range.

In using these measures for more advanced statistical procedures, either the variance or the standard deviation will be specified for use. While they certainly are not the same measure, they are very closely related. We find the value of the standard deviation as the positive square root of the variance. We will, therefore, focus our discussion on the variance.

We mentioned above that the variance was based upon the distance of each individual point from the mean. For data point i this deviation is $x_i - \overline{X}$. You may notice that if we were to sum the deviations for all of the data points, the sum would be zero. We, therefore, cannot use the measure "average deviation from the mean;" its value would be zero. Since we would like to consider any deviation from the mean as positive, we could square the deviations, thus assuring that they would be positive. We could also use the absolute value of deviations from the mean, instead of the square of deviations. However, this option is not as mathematically attractive and does not possess some desirable characteristics. Consult DeGroot (1975) for further discussion.

The variance is defined as the average or mean of the squares of the deviations from the mean. By utilizing the definition of the mean, we can define the variance symbolically as

$$s^2 = \frac{\sum_{i=1}^{n} (x_i - \overline{X})^2}{n}$$

You may notice a slightly different definition of variance in other sources. Using the above definition the variance is called a "biased" estimate of the true population variance. An "unbiased" estimate is found by dividing by n-1 instead of n. You can consult Walpole and Myers (1972) or DeGroot (1975) for a discussion of estimator bias.

EXAMPLE 5. What is the variance of the following set of data?

3, 6, 8, 2, 4, 3

2.3 Measures of Dispersion

SOLUTION. Following the formula for the variance, we first must calculate $\overline{X}$.

$$\overline{X} = \frac{\sum_{i=1}^{6} x_i}{6} = \frac{3+6+8+2+4+3}{6} = 4.33$$

The variance can now be found as:

$$s^2 = \frac{(3-4.33)^2+(6-4.33)^2+(8-4.33)^2+(2-4.33)^2+(4-4.33)^2+(3-4.33)^2}{6}$$

$$s^2 = 4.222$$

Computationally, this formula for s^2 is not very attractive. By rearranging terms we can find a much easier one to apply. This formula is

$$s^2 = \frac{\sum_{i=1}^{n} x_i^2 - \frac{\left(\sum_{i=1}^{n} x_i\right)^2}{n}}{n}$$

As with the previous definition, this is the "biased" estimate of the population variance.

EXAMPLE 6. Use the revised formula to find the variance of the data in Example 5.

SOLUTION. All we need do is find the sum of the values and the sum of the square of the values.

x_i	x_i^2
3	9
6	36
8	64
2	4
4	16
3	9
$\sum_{i=1}^{6} x_i = 26$	$\sum_{i=1}^{6} x_i^2 = 138$

The variance is:

$$s^2 = \frac{138 - \frac{(26)^2}{6}}{6} = 4.222$$

You can see that if the data is centered around the mean, the variance is very small. If there are many data points which are not close to the mean, the variance will be large. The larger the variance, the more scattered the data.

You should note the units of measurement of the variance. The units will be the square of the units of the data. If the data are in units of miles, the variance will be in units of square miles. For this reason the standard deviation may be a more desirable measure of the dispersion. Since the standard deviation is the square root of the variance, its units are the same as those of the original data. We label the standard deviation S:

$$s = \sqrt{s^2}$$

In a straightforward manner the standard deviation of the data in Example 5 is found to be $s = \sqrt{4.222} = 2.055$. The computer code for these measures of dispersion is presented with the other data analysis code later in this chapter.

2.4 Data Plots

The measures discussed previously are valuable in reducing a particular property of data to one number. However, it is quite often just as valuable to examine a visual representation of the data. This can be accomplished by plotting the data on a graph. This plot can show the values of the data and the frequency of occurrence. If the data consist of joint observations, they can also be represented. For instance, you may have monthly sales data over a number of years. It would be useful to examine a plot of sales over time.

One of the most important advantages of looking at the data as opposed to simply calculating statistics is examining joint data. Here joint data refers to the case when a single observation results in values for two variables. An example of this is the observation of a person's physical stature. Observing one person results in the joint measure of height and weight. We do not get an accurate picture of physical stature unless we look at the height/weight combination. Another example is measuring metal fatigue. In this case, it is important to look at the joint observations of force and time to failure.

Variables may be related in many different ways. It is important to be able to specify these relationships. In some cases, trend or seasonal patterns can be detected. There are analytic procedures for evaluating these types of patterns, but they are not presented here (see Makridakis and Wheelwright, 1978, and Montgomery and Johnson, 1976, for analytic procedures). You have not sufficiently examined a set of data unless you have examined a plot of the data.

In the computer program presented in the next section, we include a plotting routine. You will be able to specify single or joint data.

2.5 Data Analysis and Plotting Subroutines

In this section, we present a set of computer subroutines that incorporate all the data analysis procedures discussed in the chapter. There are actually three major subroutines. The first reduces the data to particular statistics. These are the measures of central tendency and measures of dispersion presented earlier. This program provides the capability to call the plotting subroutines. In these subroutines a curve may be plotted or a frequency histogram (specifying frequency of occurrence of observations) of a set of data can be generated.

2.5 Data Analysis and Plotting Subroutines

Figure 2.3 presents the subroutine for data analysis. This allows analysis of either a single variable or joint observations. The measures of central tendency and dispersion are displayed for each variable.

```
7500 REM SUBROUTINE FOR DATA ANALYSIS
7510 REM Statistics are derived in this main program,
7520 REM a separate subroutine plots the data.
7530 DEFINT I,J
7540 N=INUM
7550 PRINT "IS THIS ANALYSIS FOR ONE VARIABLE OR JOINT
      OBSERVATIONS?"
7560 INPUT "ENTER 1--ONE VARIABLE; 2--JOINT";IA
7570 IF IA=1 THEN GOTO 7750      'For the one-variable case
7580 REM    ***********RULES FOR JOINT OBSERVATIONS***********
7590 REM
7600 PRINT "THE DATA FOR THIS PROGRAM MUST BE ENTERED AS JOINT"
7610 PRINT "OBSERVATIONS IN THE VARIABLES X AND Y WITH THE"
7620 PRINT "NUMBER OF JOINT OBSERVATIONS GIVEN IN INUM. THE"
7630 PRINT "OBSERVATIONS ARE SORTED IN A SPECIALLY REVISED"
7640 PRINT "BUBBLE SORT ROUTINE THAT MAINTAINS THE INTEGRITY OF"
7650 PRINT "THE JOINT RELATIONSHIPS FOR VALUES OF JT<>3."
7660 JT=1                    'Set variable specification flag
7670 GOSUB 12710             'Call the revised bubble sort
7680 FOR I=1 TO INUM
7690 XVAR(I)=X(I)            'Set XVAR
7700 YVAR(I)=Y(I)            'Set YVAR
7710 NEXT I
7720 GOTO 7800
7730 REM    ***********RULES FOR FREQUENCY ANALYSIS***********
7740 REM
7750 PRINT "THE DATA FOR THIS PROGRAM MUST BE PRE-SORTED IN"
7760 PRINT "ASCENDING ORDER. THE SORTED VALUES MUST APPEAR"
7770 PRINT "IN THE ARRAY XVAR. THE "
7780 PRINT "NUMBER OF DATA POINTS MUST APPEAR IN INUM"
7790 JT=0                    'Set for frequency data
7800 REM In this section the data is analyzed
7810 REM According to the statistics presented
7820 REM in the chapter (mode, median, mean,
7830 REM range, standard deviation, variance)
7840 INPUT "ENTER CARRIAGE RETURN TO CONTINUE";Z$
7850 PRINT:PRINT
7860 IF JT<>2 THEN PRINT "FOR VARIABLE ";XV$
7870 IF JT=2 THEN PRINT "FOR VARIABLE ";YV$
7880 PRINT "***MEASURES OF CENTRAL TENDENCY***"
7890 PRINT
7900 REM    First find the mode
7910 J=1                    'Set mode counter
7920 JNUM=0                 'Number of data points representing the mode
7930 JCNT=2                 'Data point occurrence counter
7940 JJ=0
7950 IF JT=2 THEN GOTO 7970
```

```
7960 DIM ILOC (INUM)        'Dimension for location of mode
7970 FOR I=2 TO INUM+1
7980 IF I=INUM+1 THEN GOTO 8020
7990 IF XVAR(I)<>XVAR(I-1) THEN GOTO 8020 'For unequal values
8000 J=J+1                  'Update frequency counter
8010 GOTO 8180
8020 IF J>=JCNT THEN GOTO 8050   'Frequency less than current max
8030 J=1
8040 GOTO 8180
8050 IF J=JCNT THEN GOTO 8140    'Frequency equal to current max
8060 REM At this point, the current frequency represents the maximum
8070 REM frequency. Update all pertinent counters
8080 JNUM=1
8090 JCNT=J
8100 J=1
8110 JJ=1
8120 ILOC(JNUM)=I-1         'Set to locate mode
8130 GOTO 8180
8140 JNUM=JNUM+1            'Set for multiple modes
8150 ILOC(JNUM)=I-1
8160 J=1
8170 JJ=1
8180 NEXT I
8190 IF JJ=1 THEN GOTO 8220 'A mode does exist
8200 PRINT "MODE--NO OBSERVATION APPEARED MORE THAN ONCE"
8210 GOTO 8320
8220 IF JNUM>1 THEN GOTO 8250 'More than one mode exists
8230 PRINT "MODE--THE VALUE";XVAR(ILOC(1));"OCCURRED";JCNT;"TIMES"
8240 GOTO 8320
8250 PRINT "MODE--MULTIPLE MODES WERE FOUND"
8260 PRINT "      THEY ALL OCCURRED";JCNT;"TIMES"
8270 PRINT "      THESE VALUES ARE"
8280 FOR I-1 TO JNUM
8290 IF I=JNUM THEN PRINT XVAR(ILOC(I))
8300 IF I<>JNUM THEN PRINT XVAR(ILOC(I));", "; 'Print the
     multiple modes
8310 NEXT I
8320 REM NEXT, FIND THE MEDIAN
8330 PRINT
8340 XX=INUM/2
8350 II=INT (INUM/2)
8360 IF II=XX THEN GOTO 8390 'Integer number of data points
8370 PRINT "MEDIAN--";XVAR(II+1)
8380 GOTO 8410
8390 MED=(XVAR(II)+XVAR(II+1))/2
8400 PRINT "MEDIAN--";MED
8410 REM Now, calculate the mean
8420 PRINT
8430 SM=0                   'Set sum of terms equal to zero
8440 SS=0                   'Set sum of squared terms equal to zero
8450 FOR I=1 TO INUM
```

2.5 Data Analysis and Plotting Subroutines

```
8460 SM=SM+XVAR (I)       'Sum the terms
8470 SS=SS+XVAR(I)^2      'Square and sum the terms
8480 NEXT I
8490 MN=SM/INUM           'Find the mean
8500 PRINT "MEAN -- ";MN
8510 PRINT:PRINT:PRINT
8520 PRINT "***MEASURES OF DISPERSION***"
8530 PRINT
8540 REM First find the range
8550 RNG=XVAR(INUM)-XVAR(1) 'Calculate the range
8560 PRINT "RANGE -- ";RNG
8570 REM Now find the variance and standard deviation
8580 PRINT
8590 VAR=(SS-((SM^2)/INUM))/INUM 'Calculate the variance
8600 STD=SQR(VAR)                'Find the standard deviation
8610 PRINT "VARIANCE -- ";VAR
8620 PRINT
8630 PRINT "STANDARD DEVIATION -- ";STD
8640 PRINT:PRINT:PRINT
8650 IF IA=1 THEN GOTO 8890
8660 IF JT=2 THEN GOTO 8830
8670 XRNG=RNG
8680 JT=2                 'Reset variable specification flag
8690 XT=XVAR(1)           'Set XT equal to smallest XVAR value
8700 XM=XVAR(INUM)        'Set XM equal to largest XVAR value
8710 FOR I=1 to INUM
8720 X(I)=YVAR(I)         'Exchange XVAR and YVAR arrays, prepare
                           for sort
8730 Y(I)=XVAR(I)
8740 NEXT I
8750 REM Now call revised sorting routine to sort YVAR and
8760 REM keep the relationship between XVAR and YVAR
8770 GOSUB 12710          'Call the revised bubble sort routine
8780 FOR I=1 TO INUM
8790 XVAR(I)=X(I)         'Reassign XVAR
8800 YVAR(I)=Y(I)         'Reassign YVAR
8810 NEXT I
8820 GOTO 7800
8830 FOR I=1 TO INUM
8840 C=XVAR(I)
8850 XVAR(I)=YVAR(I)      'Exchange XVAR and YVAR arrays
8860 YVAR(I)=C
8870 NEXT I
8880 YRNG-RNG
8890 PRINT "DO YOU WANT A PLOT OF THE DATA?"
8900 INPUT "ENTER Y--YES, N--NO";B$
8910 IF B$"N" THEN GOTO 8960
8920 IF IA=1 THEN GOTO 8950
8930 GOSUB 10960
8940 GOTO 8960
```

```
8950 GOSUB 8970             'Go to the plotting subroutine
8960 RETURN
```

Figure 2.3—data analysis program

Figure 2.4 presents the frequency plotting subroutine, including a companion subroutine for plotting on your printer. It can be called from the data analysis program or from a main program by-passing the data analysis. It will plot a frequency histogram for one variable.

```
8970 REM***************************************************************
8980 REM FREQUENCY PLOT SUBROUTINE
8990 REM This is the frequency plotting subroutine.
9000 REM It can be used to plot a frequency histogram for a single
9010 REM set of data.
9020 REM Input array is XVAR and must be dimensioned
9030 REM in the main program. Also required is the number of data
9040 REM points, stored in INUM.
9050 ID=1
9060 XM=XVAR(INUM)
9070 RNG=XVAR(INUM)-XVAR(1) 'Set range
9080 PRINT "IS THIS FREQUENCY PLOT FOR INTEGER OR NONINTEGER DATA?"
9090 INPUT "ENTER D--INTEGER, C--NONINTEGER";A$
9100 INPUT "PLOT ON CRT OR PRINTER (C/P)";C$
9110 IF C$="P" THEN PRINT "PLEASE TURN ON PRINTER"
9120 IF C$="P" THEN INPUT "HIT CARRIAGE RETURN WHEN READY";WQ
9130 IF A$="D" THEN GOTO 10390
9140 REM This section is for noninteger data
9150 GOSUB 12330            'Scale the variables
9160 XW=RNG/60
9170 INC=10 'Use 10 classes to store the data
9180 REM Calculate frequency of occurrence in each class
9190 DIM IFQ(INC+1)          'Dimension frequency array
9200 FOR I=1 TO INUM
9210 IF I=1 THEN J=1
9220 IF XVAR(I)>XVAR(1)+J*(RNG/INC) THEN GOTO 9250
9230 IFQ(J)=IFQ(J)+1         'Increase the frequency by one
9240 GOTO 9270
9250 J=J+1                   'Go to the next class
9260 GOTO 9220               'Check the next class
9270 NEXT I
9280 IMF=1                   'Set maximum frequency
9290 FOR I=1 TO INC
9300 IF IFQ(I)>IMF THEN IMF=IFQ(I)   'Set maximum frequency
9310 NEXT I
9320 PRINT:PRINT:PRINT:PRINT
9330 IF C$="P" THEN LPRINT TAB(20);"FREQUENCY PLOT OF ";XV$
9340 IF C$="C" THEN PRINT TAB(20);"FREQUENCY PLOT OF ";XV$
9350 IF C$="P" THEN LPRINT
9360 IF C$="C" THEN PRINT
```

2.5 Data Analysis and Plotting Subroutines

```
9370 REM Determine frequency scale
9380 DIM IVA(15)            'Dimension label array
9390 IF IMF>15 THEN GOTO 9490
9400 IFS=INT(15/IMF)        'Set number of lines between frequencies
9410 ILN=IFS*IMF            'Set number of lines in plot
9420 J=1                    'Set start for vertical counter
9430 II=IFS
9440 FOR I=II TO ILN STEP IFS
9450 IVA(I)=J               'Set vertical axis label
9460 J=J+1
9470 NEXT I
9480 GOTO 9540
9490 IFS=INT((IMF/15)+1)    'Set number of frequencies per line
9500 ILN=INT((IMF/IFS)+1)   'Set number of lines in plot
9510 FOR I=1 TO ILN
9520 IVA(I)=I*IFS           'Set vertical axis label
9530 NEXT I
9540 IQ=ILN:IF ILN<9 THEN ILN=9
9550 FOR I=1 TO ILN
9560 IF C$="P" THEN GOSUB 10690
9570 IF C$="P" THEN GOTO 9830
9580 IF I>9 THEN GOTO 9680
9590 IF I=1 THEN PRINT "F";
9600 IF I=2 THEN PRINT "R";
9610 IF I=3 THEN PRINT "E";
9620 IF I=4 THEN PRINT "Q";
9630 IF I=5 THEN PRINT "U";
9640 IF I=6 THEN PRINT "E";
9650 IF I=7 THEN PRINT "N";
9660 IF I=8 THEN PRINT "C";
9670 IF I=9 THEN PRINT "Y";
9680 IJ=ILN-I+1:IF IJ>IQ THEN GOTO 9820
9690 IF IVA(IJ)<>0 THEN IK=IVA(IJ)
9700 IF IVA(IJ)=0 THEN PRINT TAB(9);"+";
9710 IF IVA(IJ)<>0 THEN PRINT TAB(4);
9720 IF IVA(IJ)<>0 THEN PRINT USING "####";IVA(IJ);
9730 IF IVA(IJ)<>0 THEN PRINT TAB(9);"+";
9740 FOR J=1 TO INC
9750 IF ID<>2 AND J=1 THEN PRINT TAB(9+J*6);CHR$(179);
9760 IF IFQ(J) IK THEN GOTO 9790
9770 IF ID=2 THEN PRINT TAB(9+J*IXW);CHR$(219);
9780 IF ID<>2 THEN PRINT TAB(12+J*6);CHR$(219);
9790 IF ID=2 THEN GOTO 9810
9800 IF J<>INC THEN PRINT TAB(9+(J+1)*6);CHR$(179);
9810 NEXT J
9820 PRINT
9830 NEXT I
9840 REM Print horizontal axis
9850 IF C$="P" THEN LPRINT TAB(9);"+";
9860 IF C$="C" THEN PRINT TAB(9);"+";
9870 IF ID=2 THEN GOTO 9930
```

```
9880 FOR I=0 TO INC
9890 IF C$="P" THEN LPRINT "-----+";
9900 IF C$="C" THEN PRINT "-----+";
9910 NEXT I
9920 GOTO 10030
9930 FOR I=1 TO 70
9940 IG=0                          'Set printer flag
9950 IH=INT(I/IXW)
9960 IF IH=(I/IXW) THEN IG=1
9970 IF C$="P" AND IG=1 THEN LPRINT "+";
9980 IF C$="C" AND IG=1 THEN PRINT "+";
9990 IF C$="P" AND IG=0 THEN LPRINT "-";
10000 IF C$="C" AND IG=0 THEN PRINT "-";
10010 NEXT I
10020 INC=10                       'Reset INC for integer case
10030 IF C$="P" THEN LPRINT
10040 IF C$="C" THEN PRINT
10050 REM Print axis values
10060 IF ID=2 THEN GOTO 10180
10070 IF XI%=0 THEN GOTO 10100
10080 IF C$="P" THEN LPRINT TAB(1);"(10^";XI%;")";
10090 IF C$="C" THEN PRINT TAB(1);"(10^";XI%;")";
10100 FOR I=0 TO INC
10110 IF C$="P" THEN LPRINT TAB(13+I*6);
10120 IF C$="C" THEN PRINT TAB(13+I*6);
10130 HAX=XVAR(1)+I*6*XW
10140 IF C$="P" THEN LPRINT USING "##.##";HAX/XE;
10150 IF C$="C" THEN PRINT USING "##.##";HAX/XE;
10160 NEXT I
10170 GOTO 10310
10180 JI=XVAR(1)
10190 FOR I=1 TO 70
10200 IF I=1 THEN GOTO 10250
10210 IF I<J THEN GOTO 10300
10220 IF INT((I-1)/IXW)<>(I-1)/IXW THEN GOTO 10300
10230 JI=XVAR(1)+(I-1)/IXW
10240 IF JI XVAR(INUM) THEN GOTO 10300
10250 IF C$="C" THEN PRINT TAB(5+IXW+I);
10260 IF C$="P" THEN LPRINT TAB(5+IXW+I);
10270 IF C$="C" THEN PRINT USING "####";JI;
10280 IF C$="P" THEN LPRINT USING "####";JI;
10290 J=I+5
10300 NEXT I
10310 REM Print horizontal axis title
10320 IF C$="C" THEN GOTO 10360
10330 LPRINT:LPRINT
10340 LPRINT TAB(35);XV$
10350 GOTO 10380
10360 PRINT:PRINT
10370 PRINT TAB(35);XV$
10380 GOTO 10590
```

2.5 Data Analysis and Plotting Subroutines

```
10390 REM For integer data
10400 REM If the range is>60, classes must be used
10410 IF RNG>60 THEN GOTO 9160
10420 ID=2                    'Set integer flag
10430 INC=60                  'Set number of data points
10440 IXW=INT(60/(RNG+1))     'Columns per data point
10450 DIM IFQ(75)
10460 J=1                     'Set coordinate counter
10470 FOR I=1 TO INUM
10480 IF XVAR(I)<>XVAR(1)+(J-1) THEN GOTO 10510
10490 IFQ(J)=IFQ(J)+1         'Find frequency for each coordinate
10500 GOTO 10550
10510 J=J+1
10520 IF J>61 THEN GOTO 10570
10530 IF XVAR(I)<>XVAR(1)+(J-1) THEN GOTO 10510
10540 IFQ(J)=IFQ (J)+1        'Find frequency for each coordinate
10550 NEXT I
10560 GOTO 9280               'Go to main plotting stage
10570 PRINT "PROGRAM DATA ERROR; PROGRAM TERMINATING"
10580 GOTO 10670
10590 IF C$="P" THEN GOTO 10670
10600 INPUT "WOULD YOU LIKE THE PLOT ON YOUR PRINTER(Y/N)";D$
10610 IF D$="N" THEN GOTO 10670
10620 PRINT "PLEASE TURN ON THE PRINTER"
10630 INPUT "HIT CARRIAGE RETURN WHEN READY";WQ
10640 C$="P"
10650 IF ID=2 THEN INC=60
10660 GOTO 9550
10670 RETURN
10680 REM*****************************************************************
10690 REM Subroutine for printing plot on the printer
10700 IF I>9 THEN GOTO 10800
10710 IF I=1 THEN LPRINT " F";
10720 IF I=2 THEN LPRINT " R";
10730 IF I=3 THEN LPRINT " E";
10740 IF I=4 THEN LPRINT " Q";
10750 IF I=5 THEN LPRINT " U";
10760 IF I=6 THEN LPRINT " E";
10770 IF I=7 THEN LPRINT " N";
10780 IF I=8 THEN LPRINT " C";
10790 IF I=9 THEN LPRINT " Y";
10800 IJ=ILN-I+1:IF IJ>IQ THEN GOTO 10940
10810 IF IVA(IJ)<>0 THEN IK=IVA(IJ)
10820 IF IVA(IJ)=0 THEN LPRINT TAB(9);"+";
10830 IF IVA(IJ)<>0 THEN LPRINT TAB(4);
10840 IF IVA(IJ)<>0 THEN LPRINT USING "####";IVA(IJ);
10850 IF IVA(IJ)<>0 THEN LPRINT TAB(9);"+";
10860 FOR J=1 TO INC
10870 IF ID<>2 AND J=1 THEN LPRINT TAB(9+J*6);CHR$(156);
10880 IF IFQ(J)<IK THEN GOTO 10910
10890 IF ID=2 THEN LPRINT TAB(9+J*IXW);CHR$(170);
```

```
10900 IF ID<>2 THEN LPRINT TAB(12+J*6);CHR$(170);
10910 IF ID=2 THEN GOTO 10930
10920 IF J<>INC THEN LPRINT TAB(9+(J+1)*6);CHR$(156);
10930 NEXT J
10940 LPRINT
10950 RETURN
```

Figure 2.4—frequency plot subroutine

Figure 2.5 presents the subroutine for plotting joint observations. It can also be used with or without the data analysis subroutine.

```
10960 REM****************************************************************
10970 REM SUBROUTINE FOR PLOTTING JOINT OBSERVATIONS
10980 REM Subroutine for plotting two variables
10990 REM Data consists of joint observations, XVAR and YVAR;
11000 REM the ranges of each variable, XRNG and YRNG. The
11010 REM data YVAR must be sorted in ascending order. The
11020 REM number of joint observations is INUM.
11030 REM The names of the variables are assumed to be
11040 REM given in the variables XV$ and YV$.
11050 REM The value XT must be specified as the smallest value
11060 REM of the XVAR array. This is assigned in the Data Analysis
11070 REM subroutine.
11080 REM In summary, this is what is required:
11090 REM    1. YVAR sorted in ascending order
11100 REM    2. XVAR joint observation (appropriate subscripts
11110 REM       must match)
11120 REM    4. The number of joint observations in INUM
11130 REM    5. Names for XVAR and YVAR (XV$ and YV$)
11140 REM    6. XT-smallest value in the XVAR array
11150 REM    7. XM-largest value in the XVAR array
11160 REM    8. XVAR, YVAR, X, and Y, must be dimensioned to
11170 REM       INUM in the calling program
11180 INPUT "WOULD YOU LIKE THE PLOT ON YOUR PRINTER (Y/N)";D$
11190 IF D$="Y" THEN PRINT "PLEASE TURN ON YOUR PRINTER"
11200 IF D$="Y" THEN INPUT "ENTER CARRIAGE RETURN TO BEGIN";Q$
11210 JT=3                    'Set flag for normal sorting
11220 REM These must be provided if the Data Analysis routine is not
11230 REM used to access this plotting subroutine.
11240 XSTEP=XRNG/60           'Find horizontal step range
11250 YSTEP=YRNG/15           'Find vertical step range
11260 DIM STOX(INUM)
11270 REM Rearrange XVAR and YVAR so YVAR is in descending order
11280 FOR I=1 TO INUM
11290 STOX(I)=XVAR(I)
11300 NEXT I
11310 FOR I=1 TO INUM
11320 YVAR(I)=X(INUM-I+1)     'X has YVAR values from previous sort
11330 SVAR(I)=STOX(INUM-I+1)
```

2.5 Data Analysis and Plotting Subroutines

```
11340 NEXT I
11350 GOSUB 12330              'Scale the YVAR and XVAR variables
11360 JL=LEN(YV$)              'Number of characters in YV$
11370 FOR I=0 TO 15
11380 IVAL=0                   'INITIALIZE IVAL
11390 VAX=YVAR(1)-I*YSTEP
11400 LOWER=VAX-YSTEP/2        'Set lower cell limit
11410 IF D$="Y" THEN LPRINT
11420 IF D$="N" THEN PRINT
11430 IF I>JL-1 THEN GOTO 11460
11440 IF I=0 THEN F$=LEFT$(YV$,1)
11450 IF I<>0 THEN F$=MID$(YV$,I+1,1)
11460 IF D$="N" THEN GOTO 11520
11470 IF I<JL THEN LPRINT F$;
11480 LPRINT TAB(3);
11490 LPRINT USING "###.##";VAX/YE;
11500 LPRINT TAB(10);"+";
11510 GOTO 11560
11520 IF I<JL THEN PRINT F$;
11530 PRINT TAB(3);
11540 PRINT USING "###.##"; VAX/YE;
11550 PRINT TAB(10);"+";
11560 IF I=0 THEN ICNT=1       'Set YVAR counter
11570 FOR J=ICNT TO INUM
11580 IF YVAR(J)<LOWER THEN GOTO 11610
11590 IVAL=IVAL+1
11600 NEXT
11610 IF IVAL=0 THEN GOTO 11940
11620 REM At least one value for row I
11630 IF IVAL=1 THEN GOTO 11720    'Sorting is not necessary
11640 FOR J=ICNT TO ICNT+IVAL-1
11650 X(J-ICNT+1)=XVAR(J)           'Assign X array for sort
11660 NEXT J
11670 N=IVAL
11680 GOSUB 12710                   'Call sort routine
11690 FOR J=ICNT TO ICNT+IVAL-1
11700 XVAR(J)=X(J-ICNT+1)           'Reassign XVAR
11710 NEXT J
11720 JC=ICNT
11730 JVAL=1
11740 FOR J=JC TO ICNT+IVAL-1
11750 IF J=JC THEN ICOL = CINT ((XVAR(JC)-XT)/XSTEP)
11760 IF J=JC THEN GOTO 11800
11770 JCOL=CINT((XVAR(J)-XT)/XSTEP)  'Find appropriate column
11780 IF JCOL<>ICOL THEN GOTO 11810  'Not in same column as previous
11790 JVAL=JVAL+1                    'Increase observations in column
11800 NEXT J
11810 PCHR$="*"
11820 IF JVAL=1 THEN GOTO 11870
11830 PCHR$=STR$(JVAL)
11840 IF D$="Y" THEN LPRINT TAB(14+ICOL);PCHR$;
```

```
11850 IF D$="N" THEN PRINT TAB(14+ICOL);PCHR$;
11860 GOTO 11890
11870 IF D$="Y" THEN LPRINT TAB(15+ICOL);PCHR$;
11880 IF D$="N" THEN PRINT TAB(15+ICOL);PCHR$;
11890 JC=JC+JVAL                            'Update counter for this row
11900 IF JC-ICNT+1>IVAL THEN GOTO 11920 'No more observations
                                            in this row
11910 GOTO 11730
11920 ICNT=ICNT+IVAL                        'Reset YVAR variable counter
11930 IF ICNT>INUM THEN GOTO 11950
11940 NEXT I
11950 IF D$="N" THEN GOTO 12020
11960 LPRINT
11970 IF YI%<>0 THEN LPRINT TAB(1);"(10^";YI%;")";
11980 LPRINT TAB(10);"+"
11990 LPRINTTAB(10);"+";
12000 LPRINT "----+";
12010 GOTO 12070
12020 PRINT
12030 IF YI%<>0 THEN PRINT TAB(1);"(10^";YI%;")";
12040 PRINT TAB(10);"+"
12050 PRINT TAB(10);"+";
12060 PRINT "----+";
12070 FOR I=1 TO 10
12080 IF D$="Y" THEN LPRINT "-----+";
12090 IF D$="N" THEN PRINT "-----+";
12100 NEXT I
12110 IF D$="Y" THEN GOTO 12150
12120 PRINT:PRINT
12130 IF XI%<>0 THEN PRINT TAB(3);"(10^";XI%;")";
12140 GOTO 12170
12150 LPRINT:LPRINT
12160 IF XI%<>0 THEN LPRINT TAB(3);"(10^";XI%;")";
12170 FOR I=1 TO 11
12180 HAX=XT+(I-1)*6*XSTEP         'Find horizontal axis value
12190 IF D$="N" THEN GOTO 12230
12200 LPRINT TAB(13+(I-1)*6);
12210 LPRINT USING "##.##";HAX/XE;
12220 GOTO 12250
12230 PRINT TAB(13+(I-1)*6);
12240 PRINT USING "##.##";HAX/XE;
12250 NEXT I
12260 IF D$="N" THEN GOTO 12300
12270 LPRINT:LPRINT
12280 LPRINT TAB(30);XV$
12290 GOTO 12320
12300 PRINT:PRINT
12310 PRINT TAB(30);XV$
12320 RETURN
```

Figure 2.5—joint observations plot subroutine

2.5 Data Analysis and Plotting Subroutines

Both the data analysis and the plot routines require that the data be sorted from smallest to largest. You may enter them in this fashion if you wish. However, we find it is easier to enter the data conveniently and use a sorting subroutine to order them. Figure 2.6 presents a sorting subroutine which is included in the program file. We will not discuss how it works, because this is done in Chapter 12. However, it does have one very important attribute. If the flag $JT=3$, then normal sorting of one variable from smallest to largest occurs. If the flag $JT \neq 3$, then the sort of variable X from smallest to largest also preserves the appropriate relationships of the joint observations. That is, each Y variable remains associated with its particular X variable. This is essential for plotting joint observations.

```
12710 REM*****************************************************************
12711 REM SUBROUTINE: QUICKSORT (REVISED)
12712 REM This subroutine uses the QUICKSORT procedure to sort a list
12713 REM of numbers into ascending order. The list must be
      provided to
12714 REM this subroutine in the array X. The number of elements
      in the
12715 REM list must be supplied in the variable N. The arrays LEF
      and RIT
12716 REM are used to store subsets to be partitioned.
      LEF and RIT
12717 REM contain the position of the first element and the last
      element
12718 REM of a subset to be partitioned. LS and RS maintain the
      left and
12719 REM right end positions of the subset being partitioned. L and R
12720 REM indicate the elements that have been compared with
      the pivot,
12721 REM PV. PT indicates the position in the arrays LEF and RIT
      of the
12722 REM last subset stored to be partitioned. If all the REM
      statements
12723 REM are removed from within the sort procedure, the subroutine
      will
12724 REM sort a list in less time.
12725 REM Changing the > in line 12764 to < and changing the < in
12726 REM line 12768 to > will result in the array being sorted in
12727 REM decreasing order. Set TR%=0 for first pass through.
12728 SIZ%=50
12729 IF TR%=1 THEN GOTO 12731         'Bypass dimension
12730 DIM LEF(SIZ%),RIT(SIZ%)
12731 LEF%(1)=1                         'Left end of subset
12732 RIT%(1)=N                         'Right end of subset
12733 TR%=1                             'Set to record pass
12734 REM Position in LEF and RIT of last subset stored
12735 PT%=1
12736 REM Set left side pointer
12737 L%=LEF%(PT%)
12738 REM Save starting value of left side pointer
12739 LS%=L%
```

```
12740 REM Set right side pointer
12741 R%=RIT%(PT%)
12742 REM Save starting value of right side pointer
12743 RS%=R%
12744 PT%=PT%-1
12745 REM If only 1 element in subset, do not partition
12746 IF(R%-L%)<=0 THEN 12817
12747 REM Temporarily set pivot to left element
12748 PV=X(L%):IF JT<>3 THEN PY=Y(L%)
12749 REM If 5 or less elements in subset leave pivot as is
12750 IF (R%-L%)<5  THEN 12763
12751 REM Set pivot to median
12752 MED=(L%+R%)/2
12753 IF(PV<X(R%)) AND (PV>X(MED)) THEN 12763
12754 IF(PV < X(MED)) AND (PV > X(R%)) THEN 12763
12755 IF (X(R%) < PV) AND (X(R%) > X(MED)) THEN 12761
12756 IF (X(R% < X(MED)) AND (X(R%) > PV) THEN 12761
12757 REM Set pivot=median and move left pointer to hole
12758 PV=X(MED):IF JT<>3 THEN PY=Y(MED)
12759 X(MED)=X(L%): IF JT<>3 THEN Y(MED)=Y (L%)
12760 GOTO 12763
12761 PV=X(R%): IF JT<>3 THEN PY=Y(R%)
12762 X(R%)=X (L%): IF JT<>3 THEN Y(R%)=Y (L%)
12763 IF(L% >= R%) THEN 12779       'If L=R, subset partitioned
12764 IF (PV > X(R%)) THEN 12771    'If true, swap
12765 R% = R%-1                     'Move pointer
12766 GOTO 12763
12767 IF (L%> = R%) THEN 12779      'If L=R, subset partitioned
12768 IF (PV < X(L%)) THEN 12775    'If true, swap
12769 L%=L%+1                       'Move pointer
12770 GOTO 12767
12771 X(L%)=X(R%)                   'Perform swap
12772 IF JT<>3 THEN Y(L%)=Y(R%)
12773 L%=L%+1                       'Move pointer
12774 GOTO 12767
12775 X(R%)=X(L%)                   'Perform swap
12776 IF JT<>3 THEN Y(R%)=Y(L%)
12777 R%=R%-1                       'Move pointer
12778 GOTO 12763
12779 X(L%)=PV                      'Place pivot in hole
12780 IF JT<>3 THEN Y(L%)=PY
12781 REM Determine which subset to store; largest is to be stored
12782 IF(L%-LS%) <=(RS%-L%) THEN 12800
12783 REM Right subset is smallest; if it contains more than 1
12784 REM element, store left subset
12785 IF(RS%-L%)<=1 THEN 12795
12786 LEF%(PT%+1)=LS%
12787 RIT%(PT%+1)=L%-1
12788 L%=L%+1
12789 LS%=L%
12790 R%=RS%
12791 PT%=PT%+1
```

```
12792 GOTO 12746
12793 REM Right subset contained only 1 element; partition left
12794 REM subset if it contains more than 1 element
12795 IF(L%-LS%)<=1 THEN 12817
12796 R%=L%-1
12797 RS%=L%
12798 L%=LS%
12799 GOTO 12746
12800 IF(L%-LS%)<=1 THEN 12810
12801 REM Left subset is smallest; if it contains more than 1
12802 REM element, store right subset
12803 LEF%(PT%+1)=L%+1
12804 RIT%(PT%+1)=RS%
12805 R%=L%-1
12806 RS%=R%
12807 L%=LS%
12808 PT%=PT%+1
12809 GOTO 12746
12810 IF(RS%-L%)<=1 THEN 12817
12811 REM Right subset contained more than 1 element; partition it
12812 L%=L%+1
12813 LS%=L%
12814 R%=RS%
12815 GOTO 12746
12816 REM Any more subsets to be partitioned?
12817 IP(PT%<1) THEN 12819
12818 GOTO 12737
12819 RETURN
```

Figure 2.6—sorting subroutine

An important aspect of the plotting subroutine for a frequency histogram is the choice of an integer or non-integer plot. If the non-integer option is selected, class intervals are automatically established. In this case, instead of plotting the frequency of each data point, the frequency of occurrence of data points within specified intervals is plotted. This will be the option you use most. If the integer option is selected (your data must be integer values), two possibilities exist. If the range of the data is less than 60, the frequency for each point will be plotted. If the range is greater than 60, class intervals are used.

The data analysis and plotting subroutines were not written as main programs so they could be more flexible in application. Therefore, a main program must be written in order to use them. There are a few important steps which must be taken in the calling program. They are specified in the program listing but are worthy of presentation here. First, general requirements applicable to all subroutines will be presented, then requirements specific to each subroutine.

General
 1. Data must be stored in variable XVAR (and YVAR for joint observations).
 2. Number of observations must be stored in INUM.

3. Name of variable(s) must be stored in XV$ for XVAR (and YV$ for YVAR if required).

4. Appropriate arrays must be dimensioned to INUM.

5. Define I and J as integer variables.

Single Observations: Data analysis
1. XVAR must be sorted from smallest to largest.

Joint Observations: Data analysis
1. Variables X and Y must also be dimensioned to INUM.

2. Data need not be sorted in the main program.

Frequency Plot
1. Data must be sorted from smallest to largest.

Joint Observation Plot
1. Set XT as the smallest number in the XVAR array.

2. Set XM as the largest number in the XVAR array.

3. YVAR must be sorted from smallest to largest.

4. XVAR variables must match with appropriate joint observations in YVAR (i.e., their position in the array must be the same).

We should mention the format used for the plots. In order to create an accurate yet readable plot, the PRINT USING statement was used. This statement specifies the number of spaces allowed to print a number and the position of the decimal point. Because of this, the size of the numbers which can be plotted is restricted. Restrictions on XVAR, for both plotting subroutines, allow numbers from 0.01 to 99.99 (9999 for integer frequency plots). For YVAR (joint observations) the restriction is 0.01 to 999.99. If your numbers are larger or smaller than those allowed, they will automatically be scaled to fit this format (except integer frequency plots). A subroutine is included for this purpose, and is presented in Figure 2.7.

```
12330 REM****************************************************************
12340 REM SCALING SUBROUTINE:
12350 REM This subroutine is used for scaling the numbers used in the
12360 REM plots to fit the format specifications. This subroutine
12370 REM should not be used without accessing it through the plotting
12380 REM subroutines.
12390 IF IA=1 THEN GOTO 12550       'Scale only the XVAR variable
12400 REM For YVAR scaling
12410 IF YVAR(1)<1 THEN GOTO 12470 'Scale variables up
12420 FOR I=0 TO 35
12430 YI%=I
12440 YE=10^I
12450 IF YVAR(1)/YE<100 THEN GOTO 12550
12460 NEXT I
12470 REM For YVAR less than one
12480 FOR I=3 TO 35
12490 YI%=-I
```

2.5 Data Analysis and Plotting Subroutines

```
12500 YE=10^YI%
12510 IF YVAR(1)/YE>100 THEN GOTO 12530
12520 NEXT I
12530 YE=YE*10                          'Set appropriate value of YE
12540 YI%=YI%+1
12550 REM For scaling the XVAR values
12560 IF XM<1 THEN GOTO 12620
12570 FOR I=0 TO 35
12580 XI%=I
12590 XE=10^I
12600 IF XM/XE<100 THEN GOTO 12700
12610 NEXT I
12620 REM For XVAR less than one
12630 FOR I=3 TO 35
12640 XI%=-I
12650 XE=10^XI%
12660 IF XM/XE>100 THEN GOTO 12680
12670 NEXT I
12680 XE=XE*10
12690 XI%=XI%+1
12700 RETURN
```

Figure 2.7—scaling subroutine

Since there are a limited number of rows and columns for plotting, a single row or column may actually represent many values, as though intervals were used. You must keep this in mind when interpreting the plots. When plotting joint observations different plot characters are used. If one spot on the plot represents only one observation, the character "*" is printed. If more than one observation is represented, the number of observations represented is printed.

Finally, you may elect to create your plot on the CRT or on your printer. You will notice that the characters are slightly different for each. You may use any characters you wish for the plot by altering the PRINT CHR$ statements in the program. Consult your manual for complete details.

Let us look at examples of the data analysis and plotting subroutines. We will first examine frequency plots and then analysis of joint observations. The following example illustrates the use of the data analysis and plotting routines.

EXAMPLE 7. Write a main program to use the data analysis program to analyze and plot the following data.

3.6, 4.2, 1.1, 12.5, 8.4, 6.3, 10.1, 0.8, 5.6, 6.5
7.4, 3.8, 9.2, 7.5, 8.8, 6.3, 1.0, 9.6, 10.2, 13.1

SOLUTION. The main program used to access the subroutines is listed in Figure 2.8. The statistical summary for this group of data is given below. The frequency histogram is presented in Figure 2.9.

```
                    ***MEASURES OF CENTRAL TENDENCY***
MODE- THE VALUE 6.3 OCCURRED 2 TIMES
MEDIAN- 6.95
MEAN- 6.8
                    ***MEASURES OF DISPERSION***
RANGE- 12.3
VARIANCE- 12.29001
STANDARD DEVIATION- 3.505771
```

```
310 REM MAIN PROGRAM TO READ IN SINGLE OBSERVATIONS
320 REM This program calls data analysis (and plotting subroutines
330 REM if desired). A revised bubble sort routine is used for
340 REM sorting the data.
350 DEFINT I,J
360 INPUT "ENTER THE NUMBER OF OBSERVATIONS TO BE ANALYZED";INUM
370 DIM XVAR(INUM),X(INUM)
380 INPUT "ENTER THE NAME OF THE VARIABLE";XV$
390 PRINT "ENTER THE OBSERVATIONS ONE AT A TIME"
400 FOR I=1 TO INUM
410 PRINT
420 PRINT "ENTER OBSERVATION";I;
430 INPUT X(I)
440 NEXT I
450 JT=3                   'Set flag for bubble sort routine
451 N=INUM
460 GOSUB 12710            'Call sorting routine
470 FOR I=1 TO INUM
480 XVAR(I)=X(I)           'Place sorted values in XVAR
490 NEXT I
500 GOSUB 7500
510 END
```

Figure 2.8—main program for frequency plots

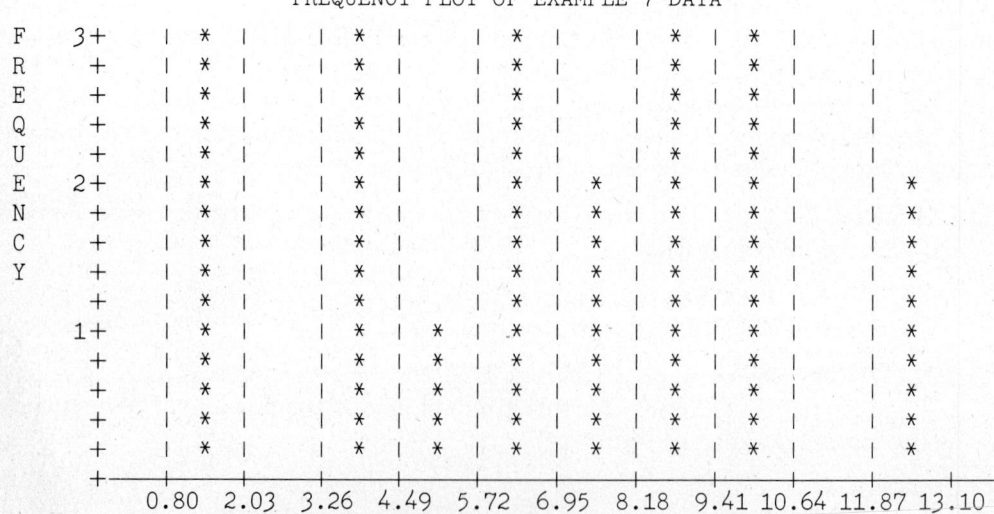

Figure 2.9—example 7, frequency histogram

EXAMPLE 8. Write a main program to analyze and plot the following joint observations

X: 3 6 9 4 5 3 1 2 8 6 7 10
Y: 4 5 10 4 7 5 1 4 10 5 9 14

SOLUTION. The main program required is presented in Figure 2.10. The data analysis output is presented below and the joint observation plot is given in Figure 2.11.

```
FOR VARIABLE X
***MEASURES OF CENTRAL TENDENCY***
MODE-MULTIPLY MODES WERE FOUND
     THEY ALL OCCURRED 2 TIMES
     THESE VALUES ARE
3 , 6
MEDIAN- 5.5
MEAN- 5.333334

***MEASURES OF DISPERSION***
RANGE- 9
VARIANCE- 7.388888
STANDARD DEVIATION- 2.718251

FOR VARIABLE Y
***MEASURES OF CENTRAL TENDENCY***
MODE-MULTIPLE MODES WERE FOUND
     THEY ALL OCCURRED 3 TIMES
     THESE VALUES ARE
4 , 5
MEDIAN- 5
MEAN- 6.5

***MEASURES OF DISPERSION***
RANGE- 13
VARIANCE- 11.91667
STANDARD DEVIATION- 3.452053

100 REM MAIN PROGRAM TO READ IN JOINT OBSERVATIONS
110 REM Program calls data analysis and plotting subroutines.
120 REM A revised bubble sort routine is used for maintaining
130 REM proper relationships between joint observations.
140 REM This is called from the data analysis subroutine.
150 DEFINT I,J
160 INPUT "ENTER THE NUMBER OF JOINT OBSERVATIONS";INUM
170 DIM XVAR(INUM),YVAR(INUM),X(INUM),Y(INUM)
180 INPUT "ENTER THE NAME OF THE FIRST VARIABLE OF THE PAIR";XV$
190 INPUT "ENTER THE NAME OF THE SECOND VARIABLE OF THE PAIR";YV$
200 PRINT "ENTER THE JOINT OBSERVATIONS"
210 FOR I=1 TO INUM
220 PRINT
230 PRINT "OBSERVATION ";I
240 PRINT "FOR ";XV$
```

```
250 INPUT X(I)
260 PRINT "FOR ";YV$;
270 INPUT Y(I)
280 NEXT I
290 TR%=0:GOSUB 7500
300 END
```

Figure 2.10—main program for plotting joint observations

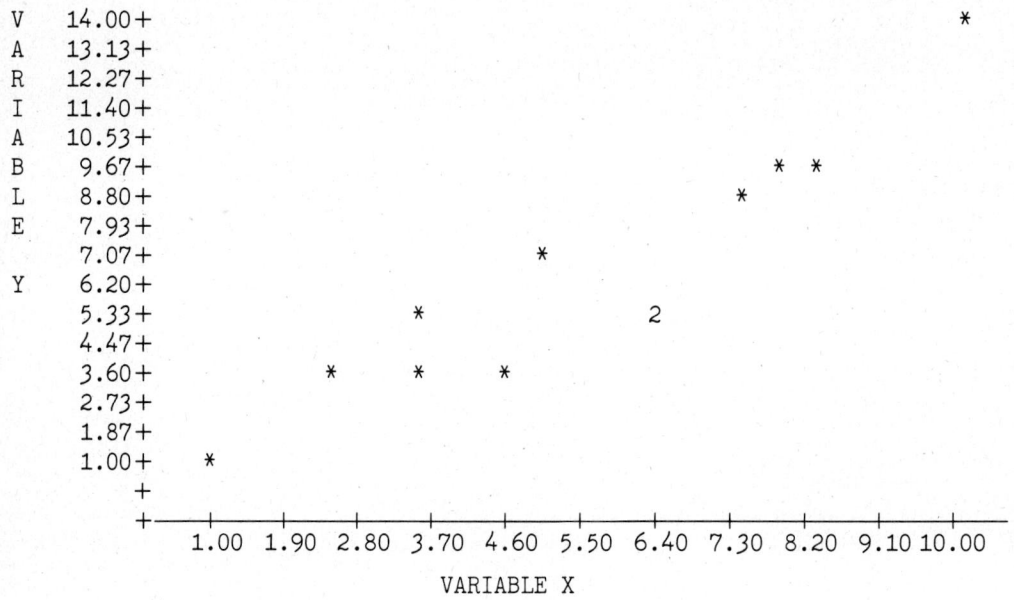

Figure 2.11—joint observation plot

2.6 Summary

This chapter has concentrated on the analysis and display of data. We are always being bombarded with data, but they are useless unless they can be reduced to a meaningful level through statistics or data plots. We have covered the most commonly used statistics and presented a plotting tool through which to examine the data visually.

We discussed measures of central tendency (mean, median, and mode) and measures of dispersion (range, variance, and standard deviation). This information, and the plot, can be used with other chapters in this book, in particular Chapters 15 (Random Numbers) and 4 (Curve Fitting).

References

DeGroot, Morris H. *Probability and Statistics*, Addison-Wesley Publishing Company, Inc., Reading, Mass., 1975.

2.6 Summary

Freund, John E. *Mathematical Statistics*, Prentice-Hall, Inc., Englewood Cliffs, N.J., 1962.
Gray, A. William and Otis M. Ulm *Elementary Probability and Statistics*, Glencoe Press, Beverly Hills, Calif., 1973.
Makridakis, Spyros and Steven C. Wheelwright *Forecasting: Methods and Applications*, John Wiley and Sons, New York, N.Y., 1978.
Montgomery, Douglas C. and Lynwood A. Johnson *Forecasting and Time Series Analysis*, McGraw-Hill Book Company, New York, N.Y., 1976
Walpole, Ronald E. and Raymond H. Myers *Probability and Statistics for Engineers and Scientists*, The Macmillan Company, New York, N.Y., 1972.

Exercises

1. Following is a collection of computer response times (in seconds) from a remote facility. Process these through the data reduction and plotting subroutines presented in this chapter.

3.2	1.0	8.5
6.1	3.7	3.0
1.3	0.6	2.8
1.2	2.1	4.6
4.6	0.9	3.3
2.3	5.3	1.1
5.1	4.4	3.2

2. Two metals are being considered for use in a new product. The most important property under consideration is tensile strength. You must make the decision as to which metal to use. Use the following test data to help support your decision. The data consist of tensile strength measurements (in pounds per square inch) from several tests.

    Metal A

9500	9632
9300	9513
9650	9481
9525	9490
9420	9691
9803	9510

    Metal B

9210	9506
10320	9005
8925	8985
9610	10145
8760	9200
10010	9600

3. Following is a set of joint data regarding percent sunlight received by a plant during the day and growth per day in centimeters. Use the plotting routine to plot this joint data.

Percent Sunlight	Growth	Percent Sunlight	Growth
25	3	85	6
40	5	50	4
30	3	30	3
50	5	75	6
95	8	10	3
60	5	30	4
80	6	15	2
40	4	70	6

    Does it appear that sunlight is the only factor relevant to plant growth?

4. Give cogent arguments for using each of the mean, median, and mode as measures of when one may be preferred over the others.
5. From problem 3, we may wish to estimate plant growth based upon the percent of sunlight received. Draw a straight line through your plot from problem 3, that line which you feel "best" represents the data. (There are formal methods for this procedure discussed in Chapter 4.) What growth would you expect for 55 percent sunlight? What is the equation of your line?

3

Matrices & Vectors

Many mathematical formulas of interest to scientists and engineers employ matrices. Matrices improve computational efficiency and provide a convenient way to represent complex relationships. This chapter details the use of matrices and vectors on the IBM Personal Computer. Many basic mathematical operations are described, including single-matrix manipulations.

3.1 Definition of a Matrix

A matrix itself does not have any mathematical meaning. It is simply a table of numbers arranged in a specific manner. The numbers are arranged in columns (or rows) of equal length. The general structure is

$$A = \begin{bmatrix} a_{11} & a_{12} & a_{13} & \cdots & a_{1n} \\ a_{21} & a_{22} & a_{23} & \cdots & a_{2n} \\ \vdots & & & & \\ a_{m1} & a_{m2} & a_{m3} & \cdots & a_{mn} \end{bmatrix}$$

The matrix A above has m rows and n columns, and is therefore classified as an m $\times$ n (read "m by n") matrix. We always list the number of rows first. If m = n, we say that the matrix is "square." Thus, the matrix B is a 3 $\times$ 2 matrix.

$$B = \begin{bmatrix} 1 & 2 \\ 3 & 6 \\ 1 & 8 \end{bmatrix}$$

Within a computer, a matrix is usually stored in an array. The number of rows and columns for each array must be specified. We do this through the DIMENSION (DIM) statement. An example of a DIM statement is

DIM A(2,4), B(5,2)

This statement specifies variables *A* and *B* as arrays. For each array variable we specify the number of rows first and the number of columns second. Thus, *A* is a 2 × 4 array and *B* is a 5 × 2 array. This dimensioning procedure ensures that the computer will allocate space for and recognize the use of the specified arrays. The computer does not require that you use all of the space allocated in the DIM statement. Matrix *A* may be only a 2 × 2. The DIM statement places only upper limits on array size.

Individual elements of array variables are referenced in a manner similar to the way the arrays are dimensioned in the DIM statement. That is, an array element is identified by its row and column position within the array. The element in the second row and third column of the array *A* is denoted A(2,3). The following is a specification of an array in computer terms.

$$A = \begin{bmatrix} 1 & 2 & 3 & 3 \\ 2 & 1 & 8 & 6 \end{bmatrix}$$

A(1,1)=1   A(1,2)=2   A(1,3)=3   A(1,4)=3

A(2,1)=2   A(2,2)=1   A(2,3)=8   A(2,4)=6

This notation is important to understand because it is used throughout the computer programs presented in this chapter. It is also necessary for reading a matrix. A subroutine for this exercise is presented in Figure 3.1.

```
13010 REM SUBROUTINE TO READ IN THE DIMENSIONS OF A MATRIX,
13020 REM the matrix itself, and allocate space through DIM
13030 INPUT "THE ROW DIMENSION OF A1 IS";IROW
13040 INPUT "THE COLUMN DIMENSION OF A1 IS";ICOL
13050 DIM A1(IROW,ICOL)             'Dimension the matrix A1
13060 PRINT "INPUT THE MATRIX A1 ONE ELEMENT AT A TIME, ROW-BY-ROW"
13070 FOR I=1 TO IROW
13080 FOR J=1 TO ICOL
13090 INPUT A1(I,J)
13100 NEXT J,I
13110 RETURN
```

Figure 3.1—matrix dimensions subroutine

3.2 Matrix Operations

There are a number of basic matrix operations that you will find useful. These operations are detailed below.

Transpose

The transpose of a matrix is a simple concept. It is accomplished by exchanging the rows and columns of a matrix. Row 1 becomes column 1, row 2 becomes column 2, etc. An m × n matrix becomes an n × m matrix when transposed. The transpose is signified by the superscript T.

3.2 Matrix Operations

$$A = \begin{bmatrix} a_{11} & a_{12} & \cdots & a_{1n} \\ a_{21} & a_{22} & \cdots & a_{2n} \\ \vdots & \vdots & & \vdots \\ a_{m1} & a_{m2} & \cdots & a_{mn} \end{bmatrix} \qquad A^T = \begin{bmatrix} a_{11} & a_{21} & \cdots & a_{m1} \\ a_{12} & a_{22} & \cdots & a_{m2} \\ \vdots & \vdots & & \vdots \\ a_{1n} & a_{2n} & \cdots & a_{mn} \end{bmatrix}$$

The matrices A and A^T have the same elements, but they are arranged in a slightly different order. Let's look at an example:

$$B = \begin{bmatrix} 3 & 6 & 1 \\ 2 & 5 & 0 \\ 2 & 1 & 0 \end{bmatrix} \qquad B^T = \begin{bmatrix} 3 & 2 & 2 \\ 6 & 5 & 1 \\ 1 & 0 & 0 \end{bmatrix}$$

We present a BASIC subroutine which performs a transpose operation in Figure 3.2. The transpose of $A1$ is stored in AT.

```
13320 REM SUBROUTINE TO TRANSPOSE A MATRIX
13330 REM Input matrix A1 to be transposed
13340 REM and its dimensions, IROW and ICOL
13350 REM Transpose is placed in matrix AT
13360 DIM AT(ICOL,IROW)
13370 FOR I=1 TO IROW
13380 FOR J=1 TO ICOL
13390 AT(J,I)=A1(I,J)
13400 NEXT J,I
13410 RETURN
```

Figure 3.2—matrix transpose subroutine

A main program which uses the transpose subroutine is given in Figure 3.3.

```
100 REM MAIN PROGRAM TO FIND THE TRANSPOSE OF A MATRIX
110 DEFINT I,J
120 GOSUB 13010       'Read in the matrix
130 GOSUB 13320       'Find the transpose, place in matrix AT
140 FOR I=1 TO ICOL
150 FOR J=1 TO IROW
160 PRINT AT(I,J);
170 NEXT J
180 PRINT
190 NEXT I
200 END
```

Figure 3.3—using the transpose subroutine

EXAMPLE 1. Find the transpose of the following matrix:

$$A1 = \begin{bmatrix} 3 & 2 \\ 6 & 5 \end{bmatrix}$$

SOLUTION.

$$AT = \begin{bmatrix} 3 & 6 \\ 2 & 5 \end{bmatrix}$$

Addition and Subtraction

Matrices may be added or subtracted in a manner very similar to scalars. In fact, matrix addition consists of the addition of corresponding scalar elements.

Not all matrices can be added (or subtracted). For some pairs of matrices, the operation is undefined. In order for the operation to be defined, the matrices must be "conformable." This means that they conform to certain specifications which allow the operation to be performed correctly. Two matrices are "conformable" for addition if they have the same number of rows and columns. That is, the matrices must be the same "size."

Once it is determined that two matrices are conformable for addition, the actual addition is a straightforward process of scalar addition. We add scalars corresponding to identical row and column positions of both matrices to yield the new element in that row and column position in the sum matrix. In general,

$$A = \begin{bmatrix} a_{11} & a_{12} & \cdots & a_{1n} \\ a_{21} & a_{22} & \cdots & a_{2n} \\ \vdots & \vdots & & \vdots \\ a_{m1} & a_{m2} & \cdots & a_{mn} \end{bmatrix} \qquad B = \begin{bmatrix} b_{11} & b_{21} & \cdots & b_{m1} \\ b_{21} & b_{22} & \cdots & b_{2n} \\ \vdots & \vdots & & \vdots \\ b_{m1} & b_{m2} & \cdots & b_{mn} \end{bmatrix}$$

$$C = A+B = \begin{bmatrix} a_{11}+b_{11} & a_{12}+b_{12} & \cdots & a_{1n}+b_{1n} \\ a_{21}+b_{21} & a_{22}+b_{22} & \cdots & a_{2n}+b_{2n} \\ \vdots & \vdots & & \vdots \\ a_{m1}+b_{m1} & a_{m2}+b_{m2} & \cdots & a_{mn}+b_{mn} \end{bmatrix}$$

3.2 Matrix Operations

We present an example of this procedure below:

$$A = \begin{bmatrix} 2 & 5 & 2 \\ 0 & 7 & 9 \end{bmatrix} \quad B = \begin{bmatrix} 6 & 10 & 12 \\ 0 & 1 & 1 \end{bmatrix}$$

$$C = A+B = \begin{bmatrix} 2+6 & 5+10 & 2+12 \\ 0+0 & 7+1 & 9+1 \end{bmatrix} \quad \begin{bmatrix} 8 & 15 & 14 \\ 0 & 8 & 10 \end{bmatrix}$$

Notice that C, the sum of A and B, will also have the same dimensions as A and B; $m \times n$. Notice also that since addition is commutative for scalars, it is also commutative for matrices.

$A+B = B+A$

The rules for matrix subtraction are the same as those for addition, with minus signs instead of plus signs. Of course, subtraction is not commutative. Adding or subtracting matrices on the computer is easy. Figure 3.4 presents a series of subroutines for adding two matrices, *A1* and *B1*. A main program to calculate the sum of two matrices is presented in Figure 3.5.

```
13430 REM SUBROUTINE ADD:
13440 REM Subroutine to add two matrices, A1 and B1, result
      stored in C1
13450 REM Their dimensions must be IROW, ICOL; JROW, JCOL,
      respectively
13460 IERR=1                            'Set no-error condition
13470 IF IROW<>JROW THEN IERR=2         'Set error condition
13480 IF ICOL<>JCOL THEN IERR=2         'Set error condition
13490 ON IERR GOSUB 13570,13530
13500 RETURN
13510 REM***********************************************************
13520 REM ERROR SUBROUTINE:
13530 REM Error subroutine for non-conformable matrices
13540 PRINT "MATRICES ARE NOT CONFORMABLE FOR THIS OPERATION"
13550 RETURN
13560 REM***********************************************************
13570 REM MATRIX ADDITION SUBROUTINE
13580 DIM C1(IROW,ICOL)
13590 FOR I=1 TO IROW
13600 FOR J=1 TO ICOL
13610 C1(I,J)=A1(I,J)+B1(I,J)
13620 NEXT J,I
13630 RETURN
```

Figure 3.4—add subroutine

```
210 REM MAIN PROGRAM TO ADD TWO MATRICES
220 DEFINT I,J
230 GOSUB 13130        'Read in both matrices
240 GOSUB 13440        'Add the matrices, store the result in matrix C1
250 FOR I=1 TO IROW
260 FOR J=1 TO ICOL
270 PRINT C1(I,J);
280 NEXT J
290 PRINT
300 NEXT I
310 END
```

Figure 3.5—using the add routine

EXAMPLE 2. Find the sum of

$$A1 = \begin{bmatrix} 6 & 2 \\ 5 & 3 \\ 1 & 2 \end{bmatrix} \text{ and } B1 = \begin{bmatrix} 7 & 7 \\ 1 & 0 \\ 3 & 0 \end{bmatrix}$$

SOLUTION.

$$C1 = A1 + B1 = \begin{bmatrix} 13 & 9 \\ 6 & 3 \\ 4 & 2 \end{bmatrix}$$

Multiplication by a Scalar

Matrices can also be used in multiplication. The simplest of these is multiplication by a scalar. All that is required in this instance is to multiply each element of the matrix by the scalar.

$$A = \begin{bmatrix} a_{11} & a_{12} & \cdots & a_{1n} \\ a_{21} & a_{22} & \cdots & a_{2n} \\ \vdots & \vdots & & \vdots \\ a_{m1} & a_{m2} & \cdots & a_{mn} \end{bmatrix} \quad gA = \begin{bmatrix} ga_{11} & ga_{12} & \cdots & ga_{1n} \\ ga_{21} & ga_{22} & \cdots & ga_{2n} \\ \vdots & \vdots & & \vdots \\ ga_{m1} & ga_{m2} & \cdots & ga_{mn} \end{bmatrix}$$

This can be represented very easily in BASIC, as you can see by the subroutine in Figure 3.6.

```
13650 REM MATRIX MULTIPLICATION (BY A SCALAR) ROUTINE:
13660 REM Subroutine to multiply matrix A1 by a scalar, S
13670 REM The result is stored in matrix D1. Dimensions of A1,
13680 REM IROW and ICOL, must be provided.
13690 DIM D1(IROW,ICOL)
13700 FOR I=1 TO IROW
13710 FOR J=1 TO ICOL
13720 D1(I,J)=S*A1(I,J)
13730 NEXT J,I
13740 RETURN
```

Figure 3.6—scalar, matrix multiplication subroutine

Matrix Multiplication

Multiplication of matrices is a very important operation. Unfortunately, it is not as easy as matrix addition. Not all matrices may be multiplied together, and, matrix multiplication does not follow a commutative law. In general, $AB = BA$. We must specify the order of multiplication.

Two matrices A and B are "conformable" for the multiplication AB if the number of columns of A equals the number of rows of B. The resulting matrix will have the same number of rows as A and the same number of columns as B. In fact, the multiplication procedure itself matches the rows of A with the columns of B.

Let the product of A and B be C. We find the element in the i row and j column of C, called the ij-element, as follows. Consider only row i of A and column j of B. Since A and B are conformable for AB, row i of A and column j of B have the same number of elements. Sum the following terms: (first element of row i) × (first element of column j), (second element of row i) × (second element of column j) ... (last element of row i) × (last element of column j). This sum is the ij element of C. We find each element of C in this manner. We illustrate this with the following example.

$$A = \begin{bmatrix} 6 & 1 \\ 7 & 3 \\ 2 & 4 \end{bmatrix} \quad B = \begin{bmatrix} 1 & 7 \\ 6 & 3 \end{bmatrix}$$

$C(1,1) = (6)(1) + (1)(6) = 12$

$C(1,2) = (6)(7) + (1)(3) = 45$

$C(2,1) = (7)(1) + (3)(6) = 25$

$C(2,2) = (7)(7) + (3)(3) = 58$

$C(3,1) = (2)(1) + (4)(6) = 26$

$C(3,2) = (2)(7) + (4)(3) = 26$

$$C = \begin{bmatrix} 12 & 45 \\ 25 & 58 \\ 26 & 26 \end{bmatrix}$$

Note that the multiplication BA is not defined because the number of columns in B is not equal to the number of rows in A. Fortunately, while matrix multiplication is tedious if performed by hand, the computer makes it easy. In fact, reading a computer program for the procedure may help in understanding it. A set of BASIC subroutines for matrix multiplication is presented in Figure 3.7.

```
13760 REM MATRIX MULTIPLICATION SUBROUTINE:
13770 REM Subroutine to multiply matrices A1 and B1 as A1 times B1
13780 REM The result is stored in matrix AB1. Row and column
       dimensions
13790 REM of A1 and B1 must be provided as IROW, ICOL; JROW, JCOL,
       RESPECTIVELY
13800 IERR=1                        'Set no-error condition
13810 IF ICOL<>JROW THEN IERR=2     'Set error condition
13820 ON IERR GOSUB 13850, 13530
13830 RETURN
13840 REM****************************************************************
13850 REM MAT MULT SUB
13860 REM access only through SUB 15650
13870 IF IEX<>0 THEN GOTO 13890     'Skip dimension for exponentiation
13880 DIM AB1(IROW,JCOL)
13890 FOR I=1 TO IROW
13900 FOR J=1 TO JCOL
13910 AB1(I,J)=0
13920 FOR K=1 TO ICOL
13930 AB1(I,J)=AB1(I,J)+A1(I,K)*B1(K,J)
13940 NEXT K,J,I
13950 RETURN
```

Figure 3.7—matrix multiplication subroutine

A main program for multiplying two matrices, $A1$ and $B1$ is presented in Figure 3.8.

```
440 REM MAIN PROGRAM TO MULTIPLY TWO MATRICES TOGETHER; A1 TIMES B1
450 DEFINT I,J
460 GOSUB 13130    'Read in both matrices
470 GOSUB 13770    'Multiply the matrices, store result in matrix AB1
480 FOR I=1 TO IROW
490 FOR J=1 TO JCOL
500 PRINT AB1(I,J);
510 NEXT J
520 PRINT
530 NEXT I
540 END
```

Figure 3.8—using matrix multiplication subroutine

3.2 Matrix Operations

EXAMPLE 3. Find *A1* times *B1*,

$$A1 = \begin{bmatrix} 2 & 1 \\ 5 & 3 \\ 1 & 4 \end{bmatrix} \quad B1 = \begin{bmatrix} 7 & 5 \\ 8 & 1 \end{bmatrix}$$

SOLUTION.

$$AB = A1 \times B1 = \begin{bmatrix} 22 & 11 \\ 59 & 28 \\ 39 & 9 \end{bmatrix}$$

Exponentiation

Quite often you may wish to multiply a matrix by itself a number of times. This process is matrix exponentiation. In order to perform this operation, or series of multiplications, the matrix must be square. Otherwise, it would not be conformable for multiplication with itself. A subroutine for this procedure is presented in Figure 3.9.

```
14640 REM MATRIX EXPONENTIATION SUBROUTINE:
14650 REM Subroutine for matrix exponentiation.
14660 REM IEX equals the power of the matrix desired.
14670 REM Input is matrix A1 with dimensions IROW=ICOL
14680 REM Output is matrix C1 with the same dimensions.
14690 DIM B1(IROW,ICOL)
14700 JROW=IROW
14710 JCOL=ICOL
14720 FOR I=1 TO IROW
14730 FOR J=1 TO ICOL
14740 B1(I,J)=A1(I,J)          'Set the matrix B1 equal to A1
14750 NEXT J,I
14760 DIM AB1(IROW,ICOL)       'Set dimension for product matrix
14770 FOR IK=1 TO IEX-1
14780 GOSUB 13770              'Multiply A1 times B1
14790 FOR I=1 TO IROW
14800 FOR J=1 TO ICOL
14810 B1(I,J)=AB1(I,J)         'Place the product matrix back in B1
14820 NEXT J,I
14830 NEXT IK
14840 RETURN
```

Figure 3.9—matrix exponentiation subroutine

We illustrate a main program for matrix exponentiation in Figure 3.10.

```
810 REM MAIN PROGRAM FOR MATRIX EXPONENTIATION
815 DEFINT I,J
820 INPUT "INPUT THE POWER OF THE MATRIX DESIRED"; IEX
830 GOSUB 13010                    'Read in the matrix A1
840 GOSUB 14650                    'Perform the exponentiation
850 PRINT "FOR A1 TO THE";IEX ,"THE RESULT IS"
860 FOR I=1 TO IROW
870 FOR J=1 TO ICOL
880 PRINT AB1(I,J);
890 NEXT J
900 PRINT
910 NEXT I
920 END
```

Figure 3.10—using the matrix exponentiation subroutine

EXAMPLE 4. Find $A1$ to the third power, where

$$A1 = \begin{bmatrix} 4 & 0 & 3 \\ 1 & 1 & 0 \\ 0 & 6 & 2 \end{bmatrix}$$

SOLUTION.

$$A1^3 = \begin{bmatrix} 82 & 126 & 84 \\ 21 & 19 & 21 \\ 42 & 42 & 26 \end{bmatrix}$$

Determinant

The determinant of a matrix is a scalar that summarizes certain information about the matrix. It is an operation involving only one matrix. Determinants exist only for square matrices. There are several methods we could use to find the determinant of a matrix. However, one is more easily programmed than the others, so that one will be used. We will use a technique called Gaussian elimination.

Gaussian elimination is a matrix reduction technique. We will use it to obtain a matrix that is upper triangular. In an upper triangular matrix, elements below and to the left of the diagonal are zero. This upper triangular matrix will be significantly different from our original matrix, but it will have the same determinant. Matrix A below is an upper triangular matrix.

$$A = \begin{bmatrix} 7 & 2 & 9 & 2 \\ 0 & 5 & 10 & 6 \\ 0 & 0 & 1 & 3 \\ 0 & 0 & 0 & 1 \end{bmatrix}$$

The determinant is found as the product of the diagonal elements. We will also see the usefulness of Gaussian elimination when we find the inverse of a matrix and solve simultaneous linear equations.

Gaussian elimination consists of the repeated application of row operations to a matrix. A row operation is the addition of a scalar multiple of one row to another row. We must try to ensure that the diagonal elements of the matrix are not zero. This may require additional row operations. If the diagonal elements cannot all be made nonzero, then the determinant is zero.

Using row operations, we must force the appropriate elements of the matrix to zero. The following example illustrates this procedure:

$$B = \begin{bmatrix} 1 & 5 & 1 \\ 3 & 6 & 3 \\ 0 & 1 & -9 \end{bmatrix}$$

1. Multiply the top row by -3 and add to the second row.

$$B_1 = \begin{bmatrix} 1 & 5 & 1 \\ 0 & -9 & 0 \\ 0 & 1 & -9 \end{bmatrix}$$

2. Multiply the second row by $1/9$ and add to the third row.

$$B_2 = \begin{bmatrix} 1 & 5 & 1 \\ 0 & -9 & 0 \\ 0 & 0 & -9 \end{bmatrix}$$

Matrix B is upper triangular, so we multiply the diagonal elements to find the determinant. This will be the same as the determinant of B.

Determinant of $B = |B_1| = |B_2| = (1)(-9)(-9) = 81$

Figure 3.11 presents BASIC subroutines to create an upper triangular matrix, and thus find the determinant of the original matrix.

```
13970 REM UPPER TRIANGULAR SUBROUTINE:
13980 REM Subroutine to create an upper triangular matrix
13990 REM Provide matrix A1 and its dimensions, IROW=ICOL.
      Result in E1
14000 IERR=1                          'Set no-error condition
14010 IF IROW<>ICOL THEN IERR=2       'Set error condition
14020 ON IERR GOSUB 14060, 13530
14030 RETURN
14040 REM***************************************************************
14050 REM UPPER TRI ROUTINE
14060 REM Upper triangular routine
14070 REM access only through SUB 13970
14080 IF INV=1 THEN ICOL=2*ICOL       'Set parameter for matrix inverse
14090 DIM E1 (IROW,ICOL)
14100 FOR I=1 TO IROW
14110 FOR J=1 TO IROW
14120 E1(I,J,)=A1(I,J)                'Assign E1 matrix to A1
14130 IF INV<>1 THEN GOTO 14160       'Skip identity augmentation
14140 REM This section is used only if a matrix is being inverted
14150 IF I=J THEN E1(I,J+IROW)=1      'Augment the identity matrix to E1
14160 NEXT J,I
14170 DET=1                           'Set determinant flag
14180 FOR I=1 TO IROW-1
14190 IF E1(I,I)=0 THEN GOSUB 14290   'Diagonal element must not be zero
14200 IF DET=0 THEN RETURN
14210 FOR J=I+1 TO IROW
14220 XM=E1(J,I)/E1(I,I)              'Multiplier to zero the column elements
14230 FOR K=I TO ICOL
14240 E1(J,K)=E1(J,K)-XM*E1(I,K)      'Calculate new elements
14250 NEXT K,J,I
14260 RETURN
14270 REM***************************************************************
14280 REM ZERO DETERMINANT DETECTION SUBROUTINE:
14290 REM Subroutine to ensure diagonal elements are nonzero during
14300 REM upper triangularization. If this is not possible,
      DET is zero
14310 FOR J=I+1 TO IROW
14320 IF E1(J,I)=0 THEN GOTO 14370
14330 FOR K=1 TO ICOL
14340 E1(I,K)=E1(I,K)+E1(J,K)         'Add rows to make diagonal nonzero
14350 NEXT K
14360 RETURN
14370 NEXT J
14380 DET=0                           'DET must be zero at this point
14390 RETURN
```

Figure 3.11—upper triangular matrix subroutine

3.2 Matrix Operations

A main program for the entire procedure is in Figure 3.12.

```
550 REM MAIN PROGRAM TO FIND THE DETERMINANT OF A MATRIX
560 DEFINT I,J
570 GOSUB 13010      'Read in the matrix
580 GOSUB 13980      'Find the determinant
590 IF DET=0 THEN GOTO 630
600 FOR I=1 TO IROW
610 DET=DET*E1(I,I)
620 NEXT I
630 PRINT "DETERMINANT IS EQUAL TO"; DET
640 END
```

Figure 3.12—using upper triangular matrix subroutine

EXAMPLE 5. Find the determinant of:

$$A1 = \begin{bmatrix} 3 & 7 & 5 \\ 9 & 0 & 1 \\ 0 & 4 & 3 \end{bmatrix}$$

SOLUTION.

$|A1| = -21$

Inverse

The inverse of a matrix serves the same purpose as the reciprocal for scalars. This is the only way a sense of division is possible for matrices. We can find inverses for square matrices only.

In order to define the inverse of a matrix, we must first define the identity matrix. An identity matrix is a square matrix with ones along its diagonal and zeros elsewhere. It is generally assigned the letter I. The following are identity matrices.

$$I = \begin{bmatrix} 1 & 0 \\ 0 & 1 \end{bmatrix} \quad I = \begin{bmatrix} 1 & 0 & 0 & 0 \\ 0 & 1 & 0 & 0 \\ 0 & 0 & 1 & 0 \\ 0 & 0 & 0 & 1 \end{bmatrix}$$

Given any matrix A, the inverse of A, denoted A^{-1}, is a matrix such that:

$AA^{-1} = A^{-1}A = I$

Of course, I must have the same dimensions as A. We present the matrix B below along with its inverse. Notice that their product is I.

$$B = \begin{bmatrix} 1 & 3 \\ 2 & 8 \end{bmatrix} \qquad B^{-1} = \begin{bmatrix} 4 & -3/2 \\ -1 & 1/2 \end{bmatrix}$$

We can find the inverse of a matrix in many ways, but we consider only the Gaussian elimination method here. We can build upon the knowledge developed in the last section. If you have not read the section on determinants yet, we suggest you do so prior to the development below.

To find the inverse, we want to perform row operations on our matrix which will turn it into an identity matrix. If we perform the same row operations on an identity matrix, the result will be the inverse of our original matrix. Therefore, we will perform the row operations required on the original matrix and the identity matrix simultaneously. When the original matrix becomes an identity, its inverse is found.

The inverse of a square matrix does not always exist. This occurs if the determinant of the matrix is zero. It is beneficial to check this condition as early as possible to avoid unnecessary computations. If the determinant is zero, then the matrix is called "singular."

We will begin by augmenting the matrix with an identity by placing the identity beside the original matrix. This is illustrated below for finding B^{-1}.

$$B = \begin{bmatrix} 1 & 3 \\ 2 & 8 \end{bmatrix} \qquad B|I = \left[\begin{array}{cc|cc} 1 & 3 & 1 & 0 \\ 2 & 8 & 0 & 1 \end{array} \right]$$

We now want to perform row operations to transform B into an identity. To do this we will first make it upper triangular, then make it lower triangular. In a lower triangular matrix, elements above and to the right of the diagonal are zero. Thus, we are left with a matrix that has only diagonal elements which may be nonzero. In addition to adding a scalar multiple of a row to another, we can also simply multiply a row by a scalar. All operations are performed over the entire $B|I$ matrix. These operations will not change the inverse. Let's see how this works for this example.

1. Multiply the first row by -2 and add to the second row

$$B|I = \left[\begin{array}{cc|cc} 1 & 3 & 1 & 0 \\ 0 & 2 & -2 & 1 \end{array} \right]$$

At this stage we see that the determinant of B is not equal to zero, so the inverse will exist.

3.2 Matrix Operations

2. Multiply the second row by $-3/2$ and add to the first row.

$$B|I_1 = \begin{bmatrix} 1 & 0 & | & 4 & -3/2 \\ 0 & 2 & | & -2 & 1 \end{bmatrix}$$

3. Multiply the second row by $1/2$.

$$B|I_2 = \begin{bmatrix} 1 & 0 & | & 4 & -3/2 \\ 0 & 1 & | & -1 & 1/2 \end{bmatrix}$$

We see that the original matrix is now an identity matrix and the original identity matrix has been replaced by B^{-1}. Therefore,

$$B^{-1} = \begin{bmatrix} 4 & -3/2 \\ -1 & 1/2 \end{bmatrix}$$

A subroutine to form a lower triangular matrix is illustrated in Figure 3.13.

```
14410 REM LOWER TRIANGULARIZATION SUBROUTINE:
14420 REM Subroutine to make a matrix lower triangular, then an
      identity.
14430 REM Find the multiplier to zero elements in the kth column
14440 REM above the diagonal.
14450 REM Input is the matrix E1 with its dimensions IROW and ICOL
14460 FOR IJ=1 TO IROW
14470 IF IJ=IROW THEN GOTO 14540
14480 IK=IROW-IJ+1
14490 FOR I=1 TO IK-1
14500 XM=E1(I,IK)/E1(IK,IK)
14510 FOR J=I+1 TO ICOL
14520 E1(I,J)=E1(I,J)-XM*E1(IK,J)
14530 NEXT J,I
14540 NEXT IJ
14550 REM Create identity by multiplying each row by the reciprocal
      of the
14560 REM diagonal element of the revised matrix
14570 FOR I=1 TO IROW
14580 DIV=E1(I,I)
14590 FOR J=1 TO ICOL
14600 E1(I,J)=E1(I,J)/DIV
14610 NEXT J,I
14620 RETURN
```

Figure 3.13—lower triangular matrix subroutine

Putting the appropriate subroutines together, the inverse of a matrix can be found using the main program in Figure 3.14.

```
650 REM MAIN PROGRAM TO FIND THE INVERSE OF A MATRIX
660 DEFINT I,J,
670 INV=1                'Set inverse flag
680 GOSUB 13010          'Read in the matrix
690 GOSUB 13980          'Make the matrix upper triangular
700 IF DET=0 THEN PRINT "MATRIX IS SINGULAR"
710 IF DET=0 THEN GOTO 800
720 GOSUB 14420          'Make E1 lower triangular
730 PRINT "THE INVERSE OF A1 IS"
740 FOR I=1 TO IROW
750 FOR J=IROW+1 TO ICOL
760 PRINT E1(I,J);
770 NEXT J
780 PRINT
790 NEXT I
800 END
```

Figure 3.14—matrix inverse program

EXAMPLE 6. Find the inverse of:

$$A1 = \begin{bmatrix} 3 & 1 & 6 \\ 0 & 2 & 5 \\ 1 & 1 & 1 \end{bmatrix}$$

SOLUTION.

$$A1^{-1} = \begin{bmatrix} 0.1875 & -0.3125 & 0.4375 \\ -0.3125 & 0.1875 & 0.9375 \\ 0.1250 & 0.1250 & -0.3750 \end{bmatrix}$$

3.3 Vectors

We will present no special programs for vectors. They can be treated as matrices with only one column. An example of a vector is:

$$\vec{a} = \begin{bmatrix} 3 \\ 2 \\ 1 \end{bmatrix}$$

Here, $\vec{a}$ is a three-element column vector, or a 3×1 matrix. All of the matrix operations we discussed earlier are also appropriate for vectors. Remember that since a vector can never be square, it cannot be multiplied by itself directly and it does not have a determinant or an inverse.

Since a vector is still an array, it must be defined in the DIM statement. For $\vec{b}$ a five-element vector and $\vec{c}$ a twenty-element vector, we need the following DIM statement:

DIM B(5), C(20)

3.4 Summary

In this chapter we have presented many commonly used matrix and vector operations. These may come in quite handy when confronted with large problems. The subroutines presented may be used in combination or separately, depending on the situation. You will see the value of the techniques presented here when examining solutions to simultaneous linear equations (Chapter 5) and linear programming (Chapter 9).

References

Hadley, G *Linear Algebra*, Addison-Wesley Publishing Company, Inc., Reading, Mass., 1961.
Stewart, G. W. *Introduction to Matrix Computations*, Academic Press, Inc., New York, N.Y., 1973.

Exercises

1. Consider the following two matrices:

$$A = \begin{bmatrix} 3 & 5 & 1 \\ 6 & 0 & 2 \\ 1 & 9 & 3 \end{bmatrix} \quad B = \begin{bmatrix} 9 & 4 & 1 \\ 0 & 0 & 6 \\ 5 & 3 & 4 \end{bmatrix}$$

Use these to illustrate that $AB \neq BA$. Develop two matrices for which $AB = BA$.

2. Write a computer program to verify the associative law for matrix multiplication: $(AB)C = A(BC)$. Be sure matrices are conformable for all necessary operations.

3. The set of simultaneous equations below:

$3x_1 + 6x_2 + 7x_3 = b_1$

$4x_1 + 2x_2 + x_3 = b_2$

$\phantom{4x_1 + {}}x_2 - x_3 = b_3$

can be written in matrix and vector notation as

$$\begin{bmatrix} 3 & 6 & 7 \\ 4 & 2 & 1 \\ 0 & 1 & -1 \end{bmatrix} \begin{bmatrix} x_1 \\ x_2 \\ x_3 \end{bmatrix} = \begin{bmatrix} b_1 \\ b_2 \\ b_3 \end{bmatrix}$$

For two different sets of (x_1, x_2, x_3) find the values of (b_1, b_2, b_3).

4. Create a square matrix A whose determinant is zero and multiply it by a square matrix B whose determinant is not zero. What is the determinant of the resulting matrix?
5. Write a computer program to illustrate that $(AB)^{-1} = B^{-1}A^{-1}$.
6. Write a computer program to illustrate that $(AB)^T = B^T A^T$.
7. What is the relationship between $|A|$ and $|kA|$, where k is any scalar?
8. Find the inverse of the following matrix.

$$A = \begin{bmatrix} 6 & 0 & 9 & 1 \\ 3 & 2 & 0 & 2 \\ 1 & 1 & 0 & 1 \\ 7 & 4 & 3 & 8 \end{bmatrix}$$

Verify that you found the inverse.

Curve Fitting with Linear Regression

4.1 Linear Regression

Linear regression is the most widely used curve fitting technique. It is based on fitting (drawing) a straight line through a series of observed data points so that the deviations of these points from this line are minimized. A straight line can be expressed as $Y=A+BX$, where A is the intercept and B is the slope. Using linear regression, we can calculate values for A and B. Then using some value for X (the independent variable) a value $\hat{Y}$ (the dependent variable) can be calculated that resides on the regression line.

Because most data will contain random variations, a particular observed data point, Y_t, may not reside on the regression line. One way to measure the amount of deviation about a regression line is to calculate the error associated with each point:

$$E_t = Y_t - \hat{Y}_t, \tag{1}$$

where Y_t is the observed data point and $\hat{Y}_t$ is the corresponding point on the fitted line. Figure 4.1 illustrates a regression line and the deviation associated with a dependent variable, Y.

Regression is based on determining the line that will minimize the sum of squares of the deviations. Thus, we want to

$$\text{minimize} \sum_{t=1}^{n} E_t^2 = \sum_{t=1}^{n} [Y_t - \hat{Y}_t]^2 \tag{2}$$

Since Y can be expressed in terms of an equation for a line, we get:

$$\sum_{t=1}^{n} E_t^2 = \sum_{t=1}^{n} [Y_t - (A+BX_t)]^2 \tag{3}$$

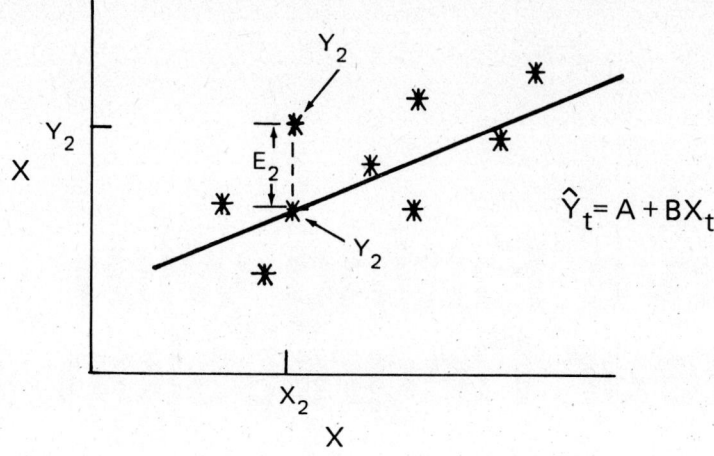

Figure 4.1—deviation from a regression line

The problem is to choose the coefficients A and B in equation 3 such that the sum of the errors squared is minimized. We can do this using differential calculus. Differentiating equation 3 with respect to A and B, we get:

$$\frac{\delta [\sum_{t=1}^{n} E_t^2]}{\delta A} = -2 \sum_{t=1}^{n} [Y_t - (A + BX_t)] \tag{4}$$

$$\frac{\delta [\sum_{t=1}^{n} E_t^2]}{\delta B} = -2 \sum [Y_t X_t - (AX_t + BX_t^2)] \tag{5}$$

When equations 4 and 5 are equated to 0, we get two simultaneous equations with two unknowns. Solving these equations gives:

$$A = \frac{\sum_{t=1}^{n} X_t^2 \sum Y_t - \left[\sum_{t=1}^{n} X_t \sum X_t Y_t\right] / n}{\sum_{t=1}^{n} X_t^2 - \sum_{t=1}^{n} X_t \sum_{t=1}^{n} X_t / n} \tag{6}$$

$$B = \frac{\sum_{t=1}^{n} X_t Y_t - \sum_{t=1}^{n} X_t \sum Y_t / n}{\sum_{t=1}^{n} X_t^2 - \sum_{t=1}^{n} X_t \sum_{t=1}^{n} X_t / n} \tag{7}$$

If these equations must be solved by hand, they appear formidable. However, these computations can easily be done by a computer.

After fitting a line through a set of data, we need some way to determine if the fit is very good. One indicator is the *coefficient of correlation*, which can have a value between -1 and $+1$. It may be expressed as

$$CC = \frac{n \sum_{t=1}^{n} X_t Y_t - \sum_{t=1}^{n} X_t \sum_{t=1}^{n} Y_t}{\sqrt{\left[n \sum_{t=1}^{n} X_t^2 - \left(\sum_{t=1}^{n} X_t\right)^2\right]\left[n \sum_{t=1}^{n} Y_t^2 - \left(\sum_{t=1}^{n} Y_t\right)^2\right]}} \quad (8)$$

A positive coefficient of correlation indicates that increases in the independent variable will result in increases in the dependent variable. The variables are directly related. If the coefficient of correlation is negative, the variables are inversely related; that is, an increase in the independent variable causes a decrease in the dependent variable. A coefficient of correlation of 0 means that the variables are not related.

Another useful indicator is the *coefficient of determination*. It is equal to the square of the coefficient of correlation:

$$CD = CC^2. \quad (9)$$

One interpretation of this indicator is that it denotes the percentage of change in the dependent variable that can be explained by changes in the regression line. A coefficient of determination of 1.0 means that 100% of the changes are explained. Consequently, this coefficient will have a value between 0 and 1.

Another useful statistic is the standard error of the estimate:

$$SE = \sum_{t=1}^{n} (Y_t - \hat{Y}_t)^2 / (n-2) \quad (10)$$

It is a measure of how closely the predicted values, $\hat{Y}_t$, compare with the actual values, Y_t. It is desirable to have this value as small as possible.

The quantity, $\sum_{t=1}^{n} (Y_t - \hat{Y}_t)^2$, in Equation 10 is known as the sum of squares about the regression. This value can be expressed in another way:

$$\sum_{t=1}^{n} (Y_t - \hat{Y}_t)^2 = \sum_{t=1}^{n} Y^2 - \left(\sum_{t=1}^{n} Y_t\right)^2/n - B\left[\sum_{t=1}^{n} X_t Y_t - \left(\sum_{t=1}^{n} X_t \sum_{t=1}^{n} Y_t\right)/n\right] \quad (11)$$

(The interested reader might refer to reference [1] to see how this expression is derived.) Although the expression on the right side of the equal sign appears complicated, it is easy to implement in a computer program. In fact, it is preferable to the expression on the left because all of the terms (Y_t, $X_t Y_t$, etc.) have been previously calculated in the process of obtaining values for A and B.

Figure 4.2 contains a program that can be used to perform linear regression where the regression line has the form

Y=A+BX.

The program in Figure 4.2 is not a subroutine, although it contains some subroutines. Consequently, the user does not have to write a main program to use this program. However, the REGRESSION subroutine starting at line number 1520 can be used by another program if data are provided in the arrays X and Y and the number of pairs of data is provided in the variable IN.

This regression program prompts you for the values for X and Y; as many as 30 pairs of data may be input. If you have more than 30 pairs of values, the dimension statement in the program for the variables X and Y must be changed. After you have input your data, you are given the opportunity to list the data, correct the data, perform the regression analysis, or quit. If you select the regression analysis option, the parameters (A and B) of the regression line, the coefficient of correlation, the coefficient of determination, and the standard error of the estimate are computed and displayed. Lastly, the input data are displayed with the respective computed values, $\hat{Y}_t$, and the error of each estimated.

```
1000 PRINT "SIMPLE LINEAR REGRESSION"
1010 REM Using linear regression this program will estimate a
1020 REM line, Y=A+BX, where X is the independent variable and
1030 REM Y is the dependent variable. If more than 30
1040 REM observations are used, the dimension statements must
1050 REM be changed. Subroutine REGRESSION may be used by other
1060 REM programs if data is provided in the arrays X and Y and
1070 REM the number of observations is provided in variable IN.
1080 DEFINT I
1090 DIM X(30), Y(30)
1100 PRINT:INPUT "INPUT NUMBER OF DATA POINTS"; IN
1110 IF IN<=0 THEN 1100
1120 IF IN<4 THEN PRINT "NO. MUST BE > 3":GOTO 1100
1130 PRINT "INPUT DATA IN PAIRS:X,Y"
1140 PRINT "WHERE X IS INDEPENDENT VARIABLE, Y IS DEPENDENT
      VARIABLE"
1150 FOR I=1 TO IN
1160 PRINT "INPUT X,Y FOR POINT"; I;
1170 INPUT X(I), Y(I)
1180 NEXT I
1190 PRINT:PRINT "AVAILABLE OPTIONS:"
1200 PRINT TAB(7) "1-LIST INPUT DATA"
1210 PRINT TAB(7) "2-MODIFY INPUT DATA"
1220 PRINT TAB(7) "3-PERFORM REGRESSION ANALYSIS"
1230 PRINT TAB(7) "4-QUIT"
1240 INPUT "OPTION"; IP
1250 IF(IP<1) OR (IP>4) THEN 1190
1260 IF IP=1 THEN GOSUB 1330
1270 IF IP=2 THEN GOSUB 1450
1280 IF IP=3 THEN GOSUB 1520
1290 IF IP=4 THEN 1860
1300 GOTO 1190
```

```
1310 REM*********************************************************************
1320 REM SUBROUTINE: LIST DATA
1330 PRINT:PRINT "LISTING OF DATA"
1340 PRINT "   X", "   Y"
1350 IC=1
1360 FOR I=1 TO IN
1370 IF I<>(IC*15) THEN 1400
1380 IC=IC+1
1390 PRINT:INPUT "PRESS ENTER TO CONTINUE";Y$:PRINT
1400 PRINT X(I),Y(I)
1410 NEXT I
1420 RETURN
1430 REM*********************************************************************
1440 REM SUBROUTINE: MODIFY DATA
1450 PRINT:INPUT "ENTER NUMBER OF DATA POINT TO BE MODIFIED"; ID
1460 PRINT "NEW VALUES FOR X AND Y FOR POINT"; ID;
1470 INPUT X(ID), Y(ID)
1480 INPUT "ANY MORE DATA POINTS TO BE MODIFIED (Y/N)"; Y$
1490 IF Y$="Y" THEN 1450
1500 RETURN
1510 REM*********************************************************************
1520 REM SUBROUTINE: REGRESSION
1530 SX=0:SY=0:SX2=0:SY2=0:SXY=0
1540 FOR I=1 TO IN
1550 SX=SX+X(I)                       'Sum of X
1560 SY=SY+Y(I)                       'Sum of Y
1570 SX2=SX2+X(I)^2                   'Sum of X^2
1580 SY2=SY2+Y(I)^2                   'Sum of Y^2
1590 SXY=SXY+X(I)*Y(I)                'Sum of X*Y
1600 NEXT I
1610 B=(IN*SXY-SX*SY)/(IN*SX2-SX^2)   'Slope of line
1620 A=(SY-B*SX)/IN                   'Intercept of line
1630 REM Coefficient of correlation
1640 CC=(SXY-SX*SY/IN)/(SQR((SX2-(SX^2)/IN)*(SY2-(SY^2)/IN)))
1650 CR=CC^2                          'Coefficient of determination
1660 SSE=SY2-SY^2/IN-B*(SXY-SX*SY/IN) 'Error sum of squares
1670 SE=SQR(SSE/(IN-2))               'Std deviation of estimate
1680 REM*********************************************************************
1690 REM SUBROUTINE: PRINT RESULTS
1700 PRINT:PRINT "REGRESSION EQUATION:"
1710 PRINT "Y="; A; "+";B;"X":PRINT
1720 PRINT "COEFFICIENT OF DETERMINATION="; CR
1730 PRINT "COEFFICIENT OF CORRELATION="; CC
1740 PRINT "STANDARD DEVIATION OF THE ESTIMATE=";SE
1750 PRINT:PRINT "ACTUAL VERSUS ESTIMATED VALUES"
1760 PRINT "X", "Y", "ESTIMATED Y", "ERROR"
1770 IC=1
1780 FOR I=1 TO IN
1790 IF I<>(IC*15) THEN 1820
1800 PRINT:INPUT "PRESS ENTER TO CONTINUE";Y$:PRINT
1810 IC=IC+1
```

```
1820 EY=A+B*X(I)
1830 PRINT X(I), Y(I), EY, Y(I)-EY
1840 NEXT I
1850 PRINT:INPUT "PRESS ENTER TO CONTINUE";Y$
1860 PRINT:RETURN
1870 PRINT:PRINT TAB(7) "END OF PROGRAM"
```

Figure 4.2—linear regression program

EXAMPLE 1. A company, known for high quality products, has had to dispose of unusually large amounts of products that did not pass specifications. In searching for a solution to this problem, it has been noted that the impurities in chemical A, which is used in these products, have changed. The following data have been collected:

Month	Pounds of Product Rejected/Month (Thousands of Pounds)	Average Amount of Impurities in Chemical A (Parts per Million)
1	115	10
2	249	31
3	208	17
4	374	42
5	307	36
6	299	33

From the data, develop an equation which can be used to predict the pounds of product that will be rejected if the average impurity of chemical A is known.

SOLUTION. The linear regression program in Figure 4.2 can be used to determine if a regression line with a high coefficient of determination can be obtained using the compiled data. The results of the analysis are shown in Figure 4.3. Using these results, the amount of product rejected can be predicted using the following expression:

$Y = 55.95308 + 7.196932(X)$

It appears that there is a correlation between the impurities of chemical A and the amount of product rejected.

```
LOAD"B:SIMREG
RUN
SIMPLE LINEAR REGRESSION

INPUT NUMBER OF DATA POINTS? 6
INPUT DATA IN PAIRS:X,Y
WHERE X IS INDEPENDENT VARIABLE, Y IS DEPENDENT VARIABLE
INPUT X,Y FOR POINT 1 ? 10,115
INPUT X,Y FOR POINT 2 ? 31,249
INPUT X,Y FOR POINT 3 ? 17,208
INPUT X,Y FOR POINT 4 ? 42,374
```

```
INPUT X,Y FOR POINT 5 ? 36,307
INPUT X,Y FOR POINT 6 ? 33,299

AVAILABLE OPTIONS:
  1-LIST INPUT DATA
  2-CORRECT INPUT DATA
  3-PERFORM REGRESSION ANALYSIS
  4-QUIT
OPTION? 3
REGRESSION EQUATION:
  Y = 55.95308+7.196932 X

COEFFICIENT OF DETERMINATION = .9433782
COEFFICIENT OF CORRELATION = .9712766
STANDARD DEVIATION OF THE ESTIMATE = 23.96282

ACTUAL VERSUS ESTIMATED VALUES
     X         Y       Estimated Y       Error
     10       115       127.9224       -12.92239
     31       249       279.058        -30.05798
     17       208       178.3009        29.69908
     42       374       358.2243        15.77576
     36       307       315.0426        -8.042633
     33       299       293.4519         5.548157

PRESS ENTER TO CONTINUE?

AVAILABLE OPTIONS:
  1-LIST INPUT DATA
  2-CORRECT INPUT DATA
  3-PERFORM REGRESSION ANALYSIS
  4-QUIT
OPTION? 4

END PROGRAM
```

Figure 4.3—product rejection example

4.2 Multiple Regression

The preceding discussion involved equations with one independent and one dependent variable:

$Y = A + BX$.

This approach may be extended to include several independent variables:

$$Y = A + B_1X_1 + B_2X_2 + \ldots + B_nX_n. \tag{12}$$

If more than one independent variable is involved, this technique is called multiple regression.

It is possible to apply the regression approach to an equation of the form:

$$Y = A + B_1Z + B_2Z_1^2 + B_3Z_1Z_2 + \ldots B_nf_n(Z_1, Z_2, \ldots, Z_n) \tag{13}$$

This equation can be made equivalent to (12) by the following substitutions:

$X_1 = Z_1$
$X_2 = Z_1^2$
$X_3 = Z_1 Z_2$
·
·
·
$X_n = f_n(Z_1, Z_2, \ldots, Z_n)$

When the above substitutions have been made, we say that equation (13) has been "transformed" to a linear model.

Regression can be applied to other nonlinear expressions if the proper transformations are first applied. For instance, consider

$$Y = Ae^{Bt} \tag{14}$$

A logarithmic transformation yields:

$$\ln Y = \ln A + Bt \tag{15}$$

The expression on the right represents the equation of a straight line where the intercept is ln A and the slope is B.

In order to apply regression to expressions such as equation (13), the original data must first be transformed. A line is then fit to the transformed data using regression. After values are obtained for the intercept (ln A) and slope (B), ln Y can be calculated. Then true values for Y can be obtained by taking the anti-ln of Y.

4.3 Stepwise Regression

Stepwise multiple regression is a form of multiple regression that adds one variable at a time to the "best fit" regression equation. Thus, we might get:

$Y = A + B_1 X_1$
$Y = A + B_1' X_1 + B_2' X_2$
$Y = A + B_1'' X_1 + B_2'' X_2 + B_3'' X_3$
·   ·
·   ·
·   ·

If there are n potential independent variables and we know that all n variables will be in the best fit equation, then we would not want to use a stepwise regression procedure. However, often the problem is to determine which of the independent variables should be included in the regression equation. This can be complicated by the fact that some of the independent variables may be highly correlated (that is, not independent). One possible approach is to investigate all possible combinations of a set of independent variables. This approach rapidly becomes computationally unattractive because of the many combinations of variables involved. The stepwise regression procedure is an answer to this problem.

Using a stepwise regression procedure, variables are added in the order which makes the greatest improvement in the goodness of fit (reduction of the variance).

4.3 Stepwise Regression

In an early stage, a variable may enter the regression equation; however, as other variables are added, the initial variable may be removed from the equation if its contribution is indicated to be insignificant. Consequently, the final regression equation will only include statistically significant variables.

Figure 4.4 contains a program that utilizes the stepwise linear regression procedure. We will not describe the mathematical details of this program here; these details are in references [1] and [2]. However, we will explain what the program can do and present some application examples.

```
1000 REM MULTIPLE AND STEPWISE REGRESSION PROGRAM
1010 REM This program will perform multiple regression or stepwise
1020 REM regression. Some data transformations are provided. Data
1030 REM management subroutines are included. A forecast can be made
1040 REM using the estimated regression equation. Results may be
1050 REM printed at CRT or printer. As many as 11 variables (10
1060 REM independent and 1 dependent) can be considered with each
1070 REM variable having 30 observations. A larger problem can be
1080 REM studied by changing the dimension statements.
1090 REM   IT=Total number of variables in problem
1100 REM   IV=Number of independent variables
1110 REM   IO=Number of data points including forecast
1120 REM   IH=Number of data points not including forecast
1130 REM   VNAM$(I)=Array containing names of variables
1140 REM   X(I,J)=Observed data
1150 REM   Y(I,J)=Estimated values for dependent variable
1160 REM   B(I)=Coefficients of regression equation
1170 REM   T95(I)=T-statistic values
1180 REM   R(I,J)=Used for all of the following
1190 REM      -Sum of squares and cross products
1200 REM      -Residual sums of squares and cross products
1210 REM      -Simple corelation coefficients
1220 REM      -Partial correlation coefficients
1230 REM      -An inverse matrix
1240 DEFDBL R,X,W,S
1250 DEFINT I,J,K,M,N
1260 NOBS=30              'Max. no. observations/var.
1270 NOVAR=11             'Max. no. variables
1280 DIM X(NOVAR,NOBS), WT(NOBS), WSUM(NOVAR), WMEAN(NOVAR)
1290 DIM R(NOVAR,NOVAR), SIGMA(NOVAR), B(NOVAR), SB(NOVAR)
1300 DIM VNAM$(NOVAR), T95(33), Y(NOBS)
1310 FOR I=1 TO 33        'Initialize T-statistic array
1320 READ T95(I)
1330 NEXT I
1340 DATA 12.706, 4.303, 3.182, 2.776, 2.571, 2.447, 2.365
1350 DATA 2.306, 2.262, 2.228, 2.201, 2.179, 2.160, 2.145
1360 DATA 2.131,2.120, 2.110, 2.101, 2.093, 2.086, 2.080
1370 DATA 2.074, 2.069, 2.064, 2.060, 2.056, 2.052, 2.048
1380 DATA 2.045, 2.042, 2.021, 2.000, 1.980
1390 PRINT TAB(5) "*** MULTIPLE AND STEPWISE REGRESSION PROGRAM ***"
1400 FOR I=1 TO NOBS      'Assume uniform weighting
1410 WT(I)=1!
```

```
1420 NEXT I
1430 PRINT:INPUT "LIST PROGRAM OPTIONS (Y/N)"; Y$
1440 IF Y$<> "Y" THEN 1470
1450 GOSUB 6980
1460 GOTO 1470
1470 PRINT:PRINT
1480 INPUT "HOW ENTER DATA - KEYBOARD OR DISK (K/D)";K$
1490 IF (K$<>"K") AND (K$<>"D") THEN PRINT "INVALID":
     GOTO 1480
1500 IF K$="K" THEN 1530
1510 GOSUB 6510
1520 GOTO 1890
1530 PRINT:INPUT "ENTER NAME OF DEPENDENT VARIABLE";TEMP$
1540 PRINT:INPUT "ENTER NUMBER OF INDEPENDENT VARIABLES"; IV
1550 IF(IV>0) AND (IV<NOVAR) THEN 1570
1560 PRINT "MUST BE > 0 AND < ";NOVAR:GOTO 1540
1570 VNAM$(IV+1)=TEMP$
1580 FOR I = 1 TO IV
1590 PRINT "ENTER NAME OF INDEPENDENT VARIABLE:";I;
1600 INPUT VNAM$(I)
1610 NEXT I
1620 PRINT:INPUT "FORECAST REQUIRED (Y/N)"; F$
1630 IF (F$<>"Y") AND (F$<>"N") THEN PRINT "INVALID":
     GOTO 1620
1640 IF F$="N" THEN I4=0:GOTO 1690
1650 INPUT "NUMBER OF PERIODS TO BE FORECAST";I4:PRINT
1660 PRINT "ENTER NUMBER OF DATA POINTS FOR EACH INDEPENDENT
     VARIABLE"
1670 INPUT "INCLUDING POINTS USED FOR FORECAST";IO
1680 GOTO 1700
1690 INPUT "ENTER NUMBER OF DATA POINTS FOR EACH INDEPENDENT
     VARIABLE";IO
1700 IF IO>=(IV+2+I4) THEN 1720
1710 PRINT "NUMBER OF DATA POINTS MUST BE >";IV+2+I4:GOTO 1640
1720 IF IO<=NOBS THEN 1740
1730 PRINT "NUMBER DATA POINTS MUST BE <";NOBS+1:GOTO 1640
1740 IT=IV+1:IH=IO-I4
1750 PRINT:PRINT "ENTER DATA"
1760 FOR I=1 TO IT
1780 PRINT:IF I<>IT THEN 1810
1790 PRINT IH;" VALUES REQUIRED FOR DEPENDENT VARIABLE ";VNAM$(IT)
1800 GOTO 1830
1810 PRINT IO;" VALUES REQUIRED FOR INDEPENDENT VARIABLE "; VNAM$(I)
1820 I5=IO
1830 IF I=IT THEN I5=IH
1840 FOR J=1 TO I5
1850 PRINT "VALUE FOR POINT ";J;
1860 INPUT X(I,J)
1870 NEXT J
1880 NEXT I
1890 PRINT:PRINT TAB(5) "DATA MANAGEMENT OPTIONS:"
1900 PRINT TAB(10) "1-LIST DATA"
```

```
1910 PRINT TAB(10) "2-CORRECT DATA"
1920 PRINT TAB(10) "3-ADD TO DATA"
1930 PRINT TAB(10) "4-ADD AN INDEPENDENT VARIABLE"
1940 PRINT TAB(10) "5-DELETE A VARIABLE"
1950 PRINT TAB(10) "6-PERFORM REGRESSION COMPUTATIONS"
1960 PRINT TAB(10) "7-STUDY ANOTHER MODEL"
1970 PRINT TAB(10) "8-STORE DATA ON DISK"
1980 PRINT TAB(10) "9-WEIGHT DATA"
1990 PRINT TAB(10) "10-TRANSFORM DATA"
2000 PRINT TAB(10) "11-QUIT"
2010 INPUT "OPTION";IP
2020 IF(IP<1) OR (IP>11) THEN PRINT "INVALID":GOTO 1890
2030 IF IP=1 THEN GOSUB 5330
2040 IF IP=2 THEN GOSUB 5530
2050 IF IP=3 THEN GOSUB 5850
2060 IF IP=4 THEN GOSUB 6250
2070 IF IP=5 THEN GOSUB 6090
2080 IF IP=6 THEN 2150
2090 IF IP=7 THEN 1400
2100 IF IP=8 THEN GOSUB 6370
2110 IF IP=9 THEN GOSUB 6680
2120 IF IP=10 THEN GOSUB 6740
2130 IF IP=11 THEN PRINT TAB(25) "END OF PROGRAM":END
2140 GOTO 1890
2150 PRINT:PRINT TAB(5) "AVAILABLE OPTIONS:"
2160 PRINT TAB(10) "1-MULTIPLE REGRESSION"
2170 PRINT TAB(10) "2-STEPWISE MULTIPLE REGRESSION
2180 PRINT TAB(10) "3-DATA MANAGEMENT OPTIONS"
2190 INPUT "OPTION";IC
2200 IF(IC<1) OR (IC>3) THEN PRINT "INVALID":GOTO 2150
2210 IF IC=3 THEN 1890
2220 C$="Y"
2230 IF IC=2 THEN C$="N"
2240 IR=1
2250 INPUT "PRINT SOLUTION AT EACH ITERATION (Y/N)";P$
2260 IF (P$<>"Y") AND (P$<>"N") THEN PRINT "INVALID":
     GOTO 2250
2270 IF P$="N" THEN IR=IR+1
2280 PRINT
2290 INPUT "OUTPUT TO APPEAR AT CRT OR PRINTER (C/P)";O$
2300 IF (O$<>"C") AND (O$<>"P") THEN PRINT "INVALID":
     GOTO 2290
2310 X$=STRING$(80,45)
2320 PRINT:PRINT:PRINT X$
2330 PRINT:PRINT
2340 IF C$="Y" THEN 2370
2350 PRINT TAB(10)"STEPWISE MULTIPLE REGRESSION"
2360 GOTO 2380
2370 PRINT TAB(10) "MULTIPLE REGRESSION"
2380 PRINT:PRINT
2390 IF O$="C" THEN 2460
```

```
2400 X$=STRING$(75,45)
2410 LPRINT:LPRINT:LPRINT X$:LPRINT:LPRINT
2420 IF C$="Y" THEN 2450
2430 LPRINT TAB(10) "STEPWISE MULTIPLE REGRESSION"
2440 GOTO 2460
2450 LPRINT TAB(10) "MULTIPLE REGRESSION"
2460 F1=3.29                   'F-test value, var. entering
2470 F2=3.29                   'F-test value, var. leaving
2480 TOL=.0001
2490 IDS=1
2500 ISTEP=1
2510 WDATA=0!
2520 FOR I=1 TO IH
2530 WDATA=WDATA+WT(I)         'Total weights
2540 NEXT I
2550 FOR I=1 TO IT
2560 WSUM(I)=0
2570 FOR J=1 TO IH             'Total weighted sum
2580 WSUM(I)=WSUM(I)+WT(J) * X(I,J)
2590 NEXT J
2600 NEXT I
2610 REM WEIGHTED SUMS OF SQUARES AND CROSS PRODUCTS
2620 FOR I=1 TO IT
2630 FOR J=1 TO IT
2640 R(I,J)=0!
2650 FOR K=1 TO IH
2660 R(I,J) = R(I,J)+WT(K) * X(I,K) * X(J,K)
2670 NEXT K
2680 NEXT J
2690 NEXT I
2700 FOR I=1 TO IT
2710 WMEAN(I)=WSUM(I)/WDATA    'Weighted mean
2720 NEXT I
2730 REM Weighted residual sum of squares and cross products
2740 FOR I=1 TO IT
2750 FOR J=1 TO IT
2760 R(I,J)=R(I,J)-WSUM(I) * WSUM(J)/WDATA
2770 NEXT J
2780 NEXT I
2790 FOR I=1 TO IT
2800 SIGMA(I)=R(I,I)^.5
2810 NEXT I
2820 REM Correlation coefficients
2830 FOR I=1 TO IT
2840 FOR J=1 TO IT
2850 R(I,J)=R(I,J)/(SIGMA(I) * SIGMA(J))
2860 NEXT J
2870 NEXT I
2880 FOR I=2 TO IT
2890 II=I-1
2900 FOR J=1 TO II
```

4.3 Stepwise Regression

```
2910 R(I,J)=R(J,I)
2920 NEXT J
2930 NEXT I
2940 PHI=WDATA-1!
2950 INDEX=1
2960 FOR J=1 TO IT
2970 SB(J)=0!
2980 B(J)=0!
2990 NEXT J
3000 IF (R(IT,IT)>-9.999999E-06) AND (R(IT,IT)<0) THEN R(IT,IT)=0
3010 SY=SIGMA(IT) * (R(IT,IT)/PHI)^.5      'Std. error of dep. var.
3020 IF C$<>"Y" THEN 3790
3030 IF ISTEP<=1 THEN 3160
3040 NN=ISTEP-1
3050 REM Regression coefficients
3060 FOR I=1 TO NN
3070 B(I)=R(I,IT) * SIGMA(IT)/SIGMA(I)     'Regression coefficients
3080 IF R(I,I)>0 THEN 3140
3090 PRINT:PRINT " SOLUTION CANNOT BE FOUND USING MULTIPLE"
3100 PRINT "REGRESSION. VARIABLE ";VNAM$(I);" IS A LINEAR COMBINATION"
3110 PRINT "OF THE OTHER VARIABLES. TRY THE STEPWISE REGRESSION OPTION."
3120 PRINT:GOTO 2150
3130 REM Std. error of regression coefficients
3140 SB(I)=SY * R(I,I)^.5/SIGMA(I)
3150 NEXT I
3160 IF IR>1 THEN 3480                     'Skip print
3170 B(IT)=WMEAN(IT)
3180 NN=IT-1
3190 FOR I=1 TO NN
3200 B(IT)=B(IT)-B(I) * WMEAN(I)
3210 NEXT I
3220 REM Coefficient of determination
3230 DETER=1!-R(IT,IT)
3240 IF DETER<.000001 THEN DETER=0!
3250 DEVEST=SY
3260 IF DEVEST<.000001 THEN DEVEST=0
3270 IF O$="P" THEN 3400
3280 PRINT:PRINT "REGRESSION NUMBER "; ISTEP
3290 PRINT TAB(5) VNAM$(IT); " = "; B(IT)
3300 FOR I=1 TO NN
3310 IF B(I)=0! THEN 3330
3320 PRINT " + "; B(I);VNAM$(I)
3330 NEXT I
3340 PRINT "COEFFICIENT OF DETERMINATION = "; DETER
3350 PRINT "STD DEVIATION OF ESTIMATE = "; DEVEST
3360 IF ISTEP<>(IDS*3) THEN 3390
3370 IF ISTEP=IT THEN 3480
3380 IDS=IDS+1:PRINT:INPUT "PRESS ENTER TO CONTINUE";Y$
3390 PRINT:GOTO 3480
3400 LPRINT:LPRINT "REGRESSION NUMBER ";ISTEP
```

```
3410 LPRINT TAB(5) VNAM$(IT); " = ";B(IT)
3420 FOR I=1 TO NN
3430 IF B(I)=0 THEN 3450
3440 LPRINT " + "; B(I);VNAM$(I)
3450 NEXT I
3460 LPRINT "COEFFICIENT OF DETERMINATION = ";DETER
3470 LPRINT "STD DEVIATION OF ESTIMATE = ";DEVEST
3480 ISTEP=ISTEP+1
3490 IF C$<>"Y" THEN 3620
3500 IF ISTEP<IT THEN 3600
3510 IF ISTEP>IT THEN 4150
3520 IF O$<>"C" THEN 3550
3530 PRINT:PRINT "FINAL SOLUTION"
3540 GOTO 3560
3550 LPRINT:LPRINT "FINAL SOLUTION"
3560 IY=IR
3570 IR=3
3580 IF C$<>"Y" THEN 3170
3590 IR=1
3600 K=ISTEP-1
3610 PHI=PHI-1!
3620 IF IR>2 THEN 4150
3630 REM Calculate new matrix
3640 FOR I=1 TO IT
3650 IF I=K THEN 3700
3660 FOR J=1 TO IT
3670 IF J=K THEN 3690
3680 R(I,J)=R(I,J)-R(I,K) * R(K,J)/R(K,K)
3690 NEXT J
3700 NEXT I
3710 FOR I=1 TO IT
3720 IF I=K THEN 3750
3730 R(I,K)=-R(I,K)/R(K,K)
3740 R(K,I)=R(K,I)/R(K,K)
3750 NEXT I
3760 R(K,K)=1!/R(K,K)
3770 IF C$<>"Y" THEN 2950
3780 GOTO 3000
3790 IX=0
3800 VX=0!
3810 IM=0
3820 VM=99999!
3830 IF (R(INDEX,INDEX)-TOL)>0! THEN 3980 'Check independence
3840 IF (INDEX+1)>=IT THEN 3870
3850 INDEX=INDEX+1
3860 GOTO 3830
3870 IF R(IT,IT)<=0 THEN 3520
3880 IF (VM*PHI/R(IT,IT))>=F2 THEN 3920   'Variance significant?
3890 K=IM
3900 PHI=PHI+1!
3910 GOTO 3160
3920 IF R(IT,IT)<=VX THEN 3950
```

4.3 Stepwise Regression

```
3930 REM Is variance reduction significant?
3940 IF (VX*(PHI-1!)/(R(IT,IT)-VX))<=F1 THEN 3520
3950 K=IX
3960 PHI=PHI-1!
3970 GOTO 3160
3980 V=R(INDEX,IT)*R(IT,INDEX)/R(INDEX,INDEX)
3990 IF V=0 THEN 3840          'Not add to equation
4000 IF V>0 THEN 4110           'Might add to equation
4010 REM Regression coefficient for variable
4020 B(INDEX)=R(INDEX,IT)*SIGMA(IT)/SIGMA(INDEX)
4030 REM Variable standard deviation
4040 SB(INDEX)=SY*R(INDEX,INDEX)^.5/SIGMA(INDEX)
4050 REM What variable causes the greatest variance reduction, and
4060 REM what variable causes the least variance increase?
4070 IF (V+VM)<=0 THEN 3840
4080 VM=-V
4090 IM=INDEX
4100 GOTO 3840
4110 IF(V-VX)<=0 THEN 3840
4120 VX=V
4130 IX=INDEX
4140 GOTO 3840
4150 IF O$="P" THEN 4170
4160 PRINT:INPUT "PRESS ENTER TO CONTINUE";Y$:PRINT
4170 IR=IY
4180 II=IT-1
4190 REM Calculate estimated values for dependent variable
4200 FOR J=1 TO IH
4210 Y(J)=B(IT)
4220 FOR I=1 TO II
4230 Y(J)=Y(J)+B(I)*X(I,J)
4240 NEXT I
4250 NEXT J
4260 III=II-1
4280 PRINT: INPUT "WANT TO SEE LIMITS ON PREDICTIONS (Y/N)";F$
4290 IF F$<>"Y" THEN 4500
4300 INPUT "USE PROGRAM SUPPLIED T-STATISTICS FOR LIMITS (Y/N)"; L$
4310 IF (L$<>"Y") AND (L$<>"N") THEN PRINT "INVALID":
     GOTO 4300
4320 IF L$="Y" THEN 4360
4330 INPUT "ENTER THE T-STATISTIC";T
4340 GOTO 4500
4350 REM If T-statistic not in array T95, then calculate value
4360 IF PHI>30! THEN 4400
4370 IHI=PHI
4380 T=T95(IHI)
4390 GOTO 4500
4400 IF PHI>40! THEN 4430
4410 T=T95(30)-(PHI-30!)*.0021
4420 GOTO 4500
4430 IF PHI>60 THEN 4460
4440 T=T95(31)-(PHI-40!)*.00105
```

```
4450 GOTO 4500
4460 IF PHI>120! THEN 4490
4470 T=T95(32)-(PHI-60!)*.02/60
4480 GOTO 4500
4490 T=T95(33)
4500 SYS=SIGMA(IT)^2*R(IT,IT)/PHI
4510 IF O$="P" THEN 4730
4520 PRINT:PRINT "ACTUAL VERSUS PREDICTED VALUES FOR ",VNAM$(IT)
4530 PRINT "ACTUAL", "PREDICTED", "DIFFERENCE","%DIFFERENCE"
4540 IC=1
4550 FOR J=1 TO IH
4560 IF J<>(IC*20) THEN 4590
4570 IC=IC+1
4580 PRINT: INPUT "PRESS ENTER TO CONTINUE";Y$:PRINT
4590 DIF=X(IT,J)-Y(J)
4600 IF X(IT,J)<>0 THEN 4630
4610 IF DIF<>0 THEN PDIF=999999!
4620 GOTO 4640
4630 PDIF=DIF/X(IT,J)*100!
4640 OBS=X(IT,J)
4650 PRINT OBS,Y(J),DIF, PDIF
4660 NEXT J
4670 IF F$<>"Y" THEN 5320
4680 PRINT:INPUT "PRESS ENTER TO CONTINUE";Y$:PRINT
4690 PRINT:PRINT "CONFIDENCE LIMITS ON PREDICTED VALUES"
4700 PRINT:PRINT "DEGREES OF FREEDOM = ";PHI;" T-STATISTIC = ";T
4710 PRINT "LOWER LIMIT", "PREDICTED", "UPPER LIMIT"
4720 GOTO 4870
4730 LPRINT "ACTUAL", "PREDICTED", "DIFFERENCE","%DIFFERENCE"
4740 FOR J=1 TO IH
4750 DIF=X(IT,J)-Y(J)
4760 IF X(IT,J)<>0 THEN 4790
4770 IF DIF<>0 THEN PDIF=999999!
4780 GOTO 4810
4790 OBS=X(IT,J)
4800 PDIF=DIF/X(IT,J)*100
4810 LPRINT OBS,Y(J),DIF,PDIF
4820 NEXT J
4830 IF F$<>"Y" THEN 5320
4840 LPRINT:LPRINT "CONFIDENCE LIMITS ON PREDICTED VALUES"
4850 LPRINT:LPRINT "DEGREES OF FREEDOM = ";PHI;
     " T-STATISTIC = ";T
4860 LPRINT:LPRINT "LOWER LIMIT","PREDICTED","UPPER LIMIT"
4870 XM=IH
4880 IC=1
4890 FOR I=1 TO IO
4900 IF I<>(IC*20) THEN 4940
4910 IC=IC+1
4920 IF O$<>"C" THEN 4940
4930 PRINT:INPUT "PRESS ENTER TO CONTINUE";Y$:PRINT
4940 VARY=SYS/XM
4950 Y(I) = B(IT)
```

4.3 Stepwise Regression

```
4960 IF III<=0 THEN 5170
4970 FOR J=1 TO III
4980 IF SB(J)=0 THEN 5020
4990 IF(SIGMA(J)=0!) OR (SIGMA(J)>1E+20) THEN 5020
5000 XSB=SYS*R(J,J)/(SIGMA(J)^2)
5010 GOTO 5030
5020 XSB=0!
5030 Y(I)=Y(I)+B(J)* X(J,I)
5040 DIF=X(J,I)-WMEAN(J)
5050 VARY=VARY+XSB*DIF^2
5060 KK=J+1
5070 FOR K=KK TO II
5080 IF SB(K)=0! THEN 5130
5090 IF (SIGMA(J)=0!)OR(SIGMA(J)>=1E+20)THEN 5130
5100 IF (SIGMA(K)=0!)OR(SIGMA(K)>=1E+20) THEN 5130
5110 XSB=SYS*R(J,K)/(SIGMA(J)*SIGMA(K))
5120 GOTO 5140
5130 XSB=0!
5140 VARY=VARY+2!*XSB*DIF*(X(K,I)-WMEAN(K))
5150 NEXT K
5160 NEXT J
5170 IF SB(II)<=0! THEN 5210
5180 IF (SIGMA(II)=0!) OR (SIGMA(II)>=1E+20) THEN 5210
5190 XSB=SYS*R(II,II)/SIGMA(II)^2
5200 GOTO 5220
5210 XSB=0!
5220 Y(I)=Y(I)+B(II)*X(II,I)
5230 VARY=VARY+XSB*(X(II,I)-WMEAN(II))^2
5240 SVARY=(VARY+SYS)^.5
5250 LLM=Y(I)-SVARY*T
5260 ULM=Y(I)+SVARY*T
5270 IF O$="P" THEN 5300
5280 PRINT LLM,Y(I), ULM
5290 GOTO 5310
5300 LPRINT LLM,Y(I),ULM
5310 NEXT I
5320 PRINT:PRINT:GOTO 2150
5330 REM SUBROUTINE: LIST DATA
5340 INPUT "ENTER NAME OF VARIABLE IDENTIFYING DATA TO BE LISTED";
     ID$
5350 GOSUB 5780
5360 IF IFL=0 THEN 5400
5370 INPUT "NO MATCH - TRY AGAIN (Y/N)";Y$
5380 IF Y$<>"Y" THEN RETURN
5390 GOTO 5340
5400 MM=IO
5410 IF I=IT THEN MM=IH
5420 PRINT "LIST OF DATA FOR "; VNAM$(I)
5430 IC=1
5440 FOR J=1 TO MM
5450 IF IC<>(IC*20) THEN 5480
5460 IC=IC+1
```

```
5470 PRINT:INPUT "PRESS ENTER TO CONTINUE";Y$:PRINT
5480 PRINT X(I,J)
5490 NEXT J
5500 PRINT:INPUT "LIST ADDITIONAL DATA (Y/N)"; Y$
5510 IF Y$="Y" THEN 5330
5520 RETURN
5530 REM SUBROUTINE: CORRECT DATA
5540 INPUT "ENTER VARIABLE NAME SPECIFYING DATA TO BE CHANGED";ID$
5550 GOSUB 5780
5560 IF IFL=0 THEN 5600
5570 INPUT "NO MATCH - TRY AGAIN (Y/N)";Y$
5580 IF Y$<>"Y" THEN RETURN
5590 GOTO 5540
5600 I5=I0
5610 IF I=IT THEN I5=IH
5620 INPUT "DO YOU WANT TO MODIFY ALL DATA FOR THIS VARIABLE
     (Y/N)";Y$
5630 IF Y$="Y" THEN 5700
5640 INPUT "ENTER POSITION OF DATA ELEMENT TO BE CHANGED"; J
5650 IF (J<1) OR (J>I5) THEN PRINT "INVALID NO.": GOTO 5640
5660 INPUT "ENTER DATA ELEMENT"; X(I,J)
5670 INPUT "WANT TO CHANGE OTHER DATA FOR THIS VARIABLE (Y/N)"; Y$
5680 IF Y$="Y" THEN 5640
5690 RETURN
5700 I5=I0
5710 IF I=IT THEN I5=IH
5720 PRINT I5," VALUES REQUIRED FOR ";VNAM$(I)
5730 FOR J=1 TO I5
5740 PRINT " ENTER VALUE FOR POINT ";J;
5750 INPUT X(I,J)
5760 NEXT J
5770 GOTO 5670
5780 REM SUBROUTINE: LOCATE MATCHING VARIABLE NAME
5790 IFL=0
5800 FOR I=1 TO IT
5810 IF ID$=VNAM$(I) THEN RETURN
5820 NEXT I
5830 IFL=1          'No match found
5840 RETURN
5850 REM SUBROUTINE: ADD TO DATA
5860 PRINT:INPUT "FORECAST REQUIRED (Y/N)";F$
5870 IF (F$<>"Y") AND (F$<>"N") THEN PRINT "INVALID":
     GOTO 5860
5880 IF F$="N" THEN I4=0:GOTO 5900
5890 INPUT "NUMBER OF PERIODS TO BE FORECAST ";I4:PRINT
5900 PRINT "ENTER NUMBER DATA POINTS FOR EACH INDEPENDENT"
5910 INPUT "VARIABLE INCLUDING POINTS USED FOR FORECAST ";IA
5920 IF IA>(IV+2-I4) THEN 5940
5930 PRINT "NUMBER DATA POINTS MUST BE >";IV+2-I4:GOTO 5900
5940 IF IA<=NOBS THEN I1=I0+1:I2=IA:GOTO 5960
5950 PRINT "NUMBER DATA POINTS MUST BE <";NOBS+1:GOTO 5900
5960 FOR I=1 TO IT
```

4.3 Stepwise Regression

```
5970 IF I=IT THEN I1=IH+1:I2=IA-I4
5980 IF I<>IT THEN 6010
5990 PRINT "ENTER ADDITIONAL DATA FOR DEPENDENT VARIABLE ";VNAM$(IT)
6000 GOTO 6020
6010 PRINT "ENTER ADDITIONAL DATA FOR INDEPENDENT VARIABLE
     ";VNAM$(I)
6020 FOR J=I1 TO I2
6030 PRINT "VALUE FOR POINT ";J;
6040 INPUT X(I,J)
6050 NEXT J
6060 NEXT I
6070 IO=IA:IH=IO-I4
6080 RETURN
6090 REM SUBROUTINE: DELETE AN INDEPENDENT VARIABLE
6100 INPUT "ENTER NAME OF VARIABLE TO BE DELETED";ID$
6110 GOSUB 5780
6120 IF IFL=0 THEN 6160
6130 INPUT "NO MATCH - TRY AGAIN (Y/N)";Y$
6140 IF Y$<>"Y" THEN RETURN
6150 GOTO 6100
6160 FOR J=I TO IT-1
6170 VNAM$(J)=VNAM$(J+1)
6180 FOR K=1 TO IO
6190 X(J,K)=X(J+1,K)
6200 NEXT K
6210 NEXT J
6220 IV=IV-1
6230 IT=IT-1
6240 RETURN
6250 REM SUBROUTINE: ADD AN INDEPENDENT VARIABLE
6260 VNAM$(IT+1)=VNAM$(IT)
6270 INPUT "ENTER NAME OF VARIABLE TO BE ADDED";VNAM$(IT)
6280 PRINT IO; " VALUES REQUIRED FOR ";VNAM$(IT)
6290 FOR J=1 TO IO
6300 X(IT+1,J)=X(IT,J)
6310 PRINT " VALUE FOR POINT ";J;
6320 INPUT X(IT,J)
6330 NEXT J
6340 IT=IT+1
6350 IV=IV+1
6360 RETURN
6370 REM SUBROUTINE: STORE DATA ON DISK
6380 INPUT "ENTER NAME OF DISK:FILE";NAM$
6390 OPEN NAM$ FOR OUTPUT AS #3
6400 WRITE#3, IT,IH,I4
6410 ID=IO
6420 FOR I=1 TO IT
6430 WRITE#3, VNAM$(I)
6440 IF I=IT THEN ID=IH
6450 FOR J=1 TO ID
6460 WRITE#3, X(I,J)
6470 NEXT J
```

```
6480 NEXT I
6490 CLOSE #3
6500 RETURN
6510 REM SUBROUTINE: READ DATA FROM DISK
6520 INPUT "ENTER NAME OF DISK:FILE";NAM$
6530 OPEN NAM$ FOR INPUT AS #3
6540 INPUT#3, IT,IH,I4
6550 IO=IH+I4
6560 IV=IT-1
6570 IF I4>0 THEN F$="Y"
6580 ID=IO
6590 FOR I=1 TO IT
6600 INPUT#3, VNAM$(I)
6610 IF (I=IT) THEN ID=IH
6620 FOR J=1 TO ID
6630 INPUT#3, X(I,J)
6640 NEXT J
6650 NEXT I
6660 CLOSE #3
6670 RETURN
6680 REM SUBROUTINE: WEIGHT DATA
6690 FOR I=1 TO IH
6700 PRINT "ENTER WEIGHT FOR DATA POINT ";I;
6710 INPUT WT(I)
6720 NEXT I
6730 RETURN
6740 REM SUBROUTINE: DATA TRANSFORM
6750 GOSUB 7300
6760 MM=IO
6770 FOR I=1 TO IT
6780 IF I=IT THEN MM=IH
6790 PRINT "ENTER TRANSFORM CODE FOR VARIABLE "; VNAM$(I);
6800 INPUT TC$
6810 IF TC$="N" THEN 6960
6820 IF TC$="E" THEN IK=1:GOTO 6860
6830 IF TC$="P" THEN IK=2:GOTO 6850
6840 PRINT "INVALID CODE":GOTO 6790
6850 INPUT "ENTER VALUE FOR EXPONENT (X^N) ";EX
6860 FOR J=1 TO MM            'Perform transforms
6870 ON IK GOTO 6880, 6940
6880 IF X(I,J)>0 THEN 6920
6890 PRINT "CANNOT TAKE THE LOG OF A NUMBER <=0 "
6900 PRINT "VALUE OF DATA ELEMENT INVOLVED IS ";X(I,J)
6910 PRINT: GOTO 6950
6920 X(I,J)=LOG(X(I,J))
6930 GOTO 6950
6940 X(I,J)=X(I,J)^EX
6950 NEXT J
6960 NEXT I
6970 RETURN
6980 REM SUBROUTINE: LIST PROGRAM OPTIONS
6990 PRINT "PROGRAM OPTIONS AVAILABLE:"
```

```
7000 PRINT TAB(10) "DATA INPUT:"
7010 PRINT TAB(10) "K=KEYBOARD"
7020 PRINT TAB(10) "D=DISK FILE"
7030 PRINT TAB(5) "DATA MANAGEMENT:"
7040 PRINT TAB(10) " CORRECTION OF DATA"
7050 PRINT TAB(10) " ADD TO DATA"
7060 PRINT TAB(10) " DELETION OF VARIABLES"
7070 PRINT TAB(10) " ADDITION OF VARIABLES"
7080 PRINT TAB(10) " WEIGHT DATA"
7090 PRINT TAB(10) " TRANSFORM DATA"
7100 PRINT TAB(10) " STORE DATA ON DISK"
7110 PRINT TAB(5) "OUTPUT RESULTS:"
7120 PRINT TAB(10) "C=CRT DISPLAY"
7130 PRINT TAB(10) "P=PRINTER"
7140 PRINT TAB(5) "COMPUTATIONAL:"
7150 PRINT TAB(10) " MULTIPLE REGRESSION"
7160 PRINT TAB(10) " STEPWISE MULTIPLE REGRESSION"
7170 PRINT TAB(5) "SOLUTION RESULTS:"
7180 PRINT TAB(10) "Y=SOLUTION AT EACH ITERATION"
7190 PRINT TAB(10) "N=FINAL SOLUTION ONLY"
7200 PRINT:INPUT "PRESS ENTER TO CONTINUE";Y$:PRINT
7210 PRINT TAB(5) "DATA WEIGHTING SCHEMES:"
7220 PRINT TAB(10) " UNIFORM, PROGRAM SUPPLIED"
7230 PRINT TAB(10) " NON-UNIFORM, USER SUPPLIED WEIGHTS"
7240 PRINT TAB(5) "CONFIDENCE LIMITS:"
7250 PRINT TAB(10) "Y=PROGRAM SUPPLIED 95% T-STATISTIC"
7260 PRINT TAB(10) "N=USER SUPPLIED T-STATISTIC"
7270 PRINT TAB(5) "FORECAST HORIZON:"
7280 PRINT TAB(10) "Y=FORECAST REQUIRED"
7290 PRINT TAB(10) "N=NO FORECAST"
7300 PRINT TAB(5) "TRANSFORMS AVAILABLE:"
7310 PRINT TAB(10) "N=NO TRANSFORMS"
7320 PRINT TAB(10) "E=LOG(X) TO BASE E"
7330 PRINT TAB(10) "P=POLYNOMIAL X^N"
7340 RETURN
```

Figure 4.4—stepwise linear regression program

Program Options

As execution of the program begins, the user is given an opportunity to receive a listing of the program options. Figure 4.5 contains a list of these options as they would appear on your display. A brief explanation of these options will be given and then each one will be illustrated using examples. Data can be input in two ways, from the keyboard or from a previously saved disk file. As soon as the program options are listed, the program prompts for how data are to be entered. A letter appearing to the right of an option, such as

K=KEYBOARD,

means that a user response of K at the appropriate time will denote that data are to be entered from the keyboard. Options appearing in Figure 4.5 without a letter in front are selected by a number from a list of options displayed on the screen at various times.

```
PROGRAM OPTIONS AVAILABLE:
  DATA INPUT:
    K=KEYBOARD
    D=DISK FILE
  DATA MANAGEMENT:
    CORRECTION OF DATA
    ADD TO DATA
    DELETION OF VARIABLES
    ADDITION OF VARIABLES
    WEIGHT DATA
    TRANSFORM DATA
    STORE DATA ON DISK
  OUTPUT RESULTS:
    C=CRT DISPLAY
    P=PRINTER
  COMPUTATIONAL:
    MULTIPLE REGRESSION
    STEPWISE MULTIPLE REGRESSION
  SOLUTION RESULTS:
    Y=SOLUTION AT EACH ITERATION
    N=FINAL SOLUTION ONLY
  DATA WEIGHTING SCHEMES:
    UNIFORM, PROGRAM SUPPLIED
    NON-UNIFORM, USER SUPPLIED WEIGHTS
  CONFIDENCE LIMITS:
    Y=PROGRAM SUPPLIED 95% T-STATISTIC
    N=USER SUPPLIED T-STATISTIC
  FORECAST HORIZON:
    Y=FORECAST REQUIRED
    N=NO FORECAST
  TRANSFORMS AVAILABLE:
    N=NO TRANSFORMS
    E=LOG(X) TO BASE E
    P=POLYNOMIAL X^N
```

Figure 4.5—program options

Several data management options are provided. After data is input, the user is given an opportunity to select any of these data management options. Data can be listed for any variable by specifying the variable name. Using the option CORRECT DATA, any data element or all data elements for a particular variable can be changed. Additional data elements can be added to a set of data by using the ADD TO DATA option. Thus, you might initially study a problem using nine data elements for each variable. However, in an attempt to improve the regression equation fit, you may append one or more data elements to the set of observations associated with each variable.

Any independent variable and its associated data can be removed from the problem being studied using the DELETE INDEPENDENT VARIABLE option. All data are stored in the two dimensional array X(I,J). Data for the independent variables are stored first, one variable per row. Data for the dependent variable are stored in row N+1, where N is the number of independent variables. Consequently, if variable N is deleted from the problem, data for the dependent variable are moved from row N+1 to row N. Similarly, an independent variable can be added to the problem using option ADD INDEPENDENT VARIABLE. Also, if you want to keep data for further analysis, you may store them in a disk file to avoid having to reenter the data from the keyboard.

Output from the regression analysis can be displayed on the CRT or at the printer. If output is displayed at the CRT, program execution is periodically stopped to give the user an opportunity to read the information being displayed before it is scrolled off the display. To continue program execution, the user is prompted to press the "ENTER" key.

Two regression computation options are provided: multiple regression and stepwise regression. Multiple regression means that all independent variables will appear in the final regression equation. However, only one variable is added at a time to the regression equation. If the user specifies that the solution is to be printed at each iteration, then the regression equation, the coefficient of determination, and the standard deviation of the estimate will be printed as each new variable is added to the regression equation. If stepwise multiple regression is selected, only the independent variables that significantly reduce the variance will appear in the regression equation.

Usually, each data element in a set of data is given equal weight. However, there may be instances when some of the elements may be more accurate, such as the more recent elements in a time series. The weighting option lets the user specify weights for each data element. It is best that the weights be normalized such that the average weight is one.

Included in the final solution are confidence limits on the predicted values. These limits are calculated using a 95% t-statistic. Given a predicted value for the dependent variable, we should expect that 95% of the time the actual value will fall within these predicted limits. If other limits are desired, the user can specify the t-statistic values that are to be employed.

As noted earlier in this chapter, some nonlinear expressions can be transformed so that linear regression can be applied. If specified, the program can perform the following data transformations:

1) Ae^{Bt} to $\ln A + Bt$
2) X^2 to X'

Once the user specifies that transformations are required, the program will prompt for the needed input. If other transformations are required (in addition to the two provided by the program), the user will have to transform that data before they are input. As each data element is transformed, it replaces the original input data; therefore, once data are transformed, they should not be transformed again during a subsequent run through the regression program.

Forecasting is a common application of linear regression. To facilitate such usage, the user can specify the number of values to be forecast. For each period to be forecast, an observed value must be entered for each independent variable. Several examples will be presented to illustrate how to use the options of the multiple and stepwise regression program. Output from the program will be presented as it appears on the monitor or printer. User responses are denoted by underlined characters.

Using the Stepwise Program

EXAMPLE 2. For the past six months, the following data had been compiled at the Brady Company. Using the data from Example 1 and the multiple and stepwise regression program, forecast the sales for period seven.

Month	Orders on Hand at Beginning of Month	Sales for the Month
1	$3550	$8350
2	3600	8730
3	3250	8120
4	3350	8450
5	3000	7900
6	3590	8650

At the beginning of the seventh month, the orders on hand had a value of $3,700. What would your sales forecast be for the seventh month?

SOLUTION. Because the multiple and stepwise regression program provides several options, the entire output from running the program will be presented here. We will intersperse some comments within this output to assist in explaining how to use this program.

The multiple and stepwise program is stored in the file STEPWISE: therefore, we start by loading this file.

```
LOAD "B:STEPWISE"
RUN
**MULTIPLE AND STEPWISE REGRESSION PROGRAM**

LIST PROGRAM OPTIONS (Y/N)?Y
PROGRAM OPTIONS AVAILABLE:
  DATA INPUT:
    K=KEYBOARD
    D=DISK FILE
  DATA MANAGEMENT:
    CORRECTION OF DATA
    ADD TO DATA
    DELETION OF VARIABLES
```

```
      ADDITION OF VARIABLES
      WEIGHT DATA
      TRANSFORM DATA
      STORE DATA ON DISK
   OUTPUT RESULTS:
      C=CRT DISPLAY
      P=PRINTER
   COMPUTATIONAL:
      MULTIPLE REGRESSION
      STEPWISE MULTIPLE REGRESSION
   SOLUTION RESULTS:
      Y=SOLUTION AT EACH ITERATION
      N=FINAL SOLUTION ONLY
PRESS ENTER TO CONTINUE?

   DATA WEIGHTING SCHEMES:
      UNIFORM, PROGRAM SUPPLIED
      NON-UNIFORM, USER SUPPLIED WEIGHTS
   CONFIDENCE LIMITS:
      Y=PROGRAM SUPPLIED 95% T-STATISTIC
      N=USER SUPPLIED T-STATISTIC
   FORECAST HORIZON:
      Y=FORECAST REQUIRED
      N=NO FORECAST
   TRANSFORMS AVAILABLE:
      N=NO TRANSFORMS
      E=LOG(X) TO BASE E
      P=POLYNOMIAL X^N
```

This list of options describes the capabilities of the program. Note the "PRESS ENTER TO CONTINUE" phase; this stops program execution so that the user can read the displayed information. Also note that some of the options have a character appearing on the left. These characters are possible responses to a prompting question, and the narrative to the right of the character describes what will happen when the designated character is entered.

Proceeding with this solution, we are asked to specify how we want to enter our data. We enter a K denoting that the data will be entered from the keyboard. We also specify that the name of our dependent variable is SALES; we have 1 independent variable named ORDERS; we want a forecast for 1 period, and we have 7 data points for the independent variable. After this information is input, the program prompts for the required data.

```
HOW ENTER DATA - KEYBOARD OR DISK (K/D)? K

ENTER NAME OF DEPENDENT VARIABLE: SALES

ENTER NUMBER OF INDEPENDENT VARIABLE: 1
ENTER NAME OF INDEPENDENT VARIABLE: 1? ORDERS

FORECAST REQUIRED (Y/N)? Y
NUMBER OF PERIODS TO BE FORECAST? 7
```

ENTER DATA

7 VALUES REQUIRED FOR INDEPENDENT VARIABLE ORDERS

VALUE FOR POINT 1? 3550
VALUE FOR POINT 2? 360
VALUE FOR POINT 3? 3250
VALUE FOR POINT 4? 3350
VALUE FOR POINT 5? 3000
VALUE FOR POINT 6? 3590
VALUE FOR POINT 7? 3700

6 VALUES REQUIRED FOR DEPENDENT VARIABLE SALES

VALUE FOR POINT 1? 8350
VALUE FOR POINT 2? 8730
VALUE FOR POINT 3? 8120
VALUE FOR POINT 4? 8450
VALUE FOR POINT 5? 7900
VALUE FOR POINT 6? 8650

The independent variable requires one more data element than is required by the dependent variable. The first six data elements of the dependent and independent variables are used to develop the regression equation. The seventh independent-variable data element is used to forecast the seventh data element for the dependent variable (sales for period seven).

After the data are input, the program provides a list of data management options. We specify option 1, LIST DATA, because we think that one of the data elements for ORDERS was input incorrectly. This option proceeds to ask for the name of the variable for which we want to list data. We respond with the name ORDERS.

```
DATA MANAGEMENT OPTIONS:
    1-LIST DATA
    2-CORRECT DATA
    3-ADD TO DATA
    4-ADD AN INDEPENDENT VARIABLE
    5-DELETE A VARIABLE
    6-PERFORM REGRESSION COMPUTATIONS
    7-STUDY ANOTHER MODEL
    8-STORE DATA ON DISK
    9-WEIGHT DATA
    10-TRANSFORM DATA
    11-QUIT
```

OPTION? 1
ENTER NAME OF VARIABLE IDENTIFYING DATA TO BE LISTED? ORDERS
LIST OF DATA FOR ORDERS
 3550
 360
 3250

4.3 Stepwise Regression

```
3350
3000
3590
3700
```

Looking at the data, we can see that element two is wrong: it should be 3600. Therefore, we need to use option 2, CORRECT DATA.

```
LIST ADDITIONAL DATA (Y/N)? N

    DATA MANAGEMENT OPTIONS:
        1-LIST DATA
        2-CORRECT DATA
        3-ADD TO DATA
        4-ADD AN INDEPENDENT VARIABLE
        5-DELETE A VARIABLE
        6-PERFORM REGRESSION COMPUTATIONS
        7-STUDY ANOTHER MODEL
        8-STORE DATA ON DISK
        9-WEIGHT DATA
       10-TRANSFORM DATA
       11-QUIT

OPTION? 2
ENTER VARIABLE NAME SPECIFYING DATA TO BE CHANGED? ORDERS
DO YOU WANT TO MODIFY ALL DATA FOR THIS VARIABLE (Y/N)? N
ENTER POSITION OF DATA ELEMENT TO BE CHANGED? 2
ENTER DATA ELEMENT? 3600
```

Note that it is possible to change all of the data elements associated with a variable. However, in this case, we only need to change element two. Having changed the known incorrect data, we will list the data associated with each variable to verify that every element is correct.

```
DO YOU WANT TO CHANGE OTHER DATA FOR THIS VARIABLE (Y/N)? N

    DATA MANAGEMENT OPTIONS:
        1-LIST DATA
        2-CORRECT DATA
        3-ADD TO DATA
        4-ADD AN INDEPENDENT VARIABLE
        5-DELETE A VARIABLE
        6-PERFORM REGRESSION COMPUTATIONS
        7-STUDY ANOTHER MODEL
        8-STORE DATA ON DISK
        9-WEIGHT DATA
       10-TRANSFORM DATA
       11-QUIT

OPTION? 1
ENTER NAME OF VARIABLE IDENTIFYING DATA TO BE LISTED? ORDERS
LIST OF DATA FOR ORDERS
```

```
3550
3600
3250
3350
3000
3590
3700
```

```
LIST ADDITIONAL DATA (Y/N)? Y
ENTER NAME OF VARIABLE IDENTIFYING DATA TO BE LISTED? SALES
LIST OF DATA FOR SALES
 8350
 8730
 8120
 8450
 7900
 8650
```

All of the data are correct. Before proceeding with the regression analysis, we will store these data in a disk file. By doing this, we will avoid having to reenter these data from the keyboard if we want to use these data at a later time. Using option 8, STORE DATA ON DISK, we store the data for ORDERS and SALES in a file called BRASALES which resides in a diskette in drive B(B:BRASALES).

```
LIST ADDITIONAL DATA (Y/N)? N
    DATA MANAGEMENT OPTIONS:
        1-LIST DATA
        2-CORRECT DATA
        3-ADD TO DATA
        4-ADD AN INDEPENDENT VARIABLE
        5-DELETE A VARIABLE
        6-PERFORM REGRESSION COMPUTATIONS
        7-STUDY ANOTHER MODEL
        8-STORE DATA ON DISK
        9-WEIGHT DATA
       10-TRANSFORM DATA
       11-QUIT
OPTION? 8
ENTER NAME OF DISK:FILE? B:BRASALES
    DATA MANAGEMENT OPTIONS:
        1-LIST DATA
        2-CORRECT DATA
        3-ADD TO DATA
        4-ADD AN INDEPENDENT VARIABLE
        5-DELETE A VARIABLE
        6-PERFORM REGRESSION COMPUTATIONS
        7-STUDY ANOTHER MODEL
        8-STORE DATA ON DISK
        9-WEIGHT DATA
       10-TRANSFORM DATA
       11-QUIT
```

4.3 Stepwise Regression

We are now ready to perform the regression computations by specifying option 6, PERFORM REGRESSION COMPUTATIONS. At this point, the program provides three options. Options 1 and 2 are the regression techniques described earlier in this chapter. Option 3, DATA MANAGEMENT OPTIONS, returns control to the 11 data management options we have been using.

```
OPTION? 6
   AVAILABLE OPTIONS:
     1-MULTIPLE REGRESSION
     2-STEPWISE MULTIPLE REGRESSION
     3-DATA MANAGEMENT OPTIONS
```

Because we have only one independent variable, stepwise regression is not needed; therefore, we will choose option 1, MULTIPLE REGRESSION. We would also like to see each iteration of the regression computations displayed at our CRT.

```
OPTION? 1
PRINT SOLUTION AT EACH ITERATION (Y/N)? Y

OUTPUT TO APPEAR AT CRT OR PRINTER (C/P)? C

    MULTIPLE REGRESSION

REGRESSION NUMBER 1
   SALES = 8366.667
COEFFICIENT OF DETERMINATION = 0
STD DEVIATION OF ESTIMATE = 315.6369

FINAL SOLUTION

REGRESSION NUMBER 2
   SALES = 4251.95
   +1.213781 ORDERS
COEFFICIENT OF DETERMINATION = .8369923
STD DEVIATION OF ESTIMATE = 142.4777
```

Regression iteration number 1 represents averaging the SALES data. The final solution, regression iteration number 2, is the best fit regression equation using ORDERS to estimate SALES. The coefficient of determination is .837; thus, 83.7% of the variation in SALES can be explained by the ORDERS.

After pressing enter to continue, we are asked if we want to see a comparison of the actual and predicted values for SALES. In addition, we must specify whether the program supplied T-statistics are to be used to calculate the confidence limits on the SALES prediction. After responding to these prompts, the remaining regression results are displayed. Looking at the confidence limits on the predicted values, we can see that the forecast for period seven is 8742.939, and that the lower and upper 95% confidence limits are 8257.523 and 9228.354, respectively. Thus, we are 95% sure that the SALES for period seven will be within this range, and the expected value is 8742.939.

```
PRESS ENTER TO CONTINUE

WANT TO SEE LIMITS ON PREDICTIONS (Y/N)? Y
USE PROGRAM SUPPLIED T-STATISTICS FOR LIMITS (Y/N)? Y
```

Chapter 4—Curve Fitting with Linear Regression

```
ACTUAL VERSUS PREDICTED VALUES FOR SALES
    ACTUAL      PREDICTED    DIFFERENCE    %DIFFERENCE
    8350        8560.872     -210.8721     -2.525414
    8730        8621.561      108.4395      1.242147
    8120        8196.738      -76.73731    -.9450407
    8450        8318.115      131.8848      1.560767
    7900        7893.293        6.70752     8.490531E-02
    8650        8609.422       40.57715      .4691

PRESS ENTER TO CONTINUE?

CONFIDENCE LIMITS ON PREDICTED VALUES

DEGREES OF FREEDOM = 4  T-STATISTIC = 2.776

    LOWER LIMIT    PREDICTED    UPPER LIMIT
     8117.411      8560.872      9004.333
     8166.716      8621.561      9076.405
     7757.031      8196.738      8636.443
     7889.873      8318.115      8746.358
     7376.975     7893.293       8409.61
     8157.076      8609.422      9061.769
     8257.523      8742.939      9228.354
```

Next, we might see how these results would look if they had been displayed on the printer. The following information will appear on the CRT:

```
AVAILABLE OPTIONS:
  1-MULTIPLE REGRESSION
  2-STEPWISE MULTIPLE REGRESSION
  3-DATA MANAGEMENT OPTIONS

OPTION? 1

PRINT SOLUTION AT EACH ITERATION (Y/N)? Y
OUTPUT TO APPEAR AT CRT OR PRINTER (C/P)? P

    MULTIPLE REGRESSION

COMPARISON OF ACTUAL VS. PREDICTED DATA (Y/N)? Y
USE PROGRAM SUPPLIED T-STATISTICS FOR LIMITS (Y/N)? Y
```

Then the regression results would be printed at the printer as follows:

```
    MULTIPLE REGRESSION

REGRESSION NUMBER 1
  SALES = 8366.667
COEFFICIENT OF DETERMINATION = 0
STD DEVIATION OF ESTIMATE = 315.6369
FINAL SOLUTION

REGRESSION NUMBER 2
  SALES = 4251.95
  +1.213781 ORDERS
COEFFICIENT OF DETERMINATION = .8369923
STD DEVIATION OF ESTIMATE = 142.4777
```

ACTUAL	PREDICTED	DIFFERENCE	%DIFFERENCE
8350	8560.872	−210.8721	−2.525414
8730	8621.561	108.4395	1.242147
8120	8196.738	−76.73731	−.9450407
8450	8318.115	131.8848	1.560767
7900	7893.293	6.70752	8.490531E−02
8650	8609.422	40.57715	.4691

CONFIDENCE LIMITS ON PREDICTED VALUES

DEGREES OF FREEDOM = 4   T-STATISTIC = 2.776

LOWER LIMIT	PREDICTED	UPPER LIMIT
8117.411	8560.872	9004.333
8166.716	8621.561	9076.405
7757.031	8196.738	8636.443
7889.873	8318.115	8746.358
7376.975	7893.293	8409.61
8157.076	8609.422	9061.769
8257.523	8742.939	9228.354

Since we have a solution to Example 2, now we will use data management option 11, QUIT, to end program execution.

EXAMPLE 3. Assume that another month has passed at the Brady Company and that sales totaled $8810 for the seventh month. Orders totaling $3760 were on hand at the beginning of the eighth month. Develop a sales forecast for the eighth month using the Multiple and Stepwise Regression Program.

SOLUTION. The data used for the seventh month sales forecast were stored on a disk file named BRASALES. Therefore, it will be easy to use these data to develop a new forecast. We need only to designate that the data be input from a disk and then enter the name of the disk file containing the data. After which, we will input the newly obtained data for the seventh month.

```
RUN
   MULTIPLE AND STEPWISE REGRESSION PROGRAM
LIST PROGRAM OPTIONS (Y/N)? N
HOW ENTER DATA - KEYBOARD OR DISK (K/D)? D
ENTER NAME OF DISK:FILE? B:BRASALES

   DATA MANAGEMENT OPTIONS:
      1-LIST DATA
      2-CORRECT DATA
      3-ADD TO DATA
      4-ADD AN INDEPENDENT VARIABLE
      5-DELETE A VARIABLE
      6-PERFORM REGRESSION COMPUTATIONS
      7-STUDY ANOTHER MODEL
      8-STORE DATA ON DISK
      9-WEIGHT DATA
     10-TRANSFORM DATA
     11-QUIT
```

We need to add the newly obtained data (sales for the seventh month and orders for the eighth month) to the data that was input from the disk file. Using option 3, ADD TO DATA, we can do this.

```
OPTION? 3

FORECAST REQUIRED (Y/N)? Y
NUMBER OF PERIODS TO BE FORECAST? 1

ENTER NUMBER DATA POINTS FOR EACH INDEPENDENT
 VARIABLE INCLUDING POINTS USED FOR FORECAST: 8
ENTER ADDITIONAL DATA FOR INDEPENDENT VARIABLE ORDERS
 VALUE FOR POINT 8? 3760
ENTER ADDITIONAL DATA FOR DEPENDENT VARIABLE SALES
 VALUE FOR POINT 7? 8810

   DATA MANAGEMENT OPTIONS:
     1-LIST DATA
     2-CORRECT DATA
     3-ADD TO DATA
     4-ADD AN INDEPENDENT VARIABLE
     5-DELETE A VARIABLE
     6-PERFORM REGRESSION COMPUTATIONS
     7-STUDY ANOTHER MODEL
     8-STORE DATA ON DISK
     9-WEIGHT DATA
    10-TRANSFORM DATA
    11-QUIT
```

Having added the additional data, we perform the regression computations resulting in a forecast for the eighth month of $8841.23 total sales.

```
OPTION? 6

   AVAILABLE OPTIONS:
     1-MULTIPLE REGRESSION
     2-STEPWISE MULTIPLE REGRESSION
     3-DATA MANAGEMENT OPTIONS

OPTION? 1
PRINT SOLUTION AT EACH ITERATION (Y/N)? N

OUTPUT TO APPEAR AT CRT OR PRINTER (C/P)? C

   MULTIPLE REGRESSION

FINAL SOLUTION

REGRESSION NUMBER 2
  SALES = 4094.04
   +1.262551 ORDERS
COEFFICIENT OF DETERMINATION = .8737094
STD DEVIATION OF ESTIMATE = 129.7577

PRESS ENTER TO CONTINUE?
```

4.3 Stepwise Regression

```
WANT TO SEE LIMITS ON PREDICTIONS (Y/N)? Y
USE PROGRAM SUPPLIED T-STATISTICS FOR LIMITS (Y/N)? Y
ACTUAL VERSUS PREDICTED VALUES FOR SALES
   ACTUAL      PREDICTED     DIFFERENCE     %DIFFERENCE
    8350        8576.096     -226.0957      -2.707733
    8730        8639.222       90.77734      1.039832
    8120        8197.33       -77.33008      -.9523408
    8450        8323.585      126.415        1.496036
    7900        7881.693       18.30762       .231742
    8650        8626.598       23.40235       .2705473
    8810        8765.478       44.52246       .5053628

PRESS ENTER TO CONTINUE?

CONFIDENCE LIMITS ON PREDICTED VALUES

DEGREES OF FREEDOM =5  T-STATISTIC = 2.571

   LOWER LIMIT     PREDICTED     UPPER LIMIT
    8213.783       8576.096       8938.409
    8271.041       8639.222       9007.404
    7826.47        8197.33        8568.191
    7963.923       8323.585       8683.247
    7451.992       7881.693       8311.393
    8259.748       8626.598       8993.448
    8379.863       8765.478       9151.092
    8441.846       8841.231       9240.615

AVAILABLE OPTIONS:
  1-MULTIPLE REGRESSION
  2-STEPWISE MULTIPLE REGRESSION
  3-DATA MANAGEMENT OPTIONS

OPTION? 3
```

Looking back at the comparison of actual versus predicted values, we see that our predictions have been better for the more recent months. Consequently, we might try weighting these latter periods more heavily. When the data are weighted, we should make sure that the average weight is one. Therefore, the sum of the weights should total seven since we will be weighting seven data elements for each variable (ORDERS and SALES for periods 1 through 7). Using option 9, WEIGHT DATA, we can weight the data however we want. The user should note that these weights are retained until one of the following actions occurs: (1) program execution is terminated by specifying option 11, QUIT; (2) weights are changed using option 9, WEIGHT DATA; or (3) option 7, STUDY ANOTHER MODEL is specified.

```
DATA MANAGEMENT OPTIONS:
  1-LIST DATA
  2-CORRECT DATA
  3-ADD TO DATA
  4-ADD AN INDEPENDENT VARIABLE
  5-DELETE A VARIABLE
  6-PERFORM REGRESSION COMPUTATIONS
  7-STUDY ANOTHER MODEL
```

```
        8-STORE DATA ON DISK
        9-WEIGHT DATA
       10-TRANSFORM DATA
       11-QUIT
OPTION? 9
AVERAGE WEIGHT SHOULD BE ONE
ENTER WEIGHT FOR DATA POINT 1?  .5
ENTER WEIGHT FOR DATA POINT 2?  .5
ENTER WEIGHT FOR DATA POINT 3?  1
ENTER WEIGHT FOR DATA POINT 4?  1
ENTER WEIGHT FOR DATA POINT 5?  1
ENTER WEIGHT FOR DATA POINT 6?  1.5
ENTER WEIGHT FOR DATA POINT 7?  1.5

  DATA MANAGEMENT OPTIONS:
     1-LIST DATA
     2-CORRECT DATA
     3-ADD TO DATA
     4-ADD AN INDEPENDENT VARIABLE
     5-DELETE A VARIABLE
     6-PERFORM REGRESSION COMPUTATIONS
     7-STUDY ANOTHER MODEL
     8-STORE DATA ON DISK
     9-WEIGHT DATA
    10-TRANSFORM DATA
    11-QUIT
OPTION? 1
```

Having weighted the data, we might list them to see if any changes have occurred. We will see that nothing has been modified, because the weights are applied during the regression computations. Even then, the original data are maintained unaltered while the weighted data are temporarily stored to be used during the regression computations.

```
ENTER NAME OF VARIABLE IDENTIFYING DATA TO BE LISTED? SALES
LIST OF DATA FOR SALES
  8350
  8730
  8120
  8450
  7900
  8650
  8810
LIST ADDITIONAL DATA (Y/N)? Y
ENTER NAME OF VARIABLE IDENTIFYING DATA TO BE LISTED? ORDERS
LIST OF DATA FOR ORDERS
  3550
  3600
  3250
  3350
```

```
       3000
       3590
       3700
       3760
    LIST ADDITIONAL DATA (Y/N)? N

       DATA MANAGEMENT OPTIONS:
          1-LIST DATA
          2-CORRECT DATA
          3-ADD TO DATA
          4-ADD AN INDEPENDENT VARIABLE
          5-DELETE A VARIABLE
          6-PERFORM REGRESSION COMPUTATIONS
          7-STUDY ANOTHER MODEL
          8-STORE DATA ON DISK
          9-WEIGHT DATA
         10-TRANSFORM DATA
         11-QUIT
```

The regression computations can be performed at this point to see if the data weights have improved the fit of the regression equation. The results show that the coefficient of determination is better and that the prediction for the last four periods is better. Using the new regression equation,

SALES = 4000.563 + 1.293905(ORDERS),

we obtain a sales forecast of $8865.65. This figure is slightly higher than the forecast made using the unweighted data.

```
    OPTION? 6
        1-MULTIPLE REGRESSION
        2-STEPWISE MULTIPLE REGRESSION
        3-DATA MANAGEMENT OPTIONS

    OPTION? 1
    PRINT SOLUTION AT EACH ITERATION (Y/N)? N

    OUTPUT TO APPEAR AT CRT OR PRINTER (C/P)? C

       MULTIPLE REGRESSION

    FINAL SOLUTION

    REGRESSION NUMBER 2
       SALES = 4000.563
       +1.293905 ORDERS
    COEFFICIENT OF DETERMINATION = .924032
    STD DEVIATION OF ESTIMATE = 103.8366

    PRESS ENTER TO CONTINUE?
    WANT TO SEE LIMITS ON PREDICTIONS (Y/N)? Y
    USE PROGRAM SUPPLIED T-STATISTICS FOR LIMITS (Y/N)? Y
    ACTUAL VERSUS PREDICTED VALUES FOR SALES
```

ACTUAL	PREDICTED	DIFFERENCE	%DIFFERENCE
8350	8593.927	−243.9268	−2.921279
8730	8658.622	71.37793	.8176166
8120	8205.755	−85.75488	−1.0566095
8450	8335.146	114.8545	1.359225
7900	7882.279	17.72119	.2243189
8650	8645.684	4.316406	4.990065E−02
8810	8788.012	21.98731	.2495721

PRESS ENTER TO CONTINUE?

CONFIDENCE LIMITS ON PREDICTED VALUES

DEGREES OF FREEDOM =5 T-STATISTIC = 2.571

LOWER LIMIT	PREDICTED	UPPER LIMIT
8304.99	8593.927	8882.864
8365.598	8658.622	8951.646
7908.569	8205.755	8502.941
8046.929	8335.146	8623.362
7539.688	7882.279	8224.87
8353.597	8645.684	8937.771
8482.481	8788.012	9093.544
8550.071	8865.648	9181.224

AVAILABLE OPTIONS:
  1-MULTIPLE REGRESSION
  2-STEPWISE MULTIPLE REGRESSION
  3-DATA MANAGEMENT OPTIONS

At this point, it would be interesting to see if the same results would be obtained using the stepwise multiple regression option. After doing this, you can see that the results are the same. Remember that the data weights are still being applied.

OPTION? 2
PRINT SOLUTION AT EACH ITERATION (Y/N)? N

OUTPUT TO APPEAR AT CRT OR PRINTER (C/P)? C

  STEPWISE MULTIPLE REGRESSION

FINAL SOLUTION

REGRESSION NUMBER 2
  SALES = 4000.563
  +1.293905 ORDERS
COEFFICIENT OF DETERMINATION = .924032
STD DEVIATION OF ESTIMATE = 103.8366

PRESS ENTER TO CONTINUE?

WANT TO SEE LIMITS ON PREDICTIONS (Y/N)? N

ACTUAL	PREDICTED	DIFFERENCE	%DIFFERENCE
8350	8593.927	−243.9268	−2.921279
8730	8658.622	71.37793	.8176166

8120	8205.755	−85.75488	−1.056095
8450	8335.146	114.8545	1.359225
7900	7882.279	17.72119	.2243189
8650	8645.684	4.316406	4.990065E-02
8810	8788.012	21.98731	.2495721

```
AVAILABLE OPTIONS:
  1-MULTIPLE REGRESSION
  2-STEPWISE MULTIPLE REGRESSION
  3-DATA MANAGEMENT OPTIONS

OPTION? 3

DATA MANAGEMENT OPTIONS:
  1-LIST DATA
  2-CORRECT DATA
  3-ADD TO DATA
  4-ADD AN INDEPENDENT VARIABLE
  5-DELETE A VARIABLE
  6-PERFORM REGRESSION COMPUTATIONS
  7-STUDY ANOTHER MODEL
  8-STORE DATA ON DISK
  9-WEIGHT DATA
  10-TRANSFORM DATA
  11-QUIT

OPTION? 11
  END OF PROGRAM
```

EXAMPLE 4. Data have been collected from a laboratory experiment. The rate of growth of a bacteria culture is being studied. The population density in parts per million (D,P/M), as a function of time, is tabulated below:

Days	.5	1.0	1.5	2.0	2.5	3.0	3.5	4.0
D,P/M	.3	.7	2.0	5.4	14.8	40.3	109.7	298.1

Previous work has established that the rate of bacteria growth should conform to an equation such as:

$$\text{density} = ae^{bt}.$$

Use the multiple and stepwise regression program to develop a regression equation that relates the bacteria density to time.

SOLUTION. We can use linear regression to fit a line to this data if the data are first transformed, giving

$$\ln(\text{density}) = \ln(a) + bt$$

Because our regression program can perform a natural log transformation, we will let the computer perform the required transformations.

```
RUN
  *** MULTIPLE AND STEPWISE REGRESSION PROGRAM ***
LIST PROGRAM OPTIONS (Y/N)? N

HOW ENTER DATA - KEYBOARD OR DISK (K/D)? K
ENTER NAME OF DEPENDENT VARIABLE? BACTERIA

ENTER NUMBER OF INDEPENDENT VARIABLES? 1
ENTER NAME OF INDEPENDENT VARIABLE: 1? TIME

FORECAST REQUIRED (Y/N)? N
ENTER NUMBER OF DATA POINTS FOR EACH INDEPENDENT VARIABLE? 8
```

The raw data (untransformed) will be entered. Then using option 10, TRANSFORM DATA, we will transform the set of data named BACTERIA (the dependent variable).

```
8 VALUES REQUIRED FOR INDEPENDENT VARIABLE TIME
VALUE FOR POINT 1?   .5
VALUE FOR POINT 2?   1.
VALUE FOR POINT 3?   1.5
VALUE FOR POINT 4?   2.
VALUE FOR POINT 5?   2.5
VALUE FOR POINT 6?   3.
VALUE FOR POINT 7?   3.5
VALUE FOR POINT 8?   4.

8 VALUES REQUIRED FOR DEPENDENT VARIABLE BACTERIA
VALUE FOR POINT 1?    .3
VALUE FOR POINT 2?    .7
VALUE FOR POINT 3?   2.
VALUE FOR POINT 4?   5.4
VALUE FOR POINT 5?  14.8
VALUE FOR POINT 6?  40.3
VALUE FOR POINT 7? 109.7
VALUE FOR POINT 8? 298.1

        1-LIST DATA
        2-CORRECT DATA
        3-ADD TO DATA
        4-ADD AN INDEPENDENT VARIABLE
        5-DELETE A VARIABLE
        6-PERFORM REGRESSION COMPUTATIONS
        7-STUDY ANOTHER MODEL
        8-STORE DATA ON DISK
        9-WEIGHT DATA
```

```
            10-TRANSFORM DATA
            11-QUIT
    OPTION? 10
      TRANSFORMS AVAILABLE:
        N=NO TRANSFORM
        E=LOG(X) TO BASE E
        P=POLYNOMIAL X^N
    ENTER TRANSFORM CODE FOR VARIABLE TIME? N
    ENTER TRANSFORM CODE FOR VARIABLE BACTERIA? E
```

Having performed the transformation, we will list the data (using option 1) to see the results. You should note that the program has replaced the original data with the transformed data. Therefore, if we were to store these data in a disk file for later use, we should not transform the data again.

```
    DATA MANAGEMENT OPTIONS:
            1-LIST DATA
            2-CORRECT DATA
            3-ADD TO DATA
            4-ADD AN INDEPENDENT VARIABLE
            5-DELETE A VARIABLE
            6-PERFORM REGRESSION COMPUTATIONS
            7-STUDY ANOTHER MODEL
            8-STORE DATA ON DISK
            9-WEIGHT DATA
           10-TRANSFORM DATA
           11-QUIT
    OPTION? 1
    ENTER NAME OF VARIABLE IDENTIFYING DATA TO BE LISTED? BACTERIA
    LIST OF DATA FOR BACTERIA
    -1.203972816467285
     -.3566749691963196
      .6931470632553101
     1.686399102210999
     2.694627285003662
     3.696351528167725
     4.697749137878418
     5.697428703308106

    LIST ADDITIONAL DATA (Y/N)? N

    DATA MANAGEMENT OPTIONS:
            1-LIST DATA
            2-CORRECT DATA
            3-ADD TO DATA
            4-ADD AN INDEPENDENT VARIABLE
            5-DELETE A VARIABLE
            6-PERFORM REGRESSION COMPUTATIONS
            7-STUDY ANOTHER MODEL
            8-STORE DATA ON DISK
```

```
        9-WEIGHT DATA
       10-TRANSFORM DATA
       11-QUIT
```

We are now ready to perform the regression computations. Remember that the regression equation obtained by the program will be in terms of the transformed data:

ln(density)=ln a+bt.

```
OPTION? 6

  AVAILABLE OPTIONS:
    1-MULTIPLE REGRESSION
    2-STEPWISE MULTIPLE REGRESSION
    3-DATA MANAGEMENT OPTIONS

OPTION? 1
PRINT SOLUTION AT EACH ITERATION (Y/N)? Y

OUTPUT TO APPEAR AT CRT OR PRINTER (C/P)? C

     MULTIPLE REGRESSION

REGRESSION NUMBER 1
  BACTERIA = 2.200632
COEFFICIENT OF DETERMINATION = 0
STD DEVIATION OF ESTIMATE = 2.438164

FINAL SOLUTION

REGRESSION NUMBER 2
  BACTERIA = −2.277928
  +1.990471 TIME
COEFFICIENT OF DETERMINATION = .9997174
STD DEVIATION OF ESTIMATE = 4.427643E-02

PRESS ENTER TO CONTINUE?

WANT TO SEE LIMITS ON PREDICTIONS (Y/N)? Y
USE PROGRAM SUPPLIED T-STATISTICS FOR LIMITS (Y/N)? Y
ACTUAL VERSUS PREDICTED VALUES FOR BACTERIA
    ACTUAL      PREDICTED     DIFFERENCE      %DIFFERENCE
   −1.203973    −1.282692     7.871926E-02    −6.538292
   −.356675     −.2874566    −6.921834E-02    19.40656
    .6931471     .7077788     1.463169E-02    −2.110907
    1.686399    1.703014     −1.661503E-02    −.9852373
    2.694627    2.69825      −3.622294E-03    −.1344265
    3.696352    3.693485      2.866507E-03    7.754962E-02
    4.697749    4.68872       9.028911E-03    .1921966
    5.697429    5.683956      1.347303E-02    .2364757

PRESS ENTER TO CONTINUE?

CONFIDENCE LIMITS ON PREDICTED VALUES

DEGREES OF FREEDOM = 6  T-STATISTIC = 2.447
```

```
      -1.411648      -1.282692      -1.53736
      -.4097376      -.2874566      -.1651757
       .5901578       .7077788       .8253997
      1.587794       1.703014       1.818234
      2.583029       2.69825        2.81347
      3.575864       3.693485       3.811106
      4.566439       4.68872        4.811001
      5.555          5.683956       5.812912
   AVAILABLE OPTIONS:
       1-MULTIPLE REGRESSION
       2-STEPWISE MULTIPLE REGRESSION
       3-DATA MANAGEMENT OPTIONS
OPTION? 3

DATA MANAGEMENT OPTIONS:
       1-LIST DATA
       2-CORRECT DATA
       3-ADD TO DATA
       4-ADD AN INDEPENDENT VARIABLE
       5-DELETE A VARIABLE
       6-PERFORM REGRESSION COMPUTATIONS
       7-STUDY ANOTHER MODEL
       8-STORE DATA ON DISK
       9-WEIGHT DATA
      10-TRANSFORM DATA
      11-QUIT
OPTION? 11
       END OF PROGRAM
```

Looking at the coefficient of determination (.9997) for the final solution, we can see that a good fit was obtained. The regression equation is

BACTERIA = −2.277928 + 1.990471 (TIME)

Assume that TIME=3 days. The predicted density of the bacteria culture expressed as a natural log would be:

ln(BACTERIA) = −2.277928 + (1.990471) (3)
              = −2.277928 + 5.971413
              = 3.693485

Removing the log from the above expression, the predicted bacteria density is:

D,P/M = exp(3.693485)
       = 40.18464

This figure is very close to the observed value of 40.3. We can write the above regression equation without the natural log. Since

$e^{-2.277928}$ = .1024964,

then

D,P/M = .1024964 $e^{(1.990471)(TIME)}$

Consequently, we have developed an equation to predict the density of the bacteria culture.

EXAMPLE 5. Some experimental data have been collected and presented below in tabular form. The dependent variable is Y and X is the independent variable.

X	0	1	2	3	4	5	6	7	8	9
Y	50	65	80	100	130	160	200	245	290	345

Determine if an equation of the form:

$$Y = A + B_1 X + B_2 X^2 + B_3 X^3$$

will provide a good fit.

SOLUTION. The multiple and stepwise regression program will be utilized. Three independent variables will be used:

X1 will represent X_1
X2 will represent X^2, and
X3 will represent X^3.

```
RUN
   *** MULTIPLE AND STEPWISE REGRESSION PROGRAM ***
LIST PROGRAM OPTIONS (Y/N)? N

HOW ENTER DATA - KEYBOARD OR DISK (K/D)? K

ENTER NAME OF DEPENDENT VARIABLE? Y

ENTER NUMBER OF INDEPENDENT VARIABLES? 3
ENTER NAME OF INDEPENDENT VARIABLE: 1? X1
ENTER NAME OF INDEPENDENT VARIABLE: 2? X2
ENTER NAME OF INDEPENDENT VARIABLE: 3? X3

FORECAST REQUIRED (Y/N)? N
ENTER NUMBER OF DATA POINTS FOR EACH INDEPENDENT VARIABLE? 10

ENTER DATA
```

Since our program can perform a transformation of X, we will let the computer perform the required calculations. Therefore, the values entered for X1, X2, and X3 will be the same.

```
   10 VALUES REQUIRED FOR INDEPENDENT VARIABLE X1
   VALUE FOR POINT   1? 0
   VALUE FOR POINT   2? 1
   VALUE FOR POINT   3? 2
   VALUE FOR POINT   4? 3
   VALUE FOR POINT   5? 4
   VALUE FOR POINT   6? 5
   VALUE FOR POINT   7? 6
   VALUE FOR POINT   8? 7
```

```
VALUE FOR POINT  9? 8
VALUE FOR POINT 10? 9

10 VALUES REQUIRED FOR INDEPENDENT VARIABLE X2
VALUE FOR POINT  1? 0
VALUE FOR POINT  2? 1
VALUE FOR POINT  3? 2
VALUE FOR POINT  4? 3
VALUE FOR POINT  5? 4
VALUE FOR POINT  6? 5
VALUE FOR POINT  7? 6
VALUE FOR POINT  8? 7
VALUE FOR POINT  9? 8
VALUE FOR POINT 10? 9

10 VALUES REQUIRED FOR INDEPENDENT VARIABLE X3
VALUE FOR POINT  1? 0
VALUE FOR POINT  2? 1
VALUE FOR POINT  3? 2
VALUE FOR POINT  4? 3
VALUE FOR POINT  5? 4
VALUE FOR POINT  6? 5
VALUE FOR POINT  7? 6
VALUE FOR POINT  8? 7
VALUE FOR POINT  9? 8
VALUE FOR POINT 10? 9

10 VALUES REQUIRED FOR DEPENDENT VARIABLE Y
VALUE FOR POINT  1?  50
VALUE FOR POINT  2?  65
VALUE FOR POINT  3?  80
VALUE FOR POINT  4? 100
VALUE FOR POINT  5? 130
VALUE FOR POINT  6? 160
VALUE FOR POINT  7? 200
VALUE FOR POINT  8? 245
VALUE FOR POINT  9? 290
VALUE FOR POINT 10? 345

   DATA MANAGEMENT OPTIONS
      1-LIST DATA
      2-CORRECT DATA
      3-ADD TO DATA
      4-ADD AN INDEPENDENT VARIABLE
      5-DELETE A VARIABLE
      6-PERFORM REGRESSION COMPUTATIONS
      7-STUDY ANOTHER MODEL
      8-STORE DATA ON DISK
```

```
       9-WEIGHT DATA
      10-TRANSFORM DATA
      11-QUIT
```

Data for variables X2 and X3 will now be transformed.

```
OPTION? 10
   TRANSFORMS AVAILABLE:
     N=NO TRANSFORMS
     E=LOG(X) TO BASE E
     P=POLYNOMIAL X^N

ENTER TRANSFORM CODE FOR VARIABLE X1? N
ENTER TRANSFORM CODE FOR VARIABLE X2? P
ENTER VALUE FOR EXPONENT (X^N)? 2
ENTER TRANSFORM CODE FOR VARIABLE X3? P
ENTER VALUE FOR EXPONENT (X^N)? 3
ENTER TRANSFORM CODE FOR VARIABLE Y? N

   DATA MANAGEMENT OPTIONS:
      1-LIST DATA
      2-CORRECT DATA
      3-ADD TO DATA
      4-ADD AN INDEPENDENT VARIABLE
      5-DELETE A VARIABLE
      6-PERFORM REGRESSION COMPUTATIONS
      7-STUDY ANOTHER MODEL
      8-STORE DATA ON DISK
      9-WEIGHT DATA
     10-TRANSFORM DATA
     11-QUIT

OPTION? 6

   AVAILABLE OPTIONS:
     1-MULTIPLE REGRESSION
     2-STEPWISE MULTIPLE REGRESSION
     3-DATA MANAGEMENT OPTIONS
```

We will first try multiple regression. This time the output should be displayed at the printer.

```
OPTION? 1
PRINT SOLUTION AT EACH ITERATION (Y/N)? Y

OUTPUT TO APPEAR AT CRT OR PRINTER (C/P)? P

     MULTIPLE REGRESSION

REGRESSION NUMBER 1
   Y = 166.5
COEFFICIENT OF DETERMINATION = 0
STD DEVIATION OF ESTIMATE = 100.8588
```

REGRESSION NUMBER 2
  Y = 19.63636
  + 32.63637 X1
COEFFICIENT OF DETERMINATION = .9598144
STD DEVIATION OF ESTIMATE = 21.44497

REGRESSION NUMBER 3
  Y = 51.22728
  + 8.943182 X1
  + 2.632576 X2
COEFFICIENT OF DETERMINATION = .9997836
STD DEVIATION OF ESTIMATE = 1.682287

FINAL SOLUTION

REGRESSION NUMBER 4
  Y = 51.2028
  + 8.987956 X1
  + 2.619464 X2
  + 9.71251E-04 X3
COEFFICIENT OF DETERMINATION = .9997836
STD DEVIATION OF ESTIMATE = 1.816944

ACTUAL	PREDICTED	DIFFERENCE	%DIFFERENCE
50	51.2028	−1.202801	−2.405602
65	62.8112	2.188805	3.367392
80	79.66434	.3356629	.4195786
100	101.7681	−1.768066	−1.768066
130	129.1282	8.717804	.6706003
160	161.7506	−1.75058	−1.094112
200	199.641	.358963	.1794815
245	242.8054	2.194626	.8957656
290	291.2494	−1.24942	−.4308346
345	344.979	2.096558E-02	6.076979E-03

CONFIDENCE LIMITS ON PREDICTED VALUES

DEGREES OF FREEDOM = 6 T-STATISTIC = 2.447

LOWER LIMIT	PREDICTED	UPPER LIMIT
45.29852	51.2028	57.20709
57.73873	62.8112	67.88366
74.5444	79.66434	84.78428
96.68426	101.7681	106.8519
124.1753	129.1282	134.0812
156.7976	161.7506	166.7036
194.5572	199.641	204.7249
237.6855	242.8054	247.9253
286.177	291.2494	296.3219
338.9748	344.979	350.9833

Look at the final solution and regression number 3. There is no difference in the coefficient of determination. The last variable added to the equation, X3, may not be needed. We will try the stepwise multiple regression option to see if this reasoning is correct.

```
    AVAILABLE OPTIONS:
      1-MULTIPLE REGRESSION
      2-STEPWISE MULTIPLE REGRESSION
      3-DATA MANAGEMENT OPTIONS
OPTION? 2
PRINT SOLUTION AT EACH ITERATION (Y/N)? Y

OUTPUT TO APPEAR AT CRT OR PRINTER (C/P)? P
     STEPWISE MULTIPLE REGRESSION

REGRESSION NUMBER 1
   Y = 166.5
COEFFICIENT OF DETERMINATION = 0
STD DEVIATION OF ESTIMATE = 100.8588

REGRESSION NUMBER 2
   Y = 65.2253
   + 3.553498 X2
COEFFICIENT OF DETERMINATION = .994506
STD DEVIATION OF ESTIMATE = 7.929282

FINAL SOLUTION

REGRESSION NUMBER 3
   Y = 51.22728
   + 8.943182 X1
   + 2.632576 X2
COEFFICIENT OF DETERMINATION = .9997836
STD DEVIATION OF ESTIMATE = 1.682287
```

ACTUAL	PREDICTED	DIFFERENCE	%DIFFERENCE
50	51.22728	−1.227276	−2.454552
65	62.80304	2.196964	3.379945
80	79.64395	.3560562	.4450703
100	101.75	−1.750008	−1.750008
130	129.1212	.8787842	.6759878
160	161.7576	−1.757584	−1.09849
200	199.6591	.3409119	.1704559
245	242.8258	2.17424	.887445
290	291.2576	−1.257599	−.4336548
345	344.9546	4.544068E-02	1.317121E-02

CONFIDENCE LIMITS ON PREDICTED VALUES

DEGREES OF FREEDOM = 7 T-STATISTIC = 2.365

LOWER LIMIT	PREDICTED	UPPER LIMIT
46.16618	51.22728	56.28837
58.30389	62.80304	67.30219
75.31597	79.64395	83.97191

97.39992	101.75	106.1001
124.7191	129.1212	133.5234
157.3554	161.7576	166.1597
195.309	199.6591	204.0092
238.4978	242.8258	247.1537
286.7585	291.2576	295.7568
339.8935	344.9546	350.0157

We were right. The final results of the stepwise regression computations left X3 out of the regression equation. Remember that the stepwise procedure will not include an independent variable in the regression unless it significantly reduces the variance of the results. In most cases, it is not so obvious when a variable should be included in the final equation. Consequently, a stepwise regression program can be very useful.

EXAMPLE 6. Since the stepwise regression procedure will not include variables in the final solution that are not significant, the temptation is to use only this procedure. However, there are times when the multiple regression procedure can produce better results. Consider the following data, where Y is the dependent variable.

X	1	2	3	4	5	6	7	8
Y	2	9	14	16	19	18	13	6

Determine if an equation of the form

$$Y = A + B_1 X + B_2 X^2$$

will provide a good fit to this data.

SOLUTION. We will show only the results after having entered the data. The data associated with X2 was transformed, and then the stepwise regression option was specified.

```
OPTION? 10
  TRANSFORMS AVAILABLE:
    N=NO TRANSFORM
    E=LOG(X) TO BASE E
    P=POLYNOMIAL X^N
  ENTER TRANSFORM CODE FOR VARIABLE X1? N
  ENTER TRANSFORM CODE FOR VARIABLE X2? P
  ENTER VALUE FOR EXPONENT (X^N)? 2
  ENTER TRANSFORM CODE FOR VARIABLE Y? N

  DATA MANAGEMENT OPTIONS:
       1-LIST DATA
       2-CORRECT DATA
       3-ADD TO DATA
       4-ADD AN INDEPENDENT VARIABLE
       5-DELETE A VARIABLE
       6-PERFORM REGRESSION COMPUTATIONS
       7-STUDY ANOTHER MODEL
       8-STORE DATA ON DISK
```

```
            9-WEIGHT DATA
           10-TRANSFORM DATA
           11-QUIT

   OPTION? 6

      AVAILABLE OPTIONS:
        1-MULTIPLE REGRESSION
        2-STEPWISE MULTIPLE REGRESSION
        3-DATA MANAGEMENT OPTIONS

   OPTION? 2
   PRINT SOLUTION AT EACH ITERATION (Y/N)? Y

   OUTPUT TO APPEAR AT CRT OR PRINTER (C/P)? C

        STEPWISE MULTIPLE REGRESSION

   FINAL SOLUTION

   REGRESSION NUMBER 1
      Y = 12.125
   COEFFICIENT OF DETERMINATION = 0
   STD DEVIATION OF ESTIMATE = 5.986593

   PRESS ENTER TO CONTINUE

   WANT TO SEE LIMITS ON PREDICTIONS (Y/N)? Y
   USE PROGRAM SUPPLIED T-STATISTICS FOR LIMITS (Y/N)? Y
   ACTUAL VERSUS PREDICTED VALUES FOR Y
        ACTUAL       PREDICTED      DIFFERENCE      %DIFFERENCE
          2            12.125         -10.125         -506.25
          9            12.125          -3.125          -34.72222
         14            12.125           1.875           13.39286
         16            12.125           3.875           24.21875
         19            12.125           6.875           36.18421
         18            12.125           5.875           32.63889
         13            12.125            .875            6.730769
          6            12.125          -6.125         -102.0833

   PRESS ENTER TO CONTINUE?

   CONFIDENCE LIMITS ON PREDICTED VALUES

   DEGREES OF FREEDOM = 7  T-STATISTIC = 2.365

        -2.892135      12.125         27.14214
        -2.892135      12.125         27.14214
        -2.892135      12.125         27.14214
        -2.892135      12.125         27.14214
        -2.892135      12.125         27.14214
        -2.892135      12.125         27.14214
        -2.892135      12.125         27.14214
        -2.892135      12.125         27.14214
```

4.3 Stepwise Regression

The results show that there is no correlation. Consequently, the average value for the dependent variable is used to predict any other value for Y. Now, we will try the multiple regression option.

```
    AVAILABLE OPTIONS:
      1-MULTIPLE REGRESSION
      2-STEPWISE MULTIPLE REGRESSION
      3-DATA MANAGEMENT OPTIONS
 OPTION? 1
 PRINT SOLUTION AT EACH ITERATION (Y/N)? Y

 OUTPUT TO APPEAR AT CRT OR PRINTER (C/P)? C

 REGRESSION NUMBER 1
    Y = 12.125
 COEFFICIENT OF DETERMINATION = 0
 STD DEVIATION OF ESTIMATE = 5.986593

 REGRESSION NUMBER 2
    Y = 8.75
     + 1.75 X1
 COEFFICIENT OF DETERMINATION = 9.417029E-02
 STD DEVIATION OF ESTIMATE = 6.154267

 FINAL SOLUTION

 REGRESSION NUMBER 3
    Y = -8.482143
     +  11.08929 X1
     +  -1.14881 X2

 COEFFICIENT OF DETERMINATION = .9779581
 STD DEVIATION OF ESTIMATE = 1.051643

 PRESS ENTER TO CONTINUE?

 WANT TO SEE LIMITS ON PREDICTIONS (Y/N)? Y
 USE PROGRAM SUPPLIED T-STATISTICS FOR LIMITS (Y/N)? Y
 ACTUAL VERSUS PREDICTED VALUES FOR?

     ACTUAL      PREDICTED       DIFFERENCE      %DIFFERENCE
       2         1.458333         .5416671        27.08336
       9         9.101191        -.1011906        -1.12434
      14        14.44643         -.4464293        -3.18878
      16        17.49405        -1.494047         -9.337795
      19        18.24405          .7559528         3.978699
      18        16.69643         1.30357           7.242055
      13        12.85119          .1488075         1.144673
       6         6.708336        -.7083359        -11.8056

 PRESS ENTER TO CONTINUE?

 CONFIDENCE LIMITS ON PREDICTED VALUES

 DEGREES OF FREEDOM = 5    T-STATISTIC = 2.571
```

−2.075587	1.458333	4.992253
6.042504	9.101191	12.15988
11.44519	14.44643	17.44767
14.43536	17.49405	20.55273
15.18536	18.24405	21.30273
13.69519	16.69643	19.69767
9.792506	12.85119	15.90988
3.164417	6.708336	10.24226

Notice what happened. Using X1 and X2 together results in a regression equation having a very good fit. The coefficient of determination has a value of .9779. However, using X1 or X2 alone would result in equations having coefficients of determination less than .1.

EXAMPLE 7. One assumption made in linear regression is that the independent variables are independent of each other. If one of the independent variables can be expressed as a linear combination of the others, the multiple regression computations cannot be completed. However, the stepwise regression computations can be performed because this procedure will not consider dependent variables. To illustrate this, consider the following data where Y is the dependent variable and X2 is the same as X1 (X1 and X2 are not independent):

X1	1	2	3	4	5	6	7	8
X1	1	2	3	4	5	6	7	8
Y	10	20	30	40	50	60	70	80

Use the multiple and stepwise regression program to develop a regression equation using this data.

SOLUTION. Only the regression computation results will be shown since we understand how to input the data.

```
OPTION? 6

   AVAILABLE OPTIONS:
     1-MULTIPLE REGRESSION
     2-STEPWISE MULTIPLE REGRESSION
     3-DATA MANAGEMENT OPTIONS

OPTION? 1
PRINT SOLUTION AT EACH ITERATION (Y/N)? Y

OUTPUT TO APPEAR AT CRT OR PRINTER (C/P)? C

REGRESSION NUMBER 1
   Y = 45
COEFFICIENT OF DETERMINATION = 0
STD DEVIATION OF ESTIMATE = 24.4949

REGRESSION NUMBER 2
   Y = 0
   + 10 X1
COEFFICIENT OF DETERMINATION = 1
STD DEVIATION OF ESTIMATE = 0
```

FINAL SOLUTION

SOLUTION CANNOT BE FOUND USING MULTIPLE REGRESSION. VARIABLE X1 IS A
LINEAR COMBINATION OF THE OTHER VARIABLES. TRY THE STEPWISE
REGRESSION OPTION.

   AVAILABLE OPTIONS:
     1-MULTIPLE REGRESSION
     2-STEPWISE MULTIPLE REGRESSION
     3-DATA MANAGEMENT OPTIONS

Note the message which appeared. The multiple regression procedure could not find a final solution because of the dependence between X1 and X2. We will now try the stepwise regression option.

OPTION? 2
PRINT SOLUTION AT EACH ITERATION (Y/N)? Y

OUTPUT TO APPEAR AT CRT OR PRINTER (C/P)? C

    STEPWISE MULTIPLE REGRESSION

REGRESSION NUMBER 1
  Y = 45
COEFFICIENT OF DETERMINATION = 0
STD DEVIATION OF ESTIMATE = 24.4949

FINAL SOLUTION

REGRESSION NUMBER 2
  Y = 0
  + 10 X1
COEFFICIENT OF DETERMINATION = 1
STD DEVIATION OF ESTIMATE = 0

PRESS ENTER TO CONTINUE?

WANT TO SEE LIMITS ON PREDICTIONS (Y/N)? Y
USE PROGRAM SUPPLIED T-STATISTICS FOR LIMITS (Y/N)? Y
ACTUAL VERSUS PREDICTED VALUES FOR? Y

ACTUAL	PREDICTED	DIFFERENCE	%DIFFERENCE
10	10	0	0
20	20	0	0
30	30	0	0
40	40	0	0
50	50	0	0
60	60	0	0
70	70	0	0
80	80	0	0

PRESS ENTER TO CONTINUE

CONFIDENCE LIMITS ON PREDICTED VALUES

```
DEGREES OF FREEDOM =  6  T-STATISTIC = 2.447
   LOWER LIMIT    PREDICTED    UPPER LIMIT
       10            10            10
       20            20            20
       30            30            30
       40            40            40
       50            50            50
       60            60            60
       70            70            70
       80            80            80
```

The results confirm the above remarks. A solution is found by eliminating X2 from the final results.

The above examples illustrate how to use the Multiple and Stepwise Regression Program. These examples also illustrate why it is good to have both procedures (multiple linear regression and a stepwise linear regression) available.

4.4 Summary

Linear regression, a widely used curve fitting technique, was discussed in this chapter. Simple linear regression was first presented and a program that uses this technique was provided. Next, stepwise regression was discussed. Several examples were used to explain a program that incorporates two techniques: multiple linear regression and stepwise linear regression. Curve fitting is a common task; as a result, we made these programs easy to use by including data management capabilities within the programs. Space did not permit an in depth discussion of applied regression analysis; however, reference [2] provides an excellent discussion of this topic.

References

Draper, N.R., and H. Smith, *Applied Regression Analysis*, John Wiley, New York, 1966.

Ralston, Anthony and Herbert S. Wilf, *Mathematical Methods for Digital Computers Volume 1*, John Wiley, New York, 1964.

Exercises

1. A series of experiments have been performed applying various amounts of a reagent to a chemical solution to obtain a precipitate of a particular substance. The following data have been collected:

Participate	Reagent
20.8	6.7
15.2	9.2
16.4	4.3
22.5	8.1
18.9	7.9
21.6	7.1

   Fit a straight line to this data using the reagent as the independent variable.

2. A tire manufacturer has been testing a new tire design. One test involves the stopping distance (reaction plus braking distance) of a 2400 pound automobile at various speeds. From this test, some data have been collected:

Stopping Distance (Ft)	Miles Per Hour
41	15
53	25
88	35
130	45
192	55
268	65

Fit a curve to this data, then estimate the stopping distance at 60 miles per hour.

3. It is thought that the salary of an engineer or scientist is related to the number of years of experience. To test this hypothesis, the following data was collected from a random sample of employees at one company.

Annual Salary	Number of Years of Experience
$25,000	6
17,000	1
35,000	10
21,000	4
20,000	2
30,000	8

Estimate the relationship of annual salary to years of experience.

4. In the production of semiconductors, the average yield (percent of good devices) per lot increases as more lots are produced. During the past year a firm started producing a new memory device. Data describing the average yield per lot has been collected:

Average Yield Per Lot (%)	Cumulative Number of Lots Produced
20	10
36	20
49	40
60	80
68	160
74	320

Using linear regression, predict what the average yield will be for lot number 400.

5. The viscosity of a liquid varies with the temperature. Fit a curve to the following data:

Viscosity	Temperature
252	5
228	15
204	25
188	35
176	45
160	55

6. A series of tests were made and the following data was collected:

Y	X1	X2
34	32	34.3
78	36	53.4
13	30	16.1
55	33	23.9
48	31	42.6
24	27	27.1
24	29	35.3

Develop a regression equation using this data.

7. Using the following data, develop a least squares line:

Y	X1	X2	X3
71	53	25	15
57	44	30	11
43	36	40	7
89	65	15	19
32	32	50	5
66	49	27	13
54	41	32	10
72	55	20	16
38	34	45	6

Are all three independent variables needed in the regression equation?

8. Specialty Engineering is a very successful company. Historical annual earnings per share and corresponding end-of-year stock prices are given below. Management estimates that this year's earnings will be $.83 per share. Using this data, you are to predict the end-of-year stock price.

Year	Stock Price	Earnings per Share
1	$1.52	$.02
2	2.07	.04
3	1.24	.02
4	3.25	.06
5	7.72	.15
6	12.17	.37
7	20.53	.54
8	31.41	.65
9	52.98	.96

5

Solving Simultaneous Linear Equations

A set of simultaneous linear equations represents a set of conditions that must be satisfied in a particular situation. The solution is a set of variables, $x_1, x_2, \ldots, x_n$, whose values satisfy all equations simultaneously. The equations may be linear or nonlinear, although we will only examine the linear case.

We present both analytic and numerical solution procedures in this chapter. The analytic procedures, matrix inverse method and Gauss-Jordan method, are based on matrix operations which were developed in Chapter 3. A numerical iterative procedure, the Gauss-Seidel Method, will also be presented.

5.1 Simultaneous Linear Equations—General Form

$a_{11} x_1 + a_{12} x_2 + \ldots + a_{1n} x_n = b_1$

$a_{21} x + a_{22} x_2 + \ldots + a_{2n} x_n = b_2$

$\qquad \cdot \qquad\qquad\qquad \cdot \quad \cdot$

$\qquad \cdot \qquad\qquad\qquad \cdot \quad \cdot$

$\qquad \cdot \qquad\qquad\qquad \cdot \quad \cdot$

$a_{m1} x + a_{m2} x_2 + \ldots + a_{mn} x_n = b_m$

This can be represented in matrix form as:

$A\vec{x} = \vec{b}$

where

$$A = \begin{bmatrix} a_{11} & a_{12} & \cdots & a_{1n} \\ a_{21} & a_{22} & \cdots & a_{2n} \\ \cdot & & & \cdot \\ \cdot & & & \cdot \\ \cdot & & & \cdot \\ a_{m1} & a_{m2} & \cdots & a_{mn} \end{bmatrix} \quad \vec{x} = \begin{bmatrix} x_1 \\ x_2 \\ \cdot \\ \cdot \\ \cdot \\ x_n \end{bmatrix} \quad \vec{b} = \begin{bmatrix} b_1 \\ b_2 \\ \cdot \\ \cdot \\ \cdot \\ b_m \end{bmatrix}$$

If $m \neq n$, that is if the number of equations is not equal to the number of unknowns, then several situations could exist. There could be no solution to the set of equations, a unique solution, or an infinite number of solutions. One or more equations could also be "redundant," or implied by other equations. The determination of these conditions is based upon linear algebra concepts which we will not discuss here. Our discussion will be limited to the case $m=n$. If you want information about the $m \neq n$ case, see Schmidt and Davis (1981) or Hadley (1961).

Another interesting situation occurs when all the b_i, $i=1,\ldots,n$, are zero. In this case, the set of equations is said to be homogeneous. An obvious solution in this instance, called the trivial solution, is $\vec{x}=\vec{0}$. This is generally not the solution of interest. We will not examine the special case of homogeneous equations here, but you can study it further by referring to Hadley (1961).

The analytic procedures presented in this chapter use matrix concepts that were introduced in Chapter 3. You may wish to review that material before continuing in this chapter. A numerical procedure, the Gauss-Seidel method, will also be presented. Although not appropriate in all circumstances, this is a useful alternative to analytic techniques.

5.2 Matrix Inverse Method

By examining the matrix representation of the set of equations, we see that the solution for $\vec{x}$ can be found by premultiplying the right hand-side vector $\vec{b}$ by the inverse of the matrix A.

$$A\vec{x}=\vec{b}$$
$$A^{-1}A\vec{x}=A^{-1}\vec{b}$$
$$I\vec{x}=A^{-1}\vec{b}$$
$$\vec{x}=A^{-1}\vec{b}$$

The following example illustrates the solution procedure.

EXAMPLE 1. Solve the following set of simultaneous linear equations for $\vec{x}$.

$3x_1 + 2x_2 + 2x_3 = 8$

$x_2 + x_3 = 4$

$x_1 + 3x_3 = 5$

5.2 Matrix Inverse Method

SOLUTION. First, let's specify the matrix A and vector $\vec{b}$.

$$A = \begin{bmatrix} 3 & 2 & 2 \\ 0 & 1 & 1 \\ 1 & 0 & 3 \end{bmatrix} \quad \vec{b} = \begin{bmatrix} 8 \\ 4 \\ 5 \end{bmatrix}$$

We find A^{-1} to be

$$A^{-1} = \begin{bmatrix} 1/3 & -2/3 & 0 \\ 1/9 & 7/9 & -1/3 \\ -1/9 & 2/9 & 1/3 \end{bmatrix}$$

So,

$$\vec{x} = A^{-1}\vec{b} = \begin{bmatrix} 1/3 & -2/3 & 0 \\ 1/9 & 7/9 & -1/3 \\ -1/9 & 2/9 & 1/3 \end{bmatrix} \begin{bmatrix} 8 \\ 4 \\ 5 \end{bmatrix} = \begin{bmatrix} 0 \\ 7/3 \\ 5/3 \end{bmatrix}$$

or,

$$x_1 = 0 \quad x_2 = 7/3 \quad x_3 = 5/3$$

We presented subroutines in Chapter 3 for the matrix inverse operation and matrix multiplication. When used in succession, they can be employed to solve for the values of x_1, x_2, . . ., x_n. A subroutine to utilize these subroutines is given in Figure 5.1. A main program to read in the set of equations is presented in Figure 5.2.

```
18000 REM SUBROUTINE FOR SOLUTION USING MATRIX INVERSE
18010 REM This subroutine uses the matrix inverse approach
18020 REM to solve a set of simultaneous linear equations.
18030 REM This should be called from a main program which
18040 REM provides the coefficient matrix in the array A,
18050 REM the right hand side vector in array B, and the
18060 REM number of equations in IE and number of variables
18070 REM in IV (these numbers should be the same).
18080 DIM A1(IE,IV),B1(IE,1) 'Set dimensions for use in other
                              subroutines
18090 IROW=IE
18100 ICOL=IV
18110 JROW=IE
18120 JCOL=1
18130 INV=1              'Set inverse flag for use in subroutines
18140 FOR I=1 TO IE
18150 FOR J=1 TO IV
```

```
18160 A1(I,J)=A(I,J)          'Set array A1 values
18170 NEXT J,I
18180 GOSUB 18850             'Makes A1 upper triangular
18190 IF DET=0 THEN PRINT "EQUATIONS ARE NOT INDEPENDENT"
18200 IF DET=0 THEN GOTO 18350
18210 GOSUB 19410             'Makes A1 lower triangular
18220 FOR I=1 TO IE
18230 B1(I,1)=B(I)            'Set array B1 values
18240 FOR J=1 TO IV
18250 A1(I,J)=E1(I,J+IV)      'Reset A1 as the inverse of the A matrix
18260 NEXT J,I
18270 ICOL=IV                 'Reset ICOL to its original value
18280 GOSUB 13750             'From MATOP.BAS; multiply A1 times B1
18290 REM AB1 contains the solutions for the X values
18300 REM Print the results
18310 PRINT "RESULTS USING THE MATRIX INVERSION TECHNIQUE"
18320 FOR I=1 TO IV
18330 PRINT "X(";I;")= ";AB1(I,1)
18340 NEXT I
18350 RETURN
```

Figure 5.1—solving simultaneous equations, matrix inverse subroutine

```
100 REM MAIN PROGRAM TO READ IN EQUATIONS FOR
110 REM simultaneous solution.
120 REM The number of equations should equal the
130 REM number of variables
140 DEFINT I,J
150 INPUT "INPUT THE NUMBER OF VARIABLES";IV
160 INPUT "INPUT THE NUMBER OF EQUATIONS";IE
170 IF IV=IE THEN GOTO 200
180 PRINT "NUMBER OF EQUATIONS<>NUMBER OF VARIABLES: RE-ENTER"
190 GOTO 150
200 DIM B(IE),A(IE,IV)
210 PRINT:PRINT
220 PRINT "FOR EACH EQUATION, INPUT THE FOLLOWING INFORMATION"
230 PRINT
240 FOR I=1 TO IE
250 PRINT "FOR EQUATION ";I
260 INPUT "ENTER THE RIGHT HAND SIDE VALUE";B(I)
270 PRINT
280 FOR J=1 TO IV
290 PRINT "ENTER THE COEFFICIENT OF VARIABLE ";J
300 INPUT A(I,J)
310 NEXT J,I
320 REM At this time, use appropriate procedure to
330 REM solve the set of simultaneous equations
340 GOSUB 18000                    'Matrix inverse
350 END
```

Figure 5.2—matrix inverse main program

5.2 Matrix Inverse Method

EXAMPLE 2. Use the main program in Figure 5.2 to solve the following set of simultaneous linear equations.

$4x_1 + 4x_2 + x_3 = 10$

$x_1 + 2x_2 = 8$

$x_1 + x_2 + x_3 = 9$

SOLUTION.

$x_1 = -7.33 \quad x_2 = 7.676 \quad x_3 = 8.67$

Finally, you should note that this procedure works only if the inverse matrix can be found. If not, then a unique solution does not exist and this method cannot be used.

5.3 Gauss-Jordan Method

Another analytic procedure is the Gauss-Jordan method. This is very closely related to the Gaussian elimination process for finding the inverse of a matrix. We will again use elementary row operations, but the inverse matrix will not be found.

The advantage of this method over the inverse method is that if the inverse does not exist, a determination can still be made regarding the status of the solution. Let us expand upon this for a moment. If the inverse of the coefficient matrix cannot be found, then one of two cases holds; either there exists no solution or there are an infinite number of solutions. The existing situation is dependent on the rank of the coefficient matrix and the rank of the matrix created by augmenting the coefficient matrix with the right-hand side vector. The specifics of matrix "rank" and its relationship to the existence of solutions may be found in Hadley (1961), Schmidt (1974), and Stewart (1973). We will not concern ourselves with the theory here, but we will use the concepts in determining the status of the solution.

The Gauss-Jordan method begins by augmenting the coefficient matrix with the right-hand side vector. From the set of equations, we have,

$A\vec{x} = \vec{b}$

yielding the augmented matrix

$AB = (A \mid \vec{b})$

Just as in finding the inverse of a matrix, we want to perform row operations necessary to reduce A to an identity matrix. If these row operations are performed on the augmented matrix AB then they are performed on the vector $\vec{b}$ as well as A. The result of this procedure yields the solution to the set of equations. The solution is found in the modified $\vec{b}$ vector.

EXAMPLE 3. Find the solution to the following set of linear equations using the Gauss-Jordan method.

$3x_1 + 2x_2 + 2x_3 = 8$

$x_2 + x_3 = 4$

$x_1 + 3x_3 = 5$

SOLUTION. The first step in the solution process is to form the augmented matrix AB.

$$AB = \begin{bmatrix} 3 & 2 & 2 & | & 8 \\ 0 & 1 & 1 & | & 4 \\ 1 & 0 & 3 & | & 5 \end{bmatrix}$$

We now want to make the A portion of AB both upper- and lower-triangular. Step by step, we obtain:

$$AB_1 = \begin{bmatrix} 3 & 2 & 2 & | & 8 \\ 0 & 1 & 1 & | & 4 \\ 0 & -2/3 & 7/3 & | & 7/3 \end{bmatrix} \quad AB_2 = \begin{bmatrix} 3 & 2 & 2 & | & 8 \\ 0 & 1 & 1 & | & 4 \\ 0 & 0 & 3 & | & 5 \end{bmatrix}$$

$$AB_3 = \begin{bmatrix} 3 & 2 & 2 & | & 8 \\ 0 & 1 & 0 & | & 7/3 \\ 0 & 0 & 3 & | & 5 \end{bmatrix} \quad AB_4 = \begin{bmatrix} 3 & 2 & 0 & | & 14/3 \\ 0 & 1 & 0 & | & 7/3 \\ 0 & 0 & 3 & | & 5 \end{bmatrix}$$

$$AB_5 = \begin{bmatrix} 3 & 0 & 0 & | & 0 \\ 0 & 1 & 0 & | & 7/3 \\ 0 & 0 & 3 & | & 5 \end{bmatrix}$$

At this point we have effectively reduced the original equations to the following set of easily solvable equations:

$3x_1 = 0$

$x_2 = 7/3$

$3x_3 = 5$

The solution is easily seen to be:

$x_1 = 0 \quad x_2 = 7/3 \quad x_3 = 5/3$

Note that these values would be found in the revised augmented matrix if the A portion were reduced to an identity matrix. Thus,

$$AB_6 = \begin{bmatrix} 1 & 0 & 0 & | & 0 \\ 0 & 1 & 0 & | & 7/3 \\ 0 & 0 & 1 & | & 5/3 \end{bmatrix}$$

You can see that this is the same solution that was found in Example 1.

The value of this approach is evident if the inverse of A does not exist. In this case, either the equations are "inconsistent," yielding no solution, or one or more equations are "redundant," yielding an infinite number of solutions. We are able to determine which of these conditions holds using the Gauss-Jordan method.

If the inverse of A does not exist, then its determinant is zero. Using the Gauss-Jordan method, after all row operations have been completed, at least one row will exist in the A portion of AB which has only zeros. This row portrays no information. The corresponding row in the b portion of the revised AB matrix is of critical importance. If it is also zero, then no information is available from the entire row, and thus a redundant equation is represented. If it is not zero, the resulting relationship implies that zero is equal to a constant different from zero, which clearly cannot be true. In this case, the equation is inconsistent, and there exists no solution that simultaneously satisfies all equations.

This may be easier to see in an example. Example 4 below illustrates the determination process. You should notice that both conditions exist in this problem. Of course, the overriding factor is inconsistency. It need only occur once to result in a conclusion of no solution.

EXAMPLE 4. Determine whether the following set of simultaneous linear equations has a unique solution, no solution, or an infinite number of solutions.

$2x_1 + 2x_2 + 4x_3 = 12$

$4x_1 + 4x_2 + 8x_3 = 20$

$x_1 + x_2 + 2x_3 = 6$

SOLUTION.

$$AB_1 = \begin{bmatrix} 2 & 2 & 4 & | & 12 \\ 4 & 4 & 8 & | & 20 \\ 1 & 1 & 2 & | & 6 \end{bmatrix} \quad AB_2 = \begin{bmatrix} 2 & 2 & 4 & | & 12 \\ 0 & 0 & 0 & | & -4 \\ 1 & 1 & 2 & | & 6 \end{bmatrix}$$

$$AB_3 = \begin{bmatrix} 2 & 2 & 4 & | & 12 \\ 0 & 0 & 0 & | & -4 \\ 0 & 0 & 0 & | & 0 \end{bmatrix}$$

equation representation:

$2x_1 + 2x_2 + 4x_3 = 12$

$\quad\quad 0 = -4$ (inconsistent)

$\quad\quad 0 = 0$ (redundant)

CONCLUSION. Due to inconsistent equation, no solution exists.

You probably noticed another interesting development in Example 4. When the inverse of A does not exist, you cannot reduce A to an identity. Thus, the solutions are not evident from the revised AB matrix. This makes sense, since there is no solution for inconsistent equations, and it is impossible to present an infinite number of solutions.

If the determination is an infinite number of solutions, this means that a certain number of variables may be set to any arbitrary values and the equations solved for the remaining variables. If there are n equations and r of the equations are redundant, then r of the n variables may be assigned arbitrary values barring inconsistent equations. However, redundant equations will be pinpointed.

Figure 5.3 presents the subroutine for the Gauss-Jordan method. It utilizes row operations to achieve both upper- and lower-triangularization. If a solution does not exist for your set of linear equations, the program will tell you. If there are an infinite number of solutions, this will also be specified. Note that revised versions of the upper-triangular and lower-triangular subroutines from Chapter 3 are included in the Gauss-Jordan procedure in Figure 5.3. Revisions were necessary to pinpoint inconsistent or redundant equations.

```
18360 REM*************************************************************
18370 REM GAUSS-JORDAN SUBROUTINE
18380 REM This subroutine uses the Gauss-Jordan procedure to
      solve sets
18390 REM of simultaneous linear equations. It uses additional
      subroutines
18400 REM which were presented in Chapter 3 and are contained in
      the file
18410 REM CHAP6.BAS. This file must be merged prior to solution
      using this
18420 REM subroutine. Other inputs are the number of equations
      (IE), the
18430 REM number of variables (IV), the variable coefficients in
      array A,
18440 REM and the right hand side vector stored in one-dimensional
      array B.
18450 DIM A1(IE,IV),B1(IE,1),D%(IV)  'Dimension required arrays
18460 IROW=IE
18470 ICOL=IV
18480 JROW=IE
18490 JCOL=1
18500 FOR I=1 TO IE
18510 FOR J=1 TO IV
18520 A1(I,J)=A(I,J)            'Set values for matrix A1
```

```
18530 NEXT J
18540 B1(I,1)=B(I)           'Set values for matrix B1
18550 D%(I)=1
18560 NEXT I
18570 INV=2                   'Set inverse flag for upcoming subroutine
18580 GOSUB 18850             'Revised upper triangular routine
18590 GOSUB 19410             'Revised lower triangular routine
18600 REM The desired result is in the last column of E1
18610 PRINT:PRINT
18620 PRINT:PRINT
18630 PRINT "RESULTS FROM GAUSS-JORDAN REDUCTION"
18640 REM First, check for inconsistent or redundant equations.
18650 ST%=0                   'Set unique solution flag
18660 FOR I=1 TO IROW
18670 IF D%(I)<>0 THEN GOTO 18750
18680 ST%=1                   'Set redundant equations flag
18690 FOR J=IROW+1 TO ICOL
18700 IF E1(I,J)=0 THEN GOTO 18730
18710 ST%=2                   'Set inconsistent equations flag
18720 GOTO 18790
18730 NEXT J
18740 REM Normal exit means redundant equation
18750 NEXT I
18760 IF ST%=0 THEN GOTO 18810    'Unique solution
18770 PRINT "THIS SET OF EQUATIONS HAS AN INFINITE # OF SOLUTIONS"
18780 GOTO 18840
18790 PRINT "THIS SET OF EQUATIONS IS INCONSISTENT; NO SOLUTION"
18800 GOTO 18840
18810 FOR I=1 TO IV
18820 PRINT "X(";I;")= ";E1(I,IV+1)
18830 NEXT I
18840 RETURN
18850 REM*************************************************************
18860 REM UPPER TRIANGULAR SUBROUTINE (REVISED):
18870 REM  Subroutine to create an upper triangular matrix
18880 REM  Provide matrix A1 and its dimensions, IROW=ICOL. Result
      in E1
18890 IF IROW<>ICOL THEN PRINT "NUMBER OF ROWS<>NUMBER OF
      COLUMNS:ERROR"
18900 IF IROW=ICOL THEN GOSUB 18920
18910 RETURN
18920 REM*************************************************************
18930 REM UPPER TRI ROUTINE (REVISED)
18940 REM  Upper triangular routine
18950 REM  access only through SUB 18850
18960 IF INV=1 THEN ICOL=2*ICOL    'Set parameter for matrix inverse
18970 IF INV<>1 THEN ICOL=ICOL+1 'Set parameter for Gauss-Jordan
18980 DIM E1(IROW,ICOL)
18990 FOR I=1 TO IROW
19000 FOR J=1 TO IROW
19010 E1(I,J)=A1(I,J)         'Assign E1 matrix to A1
19020 IF INV<>1 THEN GOTO 19060    'Skip identity augmentation
```

```
19030 REM This section is used only if a matrix is being inverted
19040 IF I=J THEN E1(I,J+IROW)=1  'Augment the identity matrix to E1
19050 GOTO 19060
19060 NEXT J
19070 IF INV<>1 THEN E1(I,IROW+1)=B1(I,1)   'For the Gauss-Jordan
                                             procedure
19080 NEXT I
19090 DET=1                    'Set determinant flag
19100 FOR I=1 TO IROW
19110 IF I<>IROW THEN GOTO 19150
19120 IF E1(I,I)<>0 THEN GOTO 19250
19130 DT=0
19140 GOTO 19170
19150 IF E1(I,I)<>0 THEN GOTO 19200
19160 GOSUB 19290         'Diagonal element must not be zero
19170 IF DT=0 THEN DET=0
19180 IF DT=0 THEN D%(I)=0
19190 IF DT=0 THEN GOTO 19250
19200 FOR J=I+1 TO IROW
19210 XM=E1(J,I)/E1(I,I)    'Multiplier to zero the column elements
19220 FOR K=I TO ICOL
19230 E1(J,K)=E1(J,K)-XM*E1(I,K)   'Calculate new elements
19240 NEXT K,J
19250 NEXT I
19260 RETURN
19270 REM*************************************************************
19280 REM ZERO DETERMINANT DETECTION SUBROUTINE (REVISED):
19290 REM   Subroutine to ensure diagonal elements are nonzero during
19300 REM   upper-triangularization. If this is not possible, DET
      is zero
19310 DT=1
19320 FOR J=I+1 TO IROW
19330 IF E1(J,I)=0 THEN GOTO 19380
19340 FOR K=1 TO ICOL
19350 E1(I,K)=E1(I,K)+E1(J,K)     'Add rows to make diagonal nonzero
19360 NEXT K
19370 RETURN
19380 NEXT J
19390 DT=0                        'DET must be zero at this point
19400 RETURN
19410 REM*************************************************************
19420 REM LOWER TRIANGULARIZATION SUBROUTINE (REVISED):
19430 REM   Subroutine to make a matrix lower triangular, then an
      identity.
19440 REM   Find the multiplier to zero elements in the kth column
19450 REM   above the diagonal.
19460 REM   Input is the matrix E1 with its dimensions IROW and ICOL
19470 FOR IJ=1 TO IROW
19480 IF IJ=IROW THEN GOTO 19560
19490 IK=IROW-IJ+1
19500 FOR I=1 TO IK-1
19505 IF INV=1 THEN GOTO 19520
19510 IF D%(IK)=0 THEN GOTO 19560
```

5.3 Gauss-Jordan Method

```
19520 XM=E1(I,IK)/E1(IK,IK)
19530 FOR J=I+1 TO ICOL
19540 E1(I,J)=E1(I,J)-XM*E1(IK,J)
19550 NEXT J,I
19560 NEXT IJ
19570 IF DET=0 THEN GOTO 19650
19580 REM Create identity by multiplying each row by the reciprocal
      of the
19590 REM diagonal element of the revised matrix
19600 FOR I=1 TO IROW
19610 DIV=E1(I,I)
19620 FOR J=1 TO ICOL
19630 E1(I,J)=E1(I,J)/DIV
19640 NEXT J,I
19650 RETURN
```

Figure 5.3—Gauss-Jordan subroutine

EXAMPLE 5. Use the subroutine presented for the Gauss-Jordan subroutine to solve the following set of simultaneous equations.

$$3x_1 + 8x_2 - 4x_3 + x_4 = 35$$
$$2x_2 + x_3 - 3x_4 = 27$$
$$11x_1 + 5x_4 = 20$$
$$x_3 + x_4 = 9$$

SOLUTION. Using this program and the main program in Figure 5.2, the final solution is found to be:

$x_1 = 1.9676$ $x_2 = 8.3426$ $x_3 = 9.3287$ $x_4 = -0.3287$

5.4 Gauss-Seidel Method

The Gauss-Seidel method is a numerical procedure for solving simultaneous linear equations. It is iterative in nature. Because it is a numerical approach, there is no guarantee that the solution will be found.

To explain the procedure, let us look at the general form again.

$$a_{11} x_1 + a_{12} x_2 + \ldots + a_{1n} x_n = b_1$$
$$a_{21} x_1 + a_{22} x_2 + \ldots + a_{2n} x_n = b_2$$
$$\vdots$$
$$a_{n1} x_1 + a_{n2} x_2 + \ldots + a_{nn} x_n = b_n$$

Suppose we reorder these equations so that the coefficient of x in the jth equation, a_{jj}, is larger in magnitude than the remaining coefficients in that equation. We can rewrite the equations so that each represents the solution of one variable, as follows:

$$x_1 = \frac{1}{a_{11}} (b_1 - a_{12}x_2 - a_{13}x_3 - \ldots - a_{1n}x_n)$$

$$x_2 = \frac{1}{a_{22}} (b_2 - a_{21}x_1 - a_{23}x_3 - a_{24}x_4 - \ldots - a_{2n}x_n) \quad (1)$$

$$\vdots$$

$$x_n = \frac{1}{a_{nn}} (b_n - a_{n1}x_1 - a_{n2}x_2 - \ldots - a_{nn-1}x_{n-1})$$

We can develop initial estimates for each variable in turn. First, approximate x_1 by

$$x^*_1 = \frac{b_1}{a_{11}}$$

Remaining initial approximations are found as:

$$x^*_2 = \frac{1}{a_{22}} (b_2 - a_{21}x^*_1)$$

$$x^*_3 = \frac{1}{a_{33}} (b_3 - a_{31}x^*_1 - a_{32}x^*_2)$$

$$\vdots$$

$$x^*_n = \frac{1}{a_{nn}} (b_n - a_{n1}x^*_1 - a_{n2}x^*_2 - \ldots - a_{nn-1}x^*_{n-1})$$

Based upon these initial approximations, the set of equations (1) is solved for each variable. These solutions are then used as inputs to re-solve equations (1), and the iterative process continues. One important aspect of the Gauss-Seidel method, that distinguishes it from the similar Jacobi method (Hildebrand, 1974) is that in solving for each variable during each iteration, the most recently calculated values of the input variables are used. For example, when you find x_2, the input value for x_1 is the value just calculated; not the one found during the previous iteration. If we specify x_i as the value calculated during the previous iteration, and x^*_i as the new value, the iterative calculations are performed as follows:

5.4 Gauss-Seidel Method

$$x_1^* = \frac{1}{a_{11}}(b_1 - a_{12}x_2 - \ldots - a_{1n}x_n)$$

$$x_2^* = \frac{1}{a_{22}}(b_2 - a_{21}x_1^* - a_{23}x_3 - \ldots - a_{2n}x_n)$$

$$x_3^* = \frac{1}{a_{33}}(b_3 - a_{31}x_1^* - a_{32}x_2^* - a_{34}x_4 - \ldots - a_{3n}x_n) \tag{2}$$

$$x_n^* = \frac{1}{a_{nn}}(b_n - a_{n1}x_1^* - a_{n2}x_2^* - \ldots - a_{nn-1}x_{n-1}^*)$$

The iterative process continues until successive solutions are within a certain tolerance. Remember, however, that the procedure may not converge. It is therefore wise to limit the number of iterations or provide a check for divergence.

The Gauss-Seidel method is too lengthy to solve large problems by hand. However, we illustrate the procedure on a small problem in the following example.

EXAMPLE 6. Solve the following set of simultaneous linear equations using the Gauss-Seidel method.

$$x_1 + 8x_2 + 2x_3 = 4$$
$$9x_1 + x_2 - x_3 = 15$$
$$2x_2 + 6x_3 = 9$$

SOLUTION. The first step in the solution is to arrange the equations so that the large values are on the diagonal of the coefficient matrix.

$$9x_1 + x_2 - x_3 = 15$$
$$x_1 + 8x_2 + 2x_3 = 4$$
$$2x_2 + 6x_3 = 9$$

We now find the initial estimates for each variable.

$$x_1^* = \frac{b_1}{a_{11}} = \frac{15}{9} = 1.6667$$

$$x_2^* = \frac{1}{a_{22}}(b_2 - a_{21}x_1^*) = \frac{1}{8}(4 - 1.6667) = \frac{7}{24} = 0.2917$$

$$x_3^* = \frac{1}{a_{33}}(b_3 - a_{31}x_1^* - a_{32}x_2^*) = \frac{1}{6}(9 - 0 - \frac{7}{12}) = \frac{101}{72} = 1.4028$$

Using the iterative formulas (2), the following table is produced. Compare the values with those you found.

Iteration	x_1	x_2	x_3
0	1.6667		
1	1.74647	−.0140	1.5047
2	1.8354	−.1056	1.5352
3	1.8490	−.1149	1.5383
4	1.8504	−.1159	1.5386
5	1.8505	−.1160	1.5387

You should find it useful to examine conditions under which the Gauss-Seidel method is guaranteed to converge. Hildebrand (1974) presents these conditions as:

1. The coefficient matrix must not contain a submatrix with p rows and q columns whose elements are all zero, where p+q=n.
2. The magnitude of the diagonal elements of the coefficient matrix must be at least the sum of the magnitudes of the other elements in that row (equation), and is larger than the sum in at least one case.

If these conditions hold, the method will converge. However, the method may still converge even if the conditions do not hold. Thus, the conditions are sufficient but not necessary for convergence.

The convergence criteria we will use is based on comparison of successive points. If all points fall within a specified amount of their previous value, the process is terminated. Divergence is handled through a limit on the number of iterations. If 100 iterations are too restrictive, you may want to establish your own iteration limit.

The Gauss-Seidel subroutine is presented in Figure 5.4. The main program from Figure 5.2 can be used to access it. This program is used in the following example.

```
19660 REM****************************************************************
19670 REM GAUSS-SEIDEL SUBROUTINE
19680 REM This subroutine uses the Gauss-Seidel procedure to
       solve sets
19690 REM of simultaneous linear equations. Inputs provided to this
19700 REM subroutine must be the following:
19710 REM  1. A set of linear equations. The variable coefficients
19720 REM     must be stored in the two-dimensional array A and the
19730 REM     right hand side values must be stored in the
19740 REM     one-dimensional array B.
19750 REM  2. Dimensions must be provided in IE (number of equations)
19760 REM     and IV (number of variables).
19770 REM  3. The number of rows are assumed to be equal to the
19780 REM     number of variables. That is, IE=IV.
19790 REM The user must input to this subroutine the G-S convergence
19800 REM criterion, E.
19810 DIM ID(IE)
19820 INPUT "SPECIFY THE CONVERGENCE CRITERION TO BE USED";E
19830 PRINT:PRINT
19840 INPUT "WOULD YOU LIKE TO SEE RESULTS FROM ALL
       ITERATIONS(Y/N)";T$
19850 REM Find the largest coefficient in each equation
```

```
19860 FOR I=1 TO IE
19870 FOR J=1 TO IV
19880 IF J<>1 THEN GOTO 19920
19890 MAX=ABS(A(I,J))
19900 ID(I)=J                'Set largest value column
19910 GOTO 19950
19920 IF ABS(A(I,J))<MAX THEN GOTO 19950
19930 MAX=ABS(A(I,J))
19940 ID(I)=J                'Set largest value column
19950 NEXT J,I
19960 REM Switch rows to put the largest values on the diagonals
19970 FOR I=1 TO IE-1
19980 IF ID(I)=I THEN GOTO 20130
19990 FOR II=I TO IE
20000 IF ID(II)<>I THEN GOTO 20030
20010 JJ=II
20020 GOTO 20040
20030 NEXT II
20040 FOR J=1 TO IV
20050 D=A(JJ,J)
20060 A(JJ,J)=A(I,J)         'Switch rows for largest on diagonal
20070 A(I,J)=D
20080 NEXT J                 'Reset largest value column
20090 D=B(I)
20100 B(I)=B(JJ)             'Switch right hand side values
20110 B(JJ)=D
20120 ID(JJ)=ID(I)           'Reset largest value column
20130 NEXT I
20140 REM Now perform the calculations necessary for the
20150 REM Gauss-Seidel procedure.
20160 REM
20170 REM Find initial approximations
20180 FOR J=1 TO IE
20190 SUM=0
20200 IF J=1 THEN GOTO 20240
20210 FOR I=1 TO J-1
20220 SUM=SUM+X(I)
20230 NEXT I
20240 X(J)=(B(J)-SUM)/A(J,J)
20250 NEXT J
20260 REM Solve via the iterative technique
20270 IK=0                         'Set iteration counter
20280 K%=1                         'Set convergence flag
20290 IK=IK+1                      'Update iteration counter
20300 IF IK>100 THEN GOTO 20560    'No convergence after 100 iterations
20310 FOR I=1 TO IE
20320 SUM=0
20330 FOR J=1 TO IV
20340 IF I=J THEN GOTO 20360
20350 SUM=SUM+A(I,J)*X(J)
20360 NEXT J
20370 XX=X(I)
```

```
20380 X(I)=(B(I)-SUM)/A(I,I)
20390 IF ABS(XX-X(I))>E THEN K%=2 'Not yet converged
20400 NEXT I
20410 IF T$="N" THEN GOTO 20480
20420 PRINT:PRINT:PRINT
20430 PRINT "FOR ITERATION NUMBER ";IK
20440 FOR I=1 TO IV
20450 PRINT "X(";I;")= ";X(I)
20460 NEXT I
20470 INPUT "ENTER CARRIAGE RETURN TO CONTINUE";F$
20480 IF K%=2 THEN GOTO 20280
20490 REM Convergence has occurred; print results
20500 PRINT "THE GAUSS-SEIDEL PROCEDURE HAS CONVERGED"
20510 PRINT "TO THE FOLLOWING SOLUTION"
20520 FOR I=1 TO IV
20530 PRINT "X(";I;")=";X(I)
20540 NEXT I
20550 GOTO 20570
20560 PRINT "NO CONVERGENCE AFTER 100 ITERATIONS: PROCESS
      TERMINATING"
20570 RETURN
```

Figure 5.4—Gauss-Seidel subroutine

EXAMPLE 7. Use the main program in Figure 5.2 to find the solution to the following set of linear equations.

$$x_1 + 14x_2 + 3x_3 + 6x_4 = 25$$

$$10x_1 + 2x_2 - x_3 - 2x_4 = 18$$

$$-x_1 - 4x_2 + 12x_3 + 2x_4 = 31$$

$$2x_1 + 3x_2 + x_3 + 8x_4 = 14$$

SOLUTION. From the computer output, the results are:

$x_1 = 2.055$ $x_2 = 0.764$ $x_3 = 2.912$ $x_4 = 0.586$

5.5 Summary

We have presented three methods for solving simultaneous linear equations. Each has advantages and disadvantages; the particular situation will dictate which one to use. For small problems, the matrix inverse or Gauss-Jordan procedures may be preferable. However, for a large number of equations, the Gauss-Seidel method should be considered due to the number of calculations involved, provided that the conditions are met for its use. The solution to simultaneous linear equations is an important tool. We will see in Chapter 9 how they are used in solving linear programming problems.

References

Hadley, G., *Linear Algebra*. Addison-Wesley Publishing Company, Inc., Reading, Mass., 1961.

Hildebrand, F.B., *Introduction to Numerical Analysis*. McGraw-Hill Book Co., New York, N.Y., 1974.

Schmidt, J. William, *Mathematical Foundations for Management Science and Systems Analysis*. Academic Press, Inc., New York, N.Y., 1974.

Schmidt, J. William and Robert P. Davis, *Foundations of Analysis in Operations Research*. Academic Press, Inc., New York, N.Y., 1981.

Stewart, G.W., *Introduction to Matrix Computations*. Academic Press, Inc., New York, N.Y., 1973.

Exercises

1. The following set of simultaneous linear equations has an infinite number of solutions.

    $$6x_1 + 3x_3 + x_4 - 2x_5 = 7$$
    $$5x_1 - x_2 + 2x_3 + 2x_4 = 6$$
    $$ 3x_2 - x_3 + 5x_4 + 4x_5 = 8$$

    Write a computer program which presents numerous solutions by setting two variables at a time equal to arbitrary values and solving for the remaining three variables.

2. Specify whether the following sets of equations have a unique solution, no solution, or an infinite number of solutions.

    a. $3x_1 + 2x_2 = 5$
       $x_1 - x_2 = 7$
       $4x_1 = 12$

    b. $2x_1 + 2x_2 + x_3 - x_4 = 7$
       $x_1 - 3x_2 + 2x_3 - 2x_4 = 5$
       $3x_1 + x_2 - x_3 + x_4 = 6$

    c. $3x_1 + x_2 = 6$
       $9x_1 + 3x_2 = 15$

    d. $8x_1 + 7x_2 - 3x_3 = 8$
       $4x_1 - x_2 + 6x_3 = 12$

3. Modify the computer subroutines presented in this chapter to calculate the number of mathematical operations involved in solving a set of simultaneous linear equations. Compare the number of operations required for each technique as the number of equations and variables increase (base your comparisons on the same set of equations). What conclusions can you draw from this exercise?

4. For a certain chemical process we have three inputs and three desired output targets. Each ounce of input A yields three units of output 1, six units of output 2, and one unit of output 3. Each ounce of input B yields two units of output 1, four units of output 2, and no units of output 3. Finally, each ounce of input C yields one unit of output 1, five units of output 2, and two units of output 3. Our requirements are that the process result in exactly 12 units of output 1, 16 units of output 2, and 11 units of output 3. How many ounces of each type of input do we need?

5. A "sparse" matrix is one that has a large number of zero elements. What impact does a "sparse" coefficient matrix have on the solution to simultaneous linear equations? Would changes in the programs presented in this chapter be required to realize this impact?

6

Roots of Polynomials

In the course of problem solving, you may often encounter the need to find the root of a function. That is, you want to find the value of the variable that makes the function equate to zero. In simple cases, this can be accomplished easily by hand. Formulas are readily available to find the roots of polynomials of degree three or less. For polynomials of higher degree, if they cannot be factored, the procedure becomes one of trial and error unless numerical techniques are used, in general requiring a computer. This chapter presents a numerical procedure for finding the real roots of a polynomial.

6.1 Roots of Polynomials

The root of a function is that value of the variable which equates the function to zero. A function may have many roots, such as sin x. However, in our discussion, we will limit ourselves to functions which are polynomials. For simple functions, you may be able to solve for the roots by inspection. Consider the following:

$f(x) = x^2 - 3$

You can see that the roots of f(x) are $\sqrt{3}$ and $-\sqrt{3}$. For more complex functions, factoring may be required:

$f(x) = x^2 - 2x - 3 = (x-3)(x+1)$

Roots: x=3, x=−1

If factoring is not possible, all is not lost. For polynomials of degree two or three there exist formulas for the roots. Following standard forms, these are:

Quadratic

$$f(x) = ax^2 + bx + c$$
$$x = \frac{-b \pm \sqrt{b^2 - 4ac}}{2a}$$

Cubic

$$f(y) = y^3 + py^2 + qy + r$$

let

$$a = \frac{1}{3}(3q-p^2) \quad b = \frac{1}{27}(2p^3-9pq+27r)$$

$$A = \sqrt[3]{\frac{-b}{2} + \sqrt{\frac{b^2}{4} + \frac{a^3}{27}}} \quad B = \sqrt[3]{\frac{-b}{2} - \sqrt{\frac{b^2}{4} + \frac{a^3}{27}}}$$

then

$$X = A+B, \; -\frac{A+B}{2} + \frac{A-B}{2}\sqrt{-3}, \; -\frac{A+B}{2} - \frac{A-B}{2}\sqrt{-3}$$

You can see that this gets quite involved.

Notice that there are two roots for a quadratic function and three for a cubic function. This relationship holds in general. An *nth* order polynomial will have n roots, although they may not all be unique. That is, more than one root may have the same value. Also, roots may not be "real." In this case, they are called "complex" and have both a real part and an imaginary part. The imaginary part of a complex number results from the square root of a negative number. We will not discuss complex numbers here. Except in rare instances, practical applications will only require the real roots of a polynomial. In the formulas given above, both real and complex roots are possible. In the numerical technique presented, only real roots will be found.

6.2 Newton-Raphson Method

The Newton-Raphson Method for finding the roots of polynomials is a particular technique of a class of iterative procedures used with nonlinear equations. We will omit discussion of the theoretical development of the method, but will illustrate its intuitive appeal and operational characteristics. For further details, you may want to consult Hildebrand (1974).

General Concepts

Let us begin by examining the way in which the Newton-Raphson method locates a root. It is an iterative method that converges, hopefully, to a root. There are cases in which there is no convergence. We will discuss these later. The general procedure is to specify an initial starting point, x_0, and generate successive points until the root is approximated. Each successive point is found as the point where the tangent line of the function, evaluated at the previous point, intersects the x-axis. By examining Figure 6.1 you can see how this method converges to the root, r.

What about multiple roots? It is evident that the same initial starting point will result in the same root. Therefore, if multiple roots exist you must start in more than one place in order to find them. In fact, there is no guarantee that you will find all the roots. A large number of starting points may lead to the same root. This is a drawback that is not unique to Newton-Raphson. In order to reduce the

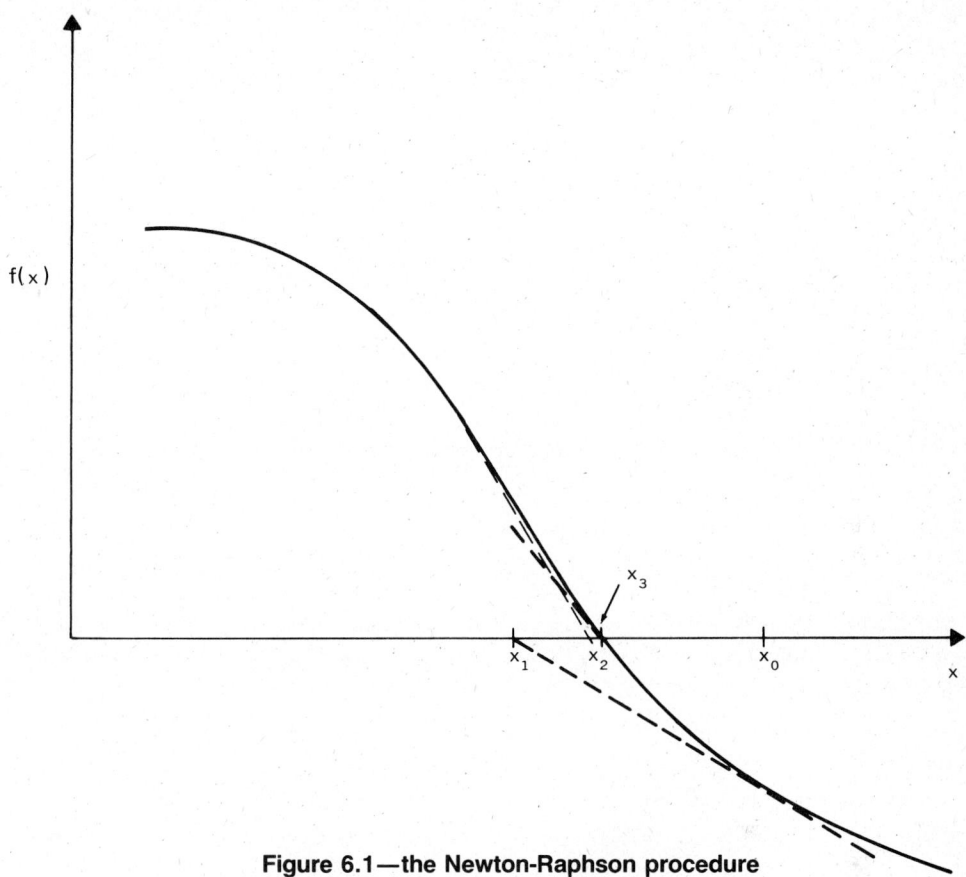

Figure 6.1—the Newton-Raphson procedure

likelihood of missing a root, a large number of initial starting values are used. This is illustrated for two roots in Figure 6.2

In examining Figures 6.1 and 6.2 you may have noticed a condition under which the method does not converge to a root or may take an extremely long time to locate a root. This unfortunate occurrence results from a point where the slope of the function is at or near zero. In this case either the tangent line never intersects the x-axis (zero slope) or the next point found is extremely far away (slope near zero). We illustrate this situation in Figure 6.3. Fortunately, we can plan ahead to avoid this situation. If, upon calculation of the slope at the current point, its value is at or near zero, we can move a particular distance from that point and try again. For many functions this is an effective procedure.

You may run into some polynomials for which this procedure simply will not work. This will be due to the shape of the function and the precise location of the root. However, in general, it is a very reliable procedure.

Operational Requirements

You know what Newton-Raphson does. We will now illustrate how it works.

The graphic illustrations presented up to this point cannot be used; if we could graph the function we would know what the roots were. The procedure used is algebraic, and uses iterative relationships to generate each successive point.

The general formula used is:

$$x_{k+1} = x_k - \frac{f(x_k)}{f'(x_k)} \tag{1}$$

where,

k = index of points

$f'(x_k)$ = derivative of the function f(x) evaluated at the point x_k (slope of the tangent line at that point)

You can see the computational problems inherent in the situation depicted in Figure 6.3. A horizontal tangent line has a slope of zero, causing the term $f(x_k)/f'(x_k)$ to be undefined. A slope near zero is almost as bad.

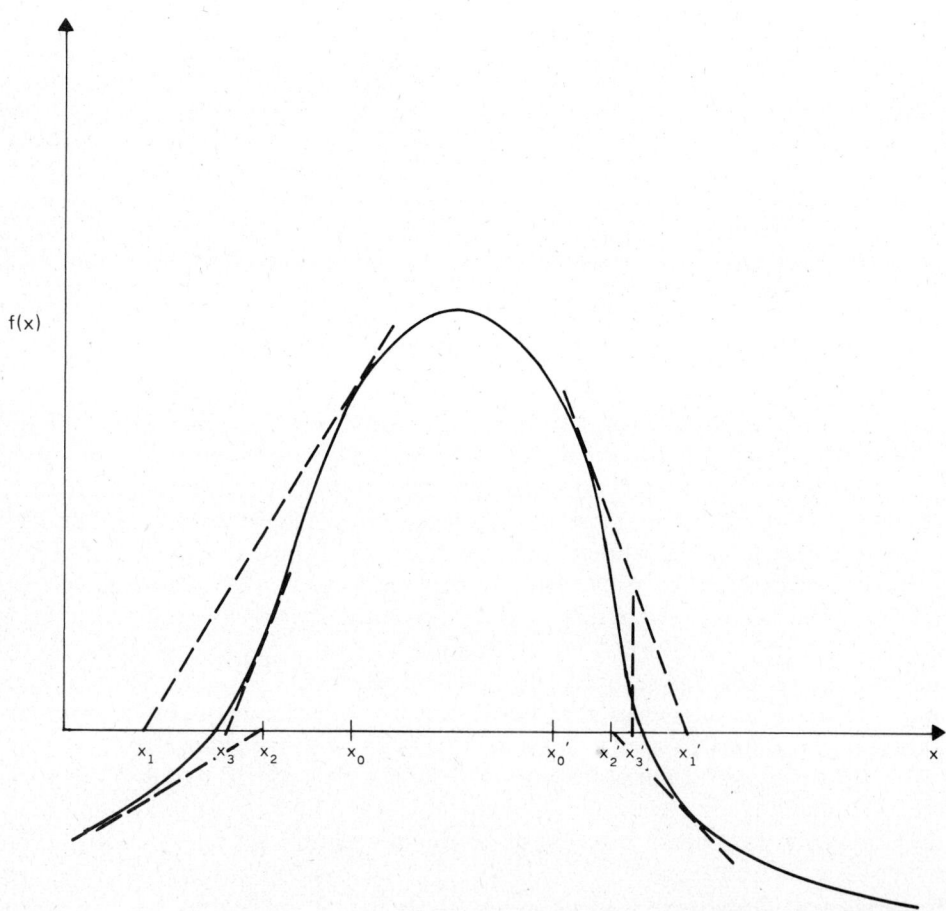

Figure 6.2—multiple starting points and roots

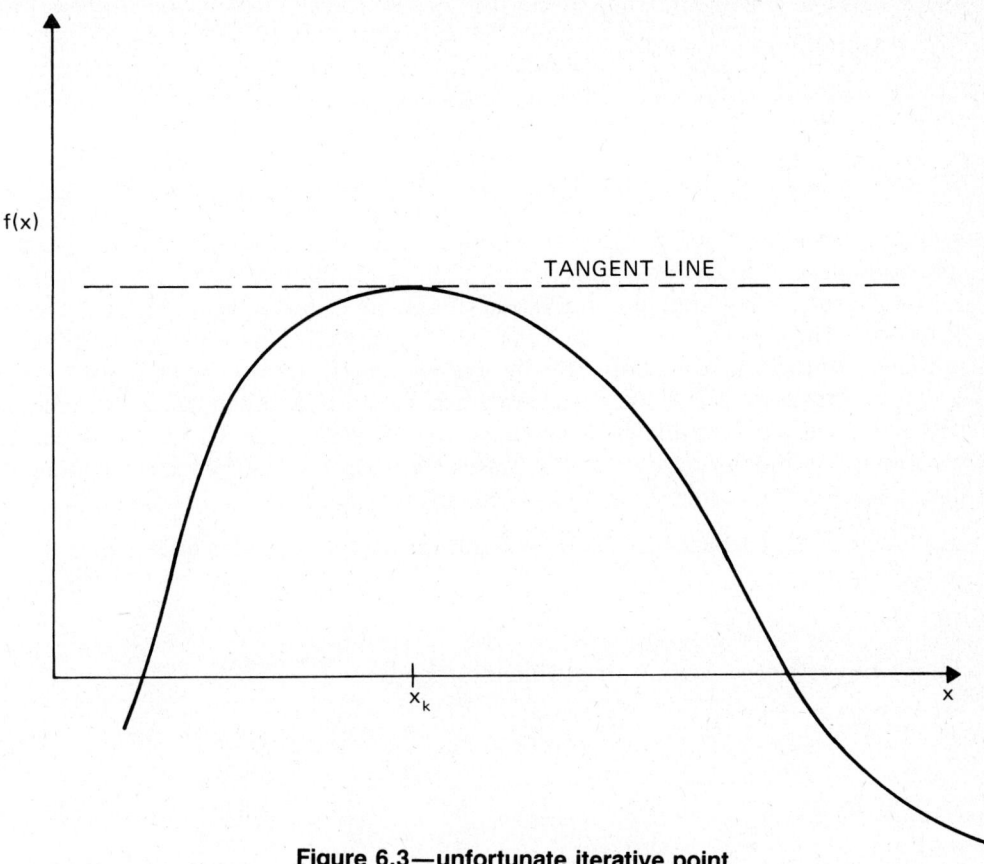

Figure 6.3—unfortunate iterative point

Does this formula make sense to you? Consider the point-slope form of an equation. The tangent line which passes through the point $f(x_k)$ has the general formula

$f(x_k) = ax_k + b$

where,

a = slope of the line
b = intercept with the f(x) axis

Now we know the slope of the line, it is simply $f'(x_k)$. Therefore, we know that

$$f(x_k) = x_k f'(x_k) + b \qquad (2)$$

Furthermore, x_{k+1} must lie on the same line. But, at x_{k+1} the function is zero. So,

$f(x_{k+1}) = x_{k+1} f'(x_k) + b$

becomes

$$0 = x_{k+1} f'(x_k) + b \qquad (3)$$

To find x_{k+1}, let us subtract (2) from (3) to get

$$x_{k+1} f'(x_k) - x_k f'(x_k) + f(x_k) = 0$$

By collecting terms and solving for x_{k+1}, we get

$$x_{k+1} = x_k - \frac{f(x_k)}{f'(x_k)}$$

which is equation (1). Therefore, this formula does indeed generate the succession of points desired in the procedure.

You have noticed that the derivative of the function is required for Newton-Raphson. Since we are limiting ourselves to polynomials you should have no difficulty with this. It is automatically included in the computer program.

The remaining issue is determining when to stop. In other words, which x is the root of the polynomial? Of course, if $f(x_k)$ is zero, then $x_{k+1} = x_k$; this can be used as a stopping criterion. In most cases the root will not be exact. A stopping rule based on the absolute value of $f(x_k)$ is appropriate in these instances.

EXAMPLE 1. Perform the Newton-Raphson method on the following polynomial.

$$f(x) = 7x^4 - 3x^3 + x^2 + x - 8$$

Stop when $|f(x)| \leq 0.001$. Start at the point $x_0 = 3$.

SOLUTION. The table below illustrates the succession of points in the solution.

k	x_k	$f(x_k)$	$f'(x_k)$	x_{k+1}
0	3	490	682	2.815
1	2.2815	153.5286	291.2470	1.7544
2	1.7544	46.9423	127.9952	1.3876
3	1.3876	13.2519	61.2616	1.1713
4	1.1713	2.8974	35.9883	1.0908
5	1.0908	0.2969	28.8130	1.0805
6	1.0805	0.0043	27.9737	1.0803
7	1.0803	0.0000		

Computer Program

The example problem above was quite tedious by hand. But based upon the exercise you can see that it is easy to program. We could use a program for which the function and its derivative must be specified in function statements. However, since the function will always be a polynomial, all that is really needed are the degree of polynomial and coefficients for each x term.

We present a subroutine for the Newton-Raphson method in Figure 6.4. This subroutine performs the necessary operations given the function, an initial starting point, and a stopping criterion. You must provide these latter items in a main program.

The construction of a main program requires considerable decision making. If you want to find all the real roots, it may be necessary to have many different

6.2 Newton-Raphson Method

starting points. This will also depend on the degree of the polynomial. In order to find all roots, you may have to run the program several times, using different starting points each time.

The main program listed in Figure 6.5 is flexible in that it allows the users to specify number of starting points and their values. You may want to revise it to meet your particular needs.

```
20600 REM ****************************************************************
20610 REM NEWTON-RAPHSON SUBROUTINE
20620 REM  Subroutine to perform the NEWTON-RAPHSON method
20630 REM  to find the real roots of a polynomial. Required inputs
20640 REM  are the degree of the polynomial (JD), coefficients for
20650 REM  all terms (stored in the vector F(I)), an initial
20660 REM  starting point (X), and a stopping criterion (Z)
20670 REM
20680 X1=X                         'Save the starting value
20690 SUM=0                        'Initialize the function
20700 PSUM=0                       'Initialize the derivative
20710 JJ=1
20720 FOR I=0 TO JD
20730   SUM=SUM+(F(I)*(X^I))       'Sum to find function value
20740   IF I=0 THEN GOTO 20760
20750   PSUM=PSUM+(F(I)*I*(X^(I-1)))    'Sum to find derivative
                                         value
20760 NEXT I
20770 IF ABS(SUM)<=Z THEN GOTO 20850    'Function meets stopping
                                         criterion
20780 IF PSUM=0 THEN GOTO 20830    'Zero slope
20790 X=X-(SUM/PSUM)               'Find the next point to evaluate
20800 JJ=JJ+1
20810 IF JJ>100 THEN GOTO 20870    'Iteration limit has been reached
20820 GOTO 20690
20830 PRINT "FOR STARTING POINT";X1;"THE POINT";X;"HAS A
      ZERO SLOPE"
20840 GOTO 20880
20850 PRINT "FOR STARTING POINT";X1;"THE ROOT IS";X
20860 GOTO 20880
20870 PRINT "THE ROOT HAS NOT BEEN FOUND AFTER 100
      ITERATIONS"
20880 RETURN
```

Figure 6.4—Newton-Raphson subroutine

```
100 REM NEWTON-RAPHSON MAIN PROGRAM:
110 REM  Main program to find the real roots of a polynomial
120 REM  using the NEWTON-RAPHSON Subroutine. The user must input
130 REM  the following data:
140 REM      JD = The degree of the polynomial
150 REM      F(I) = Coefficients of each term of the polynomial
160 REM      Z = Stopping criterion
```

```
170 REM    Also required are initial starting points, which are passed
180 REM    on to the NEWTON-RAPHSON subroutine.
190 REM
200 DEFINT I,J
210 INPUT "ENTER THE DEGREE OF THE POLYNOMIAL";JD
220 DIM F(JD)                   'Dimension the coefficient vector
230 INPUT "ENTER THE STOPPING CRITERION";Z
240 FOR I=0 TO JD
250   PRINT "ENTER THE ";I;"ORDER COEFFICIENT"
260   INPUT F(I)                'Enter the function
270 NEXT I
280 INPUT "SPECIFY A STARTING POINT";X
290 GOSUB 20600                 'Call NEWTON-RAPHSON Subroutine
300 PRINT "WOULD YOU LIKE TO TRY ANOTHER STARTING POINT?"
310 INPUT "ENTER Y--YES, N--NO";A$
320 IF A$="Y" THEN GOTO 280
330 END
```

Figure 6.5—Newton-Raphson main program

EXAMPLE 2. Use the main program in Figure 6.5 to solve for the real roots of the following function.

$f(x) = x^4 - 11x^3 - 3x^2 + x - 7$

SOLUTION. Hopefully, your choice of starting points led to the following roots:

Starting Point	Root
0	−0.9653
10	11.2634

The remaining roots of the polynomial are complex roots.

6.3 Summary

This chapter has presented a procedure for finding the roots of a polynomial. We have restricted ourselves to the determination of only the real roots. We saw that for polynomials of third order or less, closed-form analytic solutions are possible. For most higher-order polynomials, a numerical procedure is appropriate. The procedure presented here, Newton-Raphson, was shown to be a logical, readily applicable method. By varying starting points we can locate the approximate roots of most polynomials with relatively little effort.

References

Hildebrand, F. B. *Introduction to Numerical Analysis.* McGraw-Hill, Inc., New York, N.Y., 1974.

Nielsen, K. L. *Methods in Numerical Analysis.* The MacMillan Company, New York, N.Y., 1956.

Exercises

1. In optimization problems, we often must find the maximum or minimum value of a certain function. One way to locate stationary points of a function is to find the value of the variable (or set of variables) which equates the derivative of the function to zero. In the single variable case, we can use the Newton-Raphson procedure to perform the latter task. Using the method discussed above, find the real stationary points of the following function. Which is the "maximum value?"

   $f(x) = 7x^5 - 2x^3 - x^2 + 3x + 6$

2. Find the real roots of the following polynomials.

   a. $f(x) = 4x^6 - 2x^3 + 3x^2 - 19x + 12$

   b. $f(x) = 9x^4 + 3x^3 - 11x^2$

   c. $f(x) = x^5 + x^4 - 15x^3 + 6x - 3$

3. Investigate the use of the Newton-Raphson procedure to find roots of functions which are not polynomials. Is this a viable procedure? What problems are inherent in this approach?

4. Consider the following single-variable, nonlinear simultaneous equations.

   $x^2 + x = 6$

   $x^2 + 2x = 3$

   Discuss how these might be solved simultaneously using the Newton-Raphson procedure. What possible stumbling blocks are there?

5. Find all real roots of the following polynomials. Before solving, specify the minimum number of real roots you will find.

   a. $f(x) = 8x^9 - 6x^4 + 3x - 12$

   b. $f(x) = x^6 + 3x^5 - 16x^2 + 9$

   c. $f(y) = 23y^5 - y^4 - y^3 + 6y$

   d. $f(z) = 11z^6 + z^3 - 1$

6. (Advanced Topic) The Newton-Raphson procedure can also be revised to handle *all* roots of polynomials (both real and complex). Modify the Newton-Raphson subroutine so that complex roots can be found and illustrate the procedure with an example.

7

Numerical Integration

Integrating functions is a procedure used frequently in many engineering and science disciplines. When no closed-form expression can be found for an integral, we resort to numerical methods. Even if a closed-form expression is obtainable, its complexity may drive us to the ease of a computer solution.

We present and discuss two numerical integration techniques in this chapter. Each method can present a fairly accurate approximation for most functions. Those presented are the trapezoidal rule and Legendre-Gauss quadrature.

First, let us refresh our memory about integration. This will enable us to understand the reasons behind the numerical procedures better.

7.1 Basics of Integration

If you think about your experiences with integration you will probably remember that you found the integral by checking to see if its derivative was equal to the function to be integrated. In essence, you look to differentiation to integrate. You are not alone. In fact, integration is defined as the inverse of differentiation. Closed-form expressions for integrals are found by verifying that their derivatives are equal to the integrands.

It is important to distinguish between definite and indefinite integrals. Indefinite integrals do not have limits of integration, making the use of numerical methods inappropriate. Therefore, we will only be concerned with definite integrals, where the limits of integration are specified.

The integral of a function over some region is (for a single-variable function) the area under the curve generated by that function within the region. This area may be positive or negative, depending on whether the curve lies above or below the axis of the independent variable. Thus, for any function $y=f(x)$, the integral

$$\int_a^b y\,dx = \int_a^b f(x)\,dx$$

represents the area under the curve generated by y between the limits $x=a$ and $x=b$. We will use numerical procedures to calculate this area.

7.2 Numerical Methods

In order to find the area under the curve, different numerical integration procedures use a variety of approaches to divide the interval into a large number of small pieces and add the areas of each piece. The major task is to find the area of each piece. This can be accomplished in many ways, each leading to a different numerical technique. For most functions, the numerically derived result will be approximate.

In this chapter we present two different integration methods. Each has its own method of generating the shapes of the small pieces.

Trapezoidal Rule

The trapezoidal rule is one of the simplest numerical integration techniques. The general procedure is to divide the interval of integration into a large number of small subintervals, each resembling a trapezoid. We find the integral by adding the areas of all the trapezoids. This is particularly easy because the area of a trapezoid is easily found by

$$\text{area} = \tfrac{1}{2}(a+b)h$$

where a and b are the lengths of the two sides and h is the distance between them. Since each area is, in general, not a true trapezoid, the result found is only an approximation. Figure 7.1 illustrates the formation of each trapezoid.

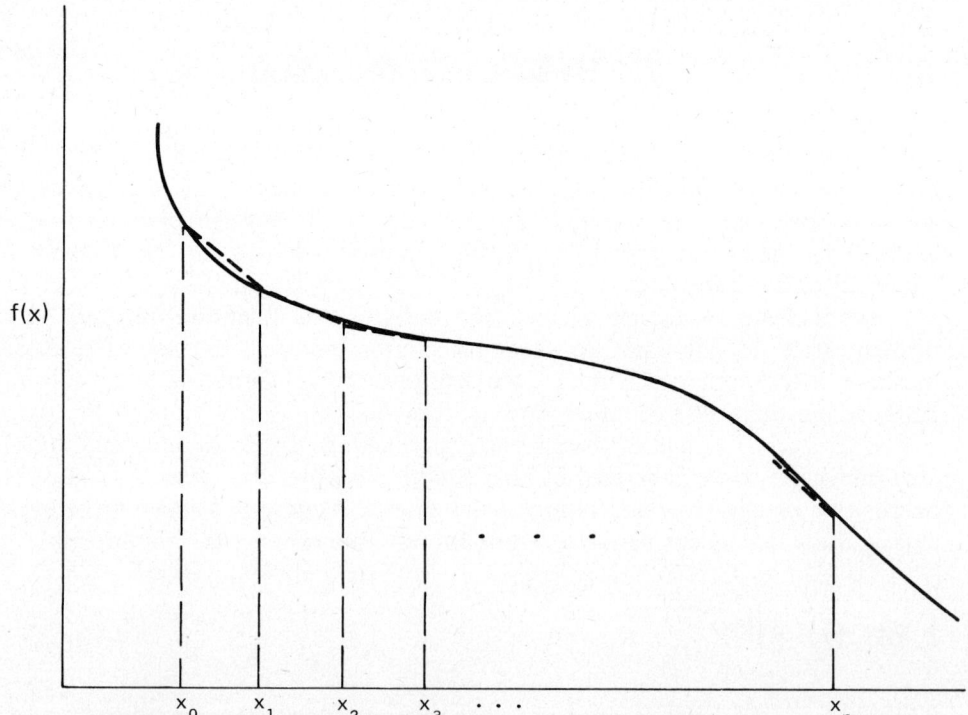

Figure 7.1—numerical integration using the trapezoidal rule

We chose points x_0, x_1, etc. so that there is a constant difference between consecutive points, say h. The lengths of the sides for each subinterval i are $y_i = f(x_i)$ and $y_{i+1} = f(x_{i+1})$. Using the formula for the area of a trapezoid, we can find the approximate area under the curve for each subinterval as:

$$A_i = \tfrac{1}{2}(y_i + y_{i+1})h$$

For n such subintervals, we have

$$\int_{x=x_0}^{x=x_n} y\,dx = A_1 + A_2 + \ldots + A_n + E$$

where E is a term representing the error resulting from the trapezoidal approximations for each subinterval. Substituting and combining terms, we find the formula easily applicable to computer application.

$$\int_{x=x_0}^{x=x_n} y\,dx = \left(\tfrac{1}{2}y_0 + y_1 + y_2 + \ldots + y_{n-1} + \tfrac{1}{2}y_n\right) + E$$

You can see that the numerical approximation will be more accurate if a large number of subintervals is used. As h gets smaller, each subinterval is more closely approximated by a trapezoid, yielding a better solution. The trade-off for this accuracy is increased computation time.

Figure 7.2 presents the BASIC subroutine for the trapezoidal rule. Notice that the number of subintervals used is an input to the subroutine. In this manner, you can control the accuracy and computation time you desire. To see how the number of subintervals affects accuracy, we developed two main programs to utilize the trapezoidal rule. Figure 7.3 presents the first. In this program the number of subintervals is an input. If the general shape of the function is known so you can properly judge the number of subintervals required, this can be the most efficient approach.

```
20900 REM************************************************************
20910 REM Subroutine to find the integral of a function using
20920 REM the trapezoidal rule. Inputs to this subroutine must be:
20930 REM     XBEG: Lower integration limit
20940 REM     XEND: Upper integration limit
20950 REM     ISUB: Number of subintervals used
20960 REM     Function must be specified in the main program FNTRAP(X)
20970 TRAP=0                    'Initialize integral
20980 FOR I=1 TO ISUB-1
20990 X=XBEG+H*I                'Find X values; not including endpoints
21000 TRAP=TRAP+FNTRAP(X)
21010 NEXT I
21020 REM CALCULATE END POINT VALUES, ADD TO TRAP
21030 TRAP=H*(TRAP+.5*(FNTRAP(XBEG)+FNTRAP(XEND)))  'Result is the
                                                    integral
21040 RETURN
```

Figure 7.2—integration trapezoidal rule subroutine

```
100 REM TRAPEZOIDAL RULE: INTEGRATION:
110 REM Main program for numerical integration using
120 REM the trapezoidal rule. Data required are the limits
130 REM of integration and the function to be integrated,
140 REM specified in line 160.
150 DEFINT I,J
160 DEF FNTRAP(X)=EXP(X)
170 INPUT "THE LOWER LIMIT OF INTEGRATION IS"; XBEG
180 INPUT "THE UPPER LIMIT OF INTEGRATION IS"; XEND
190 INPUT "THE NUMBER OF SUBINTERVALS TO BE USED IS"; ISUB
200 IF XBEG>XEND THEN GOTO 250
210 H=(XEND-XBEG)/ISUB        'Calculate equal spacing for subintervals
220 GOSUB 20900               'Call the trapezoidal rule subroutine
230 PRINT "FOR";ISUB;" SUBINTERVALS THE INTEGRAL IS";TRAP
240 GOTO 260
250 PRINT "XBEG GREATER THAN XEND; PROGRAM TERMINATING"
260 END
```

Figure 7.3—trapezoidal rule main program

The example presented in Figure 7.3 is for the function $f(x)=e^x$. A sample result for specific inputs is presented in Figure 7.4. We can compare the results with the true closed-form solution. You can see that the number of subintervals significantly affects accuracy.

Number of Subintervals	Integral
5	1.724006
10	1.719714
20	1.71864
50	1.718297
100	1.718297

Actual: 1.718281828

Figure 7.4—integration of EXP(x) over 0-1

If computation time is not a concern, you may wish to specify a desired level of accuracy. We explain this level of accuracy as follows. The numerical approximation improves as h decreases, but beyond a certain point this improvement is negligible. If we find the integral a number of times, each time doubling the number of subintervals, we will reach a point where one solution is within a small amount, say z, of the previous solution. If z is quite small, on the order of 0.0001, then you can feel confident in your solution. We present a main program for this type of analysis in Figure 7.5. Applying this to the same function examined earlier, $f(x)=EXP(x)$, we get the results in Figure 7.6. Notice how the solution converges as the number of subintervals increases.

```
570 REM****************************************************************
580 REM TRAPEZOIDAL RULE: INTEGRATION: VARIABLE SUBINTERVALS
590 REM Main program for numerical integration using the
600 REM trapezoidal rule with increasing number of subintervals
610 REM Function to be integrated must be specified in line 650
620 REM Inputs for this process are limits of integration, beginning
630 REM number of subintervals, and stopping criterion Z.
640 DEFINT I,J
650 DEF FNTRAP(X)=EXP(X)
660 INPUT "THE LOWER LIMIT OF INTEGRATION IS";XBEG
670 INPUT "THE UPPER LIMIT OF INTEGRATION IS";XEND
680 INPUT "INITIAL NUMBER OF SUBINTERVALS TO BE USED IS";ISUB
690 INPUT "SPECIFY Z, THE STOPPING CRITERION";Z
700 IST=1                    'Set initial pass flag
710 H=(XEND-XBEG)/ISUB        'Calculate equal spacing for specific ISUB
720 IF XBEG>XEND THEN GOTO 810
730 GOSUB 20900               'Call the trapezoidal rule subroutine
740 PRINT "FOR";ISUB;" SUBINTERVALS THE INTEGRAL IS";TRAP
750 IF IST=1 THEN GOTO 770
760 IF ABS(OLD-TRAP)<=Z THEN GOTO 820
770 OLD=TRAP
780 IST=0                     'Cancel initial pass flag
790 ISUB=ISUB*2               'Double the number of subintervals
800 GOTO 710
810 PRINT "XBEG IS GREATER THAN XEND: PROGRAM TERMINATING"
820 END
```

Figure 7.5—variable subintervals main program

Initial Number of Subintervals: 10
Stopping Criterion : 0.0001

Subintervals	Integral
10	1.719714
20	1.71864
40	1.718371
80	1.718304

Figure 7.6—results of one run of main program in Figure 7.5

Legendre-Gauss Quadrature

Legendre-Gauss quadrature (integration) is somewhat more theoretically complex than the trapezoidal rule and considerably more accurate. The complex nature of all quadrature formulas, there are many, resulted in lack of use until the advent of advanced computing machinery. You can find many such techniques in Hildebrand (1974).

The major difference between any quadrature method and a "traditional" numerical integration approach such as the trapezoidal rule, is that the integration procedures involve ordinates (y-values) corresponding to equally spaced abscissas (x-values). Quadrature methods present an optimal distribution of the abscissas instead of arbitrarily specifying them. In this manner, the quadrature techniques can obtain approximately the same accuracy with fewer ordinates.

In general, these abscissas will be irrational. Weights by which the corresponding ordinates are multiplied may also be irrational. It is because of this that people were reluctant to use these procedures to solve problems by hand. We will not present the theory behind the development of Legendre-Gauss quadrature. You can find many fine texts which discuss this procedure in detail. You might want to examine Hildebrand (1974), Nelson (1956), or Scheid (1968). What is important here is using this technique to perform numerical integration.

There are actually many different forms for Legendre-Gauss quadrature, depending on the number of abscissas used in the approximation. The actual abscissas and their coefficients are derived from the Legendre polynomial. The three-point quadrature formula will be used here. You can find abscissas and weights for other forms by formulas presented in Hildebrand (1974). We should note that accuracy increases as the number of abscissas increases. The three-point formula is:

$$\int_{-1}^{1} f(x)dx = \frac{1}{9}(5f(-\frac{\sqrt{15}}{5})+8f(0)+5f(\frac{\sqrt{15}}{5}))+E$$

You will notice that this formula is defined over the interval -1 to 1. A change of variables can be used to transform your integral to meet this requirement. We can include such a change in a general formula.

$$\int_{a}^{b} f(x)dx = \frac{b-a}{18}(5f(y-z)+8f(y)+5f(y+z))+E$$

where,

$y=(a+b)/2$
$z=[(b-a)/2]\sqrt{3/5}$

The accuracy of this quadrature formula is very good. In fact, the three-point formula is exact for polynomials of order five or less. You can use the formula as it is or divide a large interval into a number of subintervals, each one evaluated via Legendre-Gauss. These areas would then be summed.

We present a subroutine for Legendre-Gauss quadrature in Figure 7.7 The main program for accessing this subroutine is given in Figure 7.8 with the function $f(x)=2x^2+2x$. If you run this main program you can see that the result is exact, as was expected.

```
21050 REM*************************************************************
21060 REM LEGENDRE-GAUSS QUADRATURE SUBROUTINE:   INTEGRATION:
21070 REM Subroutine for numerical integration using
21080 REM Legendre-Gauss quadrature. Inputs are the function
21090 REM to be integrated, FNLG(X), specified in a function
      statement,
```

```
21100 REM lower integral limit A, and upper limit B. Returns LGS
21110 Y=(A+B)/2              'Calculate the interval midpoint
21120 Z=((B-A)/2)*SQR(3/5)   'Calculate Z for remaining abscissas
21130 LGS=((B-A)/18)*(5*FNLG(Y-Z)
      +8*FNLG(Y)+5*FNLG(Y+Z))            'L-G formula
21140 RETURN
```

Figure 7.7—Legendre-Gauss quadrature subroutine

```
270 REM MAIN PROGRAM FOR LEGENDRE-GAUSS QUADRATURE
280 REM Inputs are the function itself, specified in line 310
290 REM and the limits of the integration.
300 DEFINT I,J
310 DEF FNLG(X)=2*(X^2)+2*X      'Input the function to be integrated
320 INPUT "THE LOWER LIMIT OF INTEGRATION IS";A
330 INPUT "THE UPPER LIMIT OF INTEGRATION IS";B
340 IF A>B THEN GOTO 380
350 GOSUB 21050                  'Call the Legendre-Gauss subroutine
360 PRINT "THE AREA UNDER THE CURVE FROM";A;" TO";B;" IS";LGS
370 GOTO 390
380 PRINT "LOWER LIMIT GREATER THAN UPPER LIMIT: PROGRAM TERMINATING"
390 END
```

Figure 7.8—Legendre-Gauss main program

EXAMPLE. Write a main program to integrate EXP(X) over the interval 0 to 1 by using Legendre-Gauss quadrature for each of 25 subintervals of length 0.04.

SOLUTION. The main program for this procedure is given in Figure 7.9. Notice how the interval is divided into 25 subintervals. If you compare the result with that found using the trapezoidal rule, you can see how accurate the Legendre-Gauss procedure is for fewer functional evaluations. From the computer run: area under the curve is 1.718282.

```
400 REM Solution to example problem 2; Legendre-Gauss over 0-1
410 REM with subintervals of size 0.04. Function EXP(X)
420 DEFINT I,J
430 DEF FNLG(X)=EXP(X)        'Function to be evaluated
440 AREA=0                    'Initialize
450 W=SQR(3/5)                'Initialize
460 FOR I=0 to 24
470 A=.04*I
480 B=A+.04
490 K=(B-A)/2
500 Z=W*K                     'Set Z value for abscissas
510 Y=(A+B)/2                 'Calculate midpoint of subinterval
520 GOSUB 21050               'Call Legendre-Gauss subroutine
530 AREA=LGS+AREA             'Calculate the cumulative area
```

```
540 NEXT I
550 PRINT "AREA UNDER THE CURVE IS";AREA
560 END
```

Figure 7.9—main program for example problem

7.3 Summary

In this chapter we have discussed the operational aspects of numerical integration. We first briefly presented the concepts of integration. Sometimes closed-form solutions are not possible or finding them is not feasible. In these cases, numerical procedures can be implemented on your IBM Personal Computer to approximate the integral.

Two techniques were presented: the trapezoidal rule and Legendre-Gauss quadrature. The trapezoidal rule is easy to understand and works reasonably well when small subintervals are used. The Legendre-Gauss quadrature is much more complex, but also more accurate. Subroutines for each of these procedures are included. You may wish to consult the references listed below for additional integration techniques.

References

Hildebrand, F.B., *Introduction to Numerical Analysis*. McGraw-Hill Book Co., New York, N.Y., 1974.
Nelson, K.L., *Methods in Numerical Analysis*. The Macmillan Publishing Co., New York, N.Y., 1956.
Scheid, Francis, *Theory and Problems of Numerical Analysis*. Schaum's Outline Series, McGraw-Hill Book Co., New York, N.Y., 1968.

Exercises

1. Integrate the following functions over the interval 1 to 3 using the trapezoidal rule with h=0.01 and Legendre-Gauss quadrature applied over two subintervals. Compare the results. Which value of h makes the two methods roughly equivalent?

   a. $f(x) = 8x^3 - 9x + 3$

   b. $f(x) = e^{3x} + \sqrt{x}$

2. How do the integration methods presented here handle functions which dip below the x−axis? Are revisions required? Explain.

3. In statistics, the area between two reference points under the curve generated by a probability density function is equal to the probability of a point lying between the two reference points. One such density function is the "normal" density function. It is:

$$f(x) = \frac{1}{\sigma\sqrt{2\pi}} e^{-\frac{1}{2}\left(\frac{x-\mu}{\sigma}\right)^2} \quad -\infty < x < \infty$$

where

μ = mean of the distribution

7.3 Summary

σ = standard deviation of the distribution

From this information, find the following probabilities:

a. Probability $(2 \leq X \leq 5)$ for $\mu=4$, $\sigma=0.5$

b. Probability $(0.1 \leq X \leq 0.5)$ for $\mu=0.7$, $\sigma=0.3$

c. Probability $(70 \leq X \leq 85)$ for $\mu=80$, $\sigma=12$

4. Suppose we want to find only the positive area under a curve. Revise the trapezoidal rule subroutine to handle this and illustrate it on the following functions.

    a. sin x from 1 to 6

    b. $x^2 - 4x + 1$ from 0 to 5

5. Find the integral of the following functions.

    a. $f(x) = e^{x^2-2} + x^2$ over 0 to 1

    b. $f(x) = \sin x - \cos x + x^3$ over 0 to 2

    c. $f(x) = x^2 \cos x + x^3$ over 1 to 3

6. (Advanced topic) For functions of more than one variable, we are concerned with multiple integration. How would the procedures described in this chapter be adapted to address multiple integrals? Illustrate using an example problem.

Numerical Solutions to Differential Equations

When analyzing physical systems, we find that differential equations occur naturally quite frequently. The solution to many simple forms of differential equations can be easily obtained by hand. In many cases, however, a straightforward solution is not possible. As in other situations when a closed-form solution cannot be used, or a computer solution is desired, we resort to numerical methods. In this chapter we present such a procedure to obtain approximate solutions to first-order ordinary differential equations. The procedure can be extended to solve systems of differential equations.

8.1 Overview of Differential Equations

A differential equation is an equation containing derivatives of an unknown function. In general, the solution of the differential equation requires determination of the unknown function, g(x). Many important problems engineers and scientists face are formulated as differential equations. Some typical differential equations occurring in physical systems are the wave equation in fluid mechanics, the potential equation in electricity, and the diffusion equation in elasticity.

The aforementioned differential equations happen to be partial differential equations. These have terms of partial derivatives rather than ordinary derivatives. The latter case is represented by ordinary differential equations. Boyce and DiPrima (1977) present two excellent examples of ordinary differential equations representing physical systems. The first relates the charge on a condenser, Q(t), in a circuit with capacitance C, resistance R, inductance L, and impressed voltage E(t).

$$L \frac{d^2Q(t)}{dt^2} + R \frac{dQ(t)}{dt} + \frac{Q(t)}{C} - E(t) \tag{1}$$

The second represents the decay of an amount R(t) of a radioactive substance over time, where k is a known constant.

$$\frac{dR(t)}{dt} = -kR(t) \tag{2}$$

The equations (1) and (2) above are different types of differential equations. Equation (1) is a second-order differential equation and equation (2) is a first-order differential equation. The order is that of the highest derivative that appears in the equation. We will be concerned only with the first-order case.

The general form of the ordinary differential equations we will be examining is given as follows:

$$y' = f(x,y) \qquad y(x_0) = y_0 \tag{3}$$

This is called an initial value problem, since the second equation above prescribes the point (x_0, y_0) through which the integral curve passes, thus specifying an initial condition.

Solving a problem represented by equations (3) analytically cannot be reduced to a general methodology. Solution procedures are available only for special classes of equations. A determination of the solution must begin with a determination of the class of equations to which the problem of interest belongs. For some of these classes, standard solution forms are available. As an example, consider the following differential equation:

$$y' + ay = 0 \qquad \text{a is a real constant.} \tag{4}$$

The solution to this can be expressed as

$$y = ce^{-ax} \qquad \text{c an arbitrary constant.} \tag{5}$$

With the specification of an initial condition, the value of c can be assessed.

EXAMPLE 1. Solve the following initial value problem analytically.

$$y' + 5y = 0 \qquad y(0) = 2$$

SOLUTION. You can see that this differential equation is of the same form as equation (4) and the solution is therefore specified in equation (5).

$$y = ce^{-ax} = ce^{-5x}$$

To find the value of c, we apply the initial value condition

$$y(0) = ce^{-5(0)} = c = 2$$

The final result is

$$y = 2e^{-5x}$$

Solving these problems analytically is often an interesting exercise, but we are really not concerned with that here. We, and you, want to address the solution of differential equations on the computer. A discussion of solving these via numerical methods is presented in the next section.

8.2 Numerical Solutions

As we have discussed earlier, there may be several reasons to prefer a computerized numerical solution to an analytic solution. A straightforward closed-form solution may not be obtainable, an available analytic procedure may be long and complex, or we simply may not want to expend the necessary energy to determine the analytic solution. There happen to be some powerful numerical techniques available to approximate the solution to differential equations that we can use in these types of situations. These methods have become more popular as their ease of application increases with the power of the computer.

We need to make some assumptions prior to presenting the numerical procedure. First, we will be considering first-order ordinary differential equations of the form given in equations (3). We must also assume that there exists a unique solution to the problem in some interval surrounding x_0. In order to draw this conclusion, we must assume that $f(x,y)$ and its partial derivative with respect to y are continuous in a rectangular region containing (x_0, y_0). These assumptions may not always be straightforward to verify, but this will not prevent us from attempting to find a solution. We must, however, be very careful in interpreting our result. If the required assumptions do not hold, our result could be very far from the correct solution.

The numerical method we present, known as the Runge-Kutta method, represents a class of methods called one-step or starting methods. Another group of methods, called multistep or continuing methods, is discussed in Boyce and DiPrima (1977).

The Runge-Kutta method was developed initially by Carl Runge in 1895 and continued by M.W. Kutta in 1901. The procedure is based on the Taylor-series expansion of the function y. This expansion is

$$y_{n+1} = y_n + hy'_n + \frac{h^2}{2} y''_n + \frac{h^3}{6} y'''_n + \frac{h^4}{24} y^{iv}_n + \cdots \tag{6}$$

We will consider the Runge-Kutta formula found by truncating the expansion following the terms presented above. An approximation is used for calculating y using the following formula:

$$y_{n+1} = y_n + h[a_0 f(x_n, y_n) + a_1 f(x_n + b_1 h, y_n + c_1 h) + a_2 f(x_n + b_2 h, y_n + c_2 h) + a_3 f(x_n + b_3 h, y_n + c_3 h)]$$

The a's, b's, and c's are determined such that the coefficients agree with the corresponding coefficients of equation (6). Notice that this latter expression does not require the evaluation of any derivatives. This is an important advantage. There are several evaluations of $f(x,y)$, but this is no problem for computer application.

By using this type of approximation, we can formulate the Runge-Kutta formula as follows:

$$y_{n+1} = y_n + \frac{h}{6} (k_0 + 2k_1 + 2k_2 + k_3) \tag{7}$$

where

$$k_0 = f(x_n, y_n)$$
$$k_1 = f(x_n + 0.5h, y_n + 0.5hk_0)$$
$$k_2 = f(x_n + 0.5h, y_n + 0.5hk_1)$$
$$k_3 = f(x_n + h, y_n + hk_2)$$

Because of the number of terms used from the Taylor-series expansion, this is called a fourth-order Runge-Kutta formula.

Throughout the discussion so far, we have assumed a uniform step size h on the x-axis. That is, $x_n = x_0 + nh$. This step size, or spacing, has an impact on the accuracy of our approximation. In fact, the error in using the formula of equation (7) is proportional to h, and for some finite interval the formula has an accumulated error of at most a constant times h (Boyce and DiPrima, 1977). Thus, we can control our error through selection of the step size h. Remember, though, that this increased accuracy comes at the expense of increased computation time. We will not discuss the error term here, but you may find it worthwhile to consult the references at the end of the chapter in this regard.

At this time, let us examine a sample problem to see how the procedure works in actually solving a differential equation.

EXAMPLE 2. Solve the following differential equation using the Runge-Kutta formula from equation (7) at $x = 0.2$ and $h = 0.2$.

$$y' - 2xy = x \qquad y(0) = 1$$

SOLUTION. We first rewrite the differential equation in standard form.

$$y' = x + 2xy \qquad y(0) = 1$$

We can now immediately begin the solution procedure by recognizing from the initial condition that $x_0 = 0$, $y_0 = 1$. First calculate the k values.

$$k_0 = f(x_0, y_0) = f(0, 1) = 0$$
$$k_1 = f(x_0 + 0.5h, y_0 + 0.5hk_0) = f(0+0.1, 1 + (0.1)(0)) = 0.3$$
$$k_2 = f(x_0 + 0.5h, y_0 + 0.5hk_1) = f(0+0.1, 1 + (0.1)(0.3)) = 0.306$$
$$k_3 = f(x_0 + h, y_0 + hk_2) = f(0+0.2, 1 + (0.2)(0.306)) = 0.62448$$

Plugging these into equation (7) yields:

$$y = 1 + \frac{0.2}{6}(0 + (2)(0.3) + (2)(0.306) + 0.62448) = 1.061216$$

This result tells us that the approximate value of the function $y = g(x)$ (the solution to the differential equation) evaluated at $x = 0.2$ is 1.061216. It is interesting to note that this corresponds to the actual value found from the true function $g(x)$. This function is

$$y = g(x) = -1/2 + 3/2\, e^{x^2}.$$

If we evaluate this function at $x = 0.2$, we find

$$g(0.2) = -1/2 + 3/2\, e^{(0.2)^2} = 1.06121616$$

We can improve the accuracy of the approximation by decreasing the step size. In the previous example this hardly seems to be necessary due to the accuracy of the result. This will not always be the case, however. Though computation time may increase, the increase in accuracy may be worthwhile.

How do we effect an increase in the step size operationally? The Runge-Kutta method presents us with a value of x and an associated value of y=g(x). Remember that h is a uniform spacing along the x-axis. We specify successive values of x by using the following relationships:

$$x_n = x_0 + nh \tag{8}$$

$$x_n = x_{n-1} + h \tag{9}$$

In the previous example we used a value of h=0.2. Since $x_0 = 0$ in this case, we would have the following sequence of x values for successive iterations throughout the procedure.

n	x	value
0	x_0	0
1	$x_1 = 0+h$	0.2
2	$x_2 = 0.2+h$	0.4
3	$x_3 = 0.4+h$	0.6
.	.	.
.	.	.
.	.	.

Thus, if we made another pass through the formula in Example 2, starting where we left off, we would find the approximate value of y=g(x) at x=0.4. Another pass would lead to g(x) at x=0.6, and so on. By examining equation (8) we can easily determine the appropriate step size h required for a particular number of iterations to the desired point at which to evaluate g(x).

For example, suppose $x_0=1$ and we are interested in the solution of a differential equation at the point $x_n=5$. If we wanted to use 20 iterations through the Runge-Kutta formula, we would solve equation (8) for h to find the step size.

$$h = \frac{x_n - x_0}{n} = \frac{5-1}{20} = 0.2$$

When selecting the step size, it is important to remember that the desired point of evaluation, x_n, must be an integer multiple of h from the initial value x_0.

Now you know how the Runge-Kutta method works. In the next section, we will examine the computer application of this procedure.

8.3 Application of the Runge-Kutta Method

Through Example 2 you can see that application of the Runge-Kutta method is not complex and should be easily performed by your IBM Personal Computer. The repeated functional evaluations make this a natural for computer implementation.

162 Chapter 8—Numerical Solutions to Differential Equations

The general solution procedure is specified below. The computer program follows these same steps.

1. Specify f(x,y) from the problem statement.
2. Choose an appropriate step size h for the problem.
3. Begin the procedure at x_0 and y_0, which are specified as the initial condition of the problem.
4. Calculate k_0, k_1, k_2, and k_3 terms based upon current values for x_n and y_n.
5. Find the value of $y_{n+1} = g(x_{n+1})$ from equation (7).
6. If more function evaluations are desired, return to step 4; otherwise stop.

This procedure is not difficult to program. A subroutine to perform the Runge-Kutta method is presented in Figure 8.1. We can easily write a main program to obtain input information for the Runge-Kutta subroutine. This is illustrated in the following example.

```
21200 REM***************************************************************
21210 REM RUNGE-KUTTA: FIXED STEPSIZE
21220 REM Subroutine to perform the Runge-Kutta method for
21230 REM solving differential equations. Input to the
21240 REM subroutine must be:
21250 REM   1. Stepsize to be used: H
21260 REM   2. Initial condition value for X: R
21270 REM   3. Initial condition value for Y(X): S
21280 REM   4. Stopping point along x-axis: MAX
21290 REM   5. Function specified in FNDE
21300 REM
21310 REM Set double precision variables
21320 DEFDBL X,Y, K
21330 DEFINT I,J
21340 L=INT((MAX-R)/H)       'Number of required evaluation points
21350 DIM X(L+1),Y(L+1)           'Set dimensions
21360 I=0
21370 X(0)=R                  'Set initial condition
21380 Y(0)=S                  'Set initial condition
21390 PRINT:PRINT:PRINT
21400 PRINT "RUNGE-KUTTA PROCEDURE FOR STEPSIZE= ";H
21410 PRINT
21420 PRINT "STEP X           X      Y=G(X)"
21430 PRINT "**********************************"
21440 PRINT TAB(2);I;
21450 PRINT TAB(10);
21460 PRINT USING "#####.#####";X(I);
21470 PRINT TAB(23);
21480 PRINT USING "#####.#####";Y(I);
21490 X(I+1)=X(I)+H
21500 IF X(I+1) > MAX THEN GOTO 21580
21510 K=FNDE(X(I), Y(I))
21520 K1=FNDE(X(I)+.5*H,Y(I)+.5*H*K)
21530 K2=FNDE(X(I)+.5*H,Y(I)+.5*H*K1)
21540 K3=FNDE(X(I)+H,Y(I)+H*K2)
21550 Y(I+1)=Y(I)+(H/6)*(K+2*K1+2*K2+K3)
```

```
21560 I=I+1
21570 GOTO 21440
21580 RETURN
```

Figure 8.1—Runge-Kutta subroutine

EXAMPLE 3. Write a main program to utilize the subroutine in Figure 8.1 and use it to solve the following differential equation at x=0.2, 0.4, and 0.5.

$$y' + 2xy - 3x = 0 \qquad y(0) = 1$$

SOLUTION. A main program to collect input data for a differential equation is presented in Figure 8.2. Notice that the function must be specified within the program. Inputs include the step size h, the initial conditions, and the last x value to be evaluated.

```
100 REM RUNGE-KUTTA MAIN PROGRAM
110 REM Main program to utilize the Runge-Kutta
120 REM subroutine. Access to the fixed stepsize
130 REM subroutine or the revised (variable stepsize)
140 REM subroutine is possible. The differential
150 REM equation to be solved must be input in the
160 REM function statement in line 290.
170 INPUT "ENTER THE DESIRED OR INITIAL STEPSIZE";H
180 PRINT:PRINT
190 PRINT "ENTER THE INITIAL CONDITIONS FOR Y(X0)=Y0"
200 PRINT
210 INPUT "ENTER THE INITIAL VALUE X0";R
220 INPUT "ENTER THE INITIAL VALUE Y0";S
230 PRINT:PRINT
240 PRINT "SPECIFY THE LAST X-VALUE AT WHICH TO EVALUATE G(X)"
250 PRINT "NOTE: IF THIS VALUE IS NOT AN INTEGER MULTIPLE"
260 PRINT "OF H FROM X0, IT WILL NOT BE EVALUATED"
270 INPUT MAX
280 REM Input the differential equation to be solved
290 DEF FNDE(X,Y)=3*X-2*X*Y
300 GOSUB 21200          'Fixed stepsize subroutine
310 END
```

Figure 8.2—Runge-Kutta main program

From the problem statement we must obtain evaluations for x=0.2, 0.4, and 0.5. Since there are different intervals between these points we could proceed in many different ways. Using a step size of h=0.1, we would find g(x) for x=0.1, 0.2, 0.3, 0.4, 0.5, which would meet the requirements. We could also use a step size of h=0.2 for x=0.2 and 0.4, and a step size of 0.5 for x=0.5. Using the main program in these two cases, we get the results presented below.

Step Size	X	Y=G(X)
0.1	0	1
	0.1	1.00498
	0.2	1.01961
	0.3	1.04303
	0.4	1.07393
	0.5	1.11060
0.5	0	1
	0.5	1.11068

Using data found from the Runge-Kutta method for various values of x, one could approximate the function y=g(x) within that range. Chapter 4 discusses some curve fitting techniques which might be used.

The Runge-Kutta procedure can be made more reliable and efficient through automatic adjustment of the step size during computation. You may desire a certain accuracy in the approximation during each step. By specifying this desired accuracy, you allow the program to adjust the step size to ensure that the accuracy is being maintained. If the error is too large at any point, the step size is automatically decreased in order to increase accuracy. If the error is extremely small, the step size can be increased in order to improve efficiency. This type of procedure is included in the subroutine presented in Figure 8.3.

```
21590 REM*****************************************************************
21600 REM RUNGE-KUTTA: VARYING STEPSIZE
21610 REM Subroutine to perform the Runge-Kutta method for
21620 REM solving differential equations. Input to the
21630 REM subroutine must be:
21640 REM  1. Initial stepsize to be used: H
21650 REM  2. Initial condition value for X: R
21660 REM  3. Initial condition value for Y(X): S
21670 REM  4. Stopping point along X-axis: MAX
21680 REM  5. Function specified in FNDE in main program
21690 REM
21700 REM This subroutine will ask for the specification of a
21710 REM desired tolerance at each step, T
21720 REM
21730 REM Set double precision variables
21740 DEFDBL X,Y,K
21750 DEFINT I,J
21760 INPUT "ENTER DESIRED TOLERANCE AT EACH STEP";T
21770 L=INT((MAX-R)/H)         'Number of required evaluation points
21780 DIM X(L+1),Y(L+1)            'Set dimensions
21790 I=0
21800 J=1                      'Set "same or half" stepsize flag
21810 X(0)=R                   'Set initial condition
21820 Y(0)=S                   'Set initial condition
21830 PRINT:PRINT:PRINT
21840 PRINT "RUNGE-KUTTA PROCEDURE FOR STEPSIZE= ";H
```

8.3 Application of the Runge-Kutta Method

```
21850 PRINT
21860 PRINT "STEP     STEPSIZE          X          Y=G(X)"
21870 PRINT "*******************************************"
21880 PRINT TAB(2);I;
21890 PRINT TAB(10);
21900 PRINT USING "#.###";H;
21910 PRINT TAB(18);
21920 PRINT USING "#####.#####";X(I);
21930 PRINT TAB(31);
21940 PRINT USING "#####.#####";Y(I);
21950 IF J=2 THEN H=H*2          'Double stepsize
21960 J=1                         'Reset flag
21970 X(I+1)=X(I)+H
21980 IF X(I+1)>MAX THEN GOTO 22270
21990 K=FNDE(X(I),Y(I))
22000 K1=FNDE(X(I)+.5*H,Y(I)+.5*H*K)
22010 K2=FNDE(X(I)+.5*H,Y(I)+.5*H*K1)
22020 K3=FNDE(X(I)+H,Y(I)+H*K2)
22030 Y(I+1)=Y(I)+(H/6)*(K+2*K1+2*K2+K3)
22040 REM Halve the stepsize to check the accuracy of the
      approximation
22050 HH=H/2
22060 YY=Y(I)
22070 XX=X(I)
22080 FOR II=1 TO 2              'Find the next Y for half the
                                  stepsize
22090 M=FNDE(XX,YY)
22100 M1=FNDE(XX+.5*HH,YY+.5*HH*M)
22110 M2=FNDE(XX+.5*HH,YY+.5*HH*M1)
22120 M3=FNDE(XX+HH,YY+H*M2)
22130 YY=YY+(HH/6)*(M+2*M1+2*M2+M3)
22140 XX=XX+HH
22150 NEXT II
22160 REM Check to see if the approximation meets the specified
      tolerance
22170 IF ABS (YY-Y(I+1))>T THEN GOTO 22220   'Decrease stepsize
22180 IF ABS(YY-Y(I+1))<T/100 THEN GOTO 22200 'Increase stepsize
22190 GOTO 22250                             'Keep current stepsize
22200 J=2                         'Set "double stepsize" flag
22210 GOTO 22250
22220 PRINT "TOLERANCE NOT MET ON THE FOLLOWING STEP: H HALVED"
22230 Y(I)=YY
22240 H=HH
22250 I=I+1
22260 GOTO 21880
22270 RETURN
```

Figure 8.3—Runge-Kutta, varying step size subroutine

This procedure accepts an initial step size from you, the user. An acceptable error must also be specified. The program calculates each point using both the specified step size, and one-half of that value, and compares them. If the difference between these values is greater than the acceptable error, the step size is cut in half. If the difference is less than the acceptable error divided by 100, the step size is doubled. In any other case, the step size remains the same.

EXAMPLE 4. Solve the differential equation in Example 3 for $x=0.1,\ldots,0.8$ using an initial step size of 0.1 and a tolerance of 0.01, using the subroutine in Figure 8.3.

$$y' + 2xy - 3x = 0 \qquad y(0) = 1$$

SOLUTION. Using the main program in Figure 8.2 to access the variable-step size subroutine, we obtain the results illustrated below. Notice how this procedure may not yield the specific results we need.

Step	Step Size	X	Y=G(X)
0	0.100	0.00000	1.00000
1	0.100	0.10000	1.00498
2	0.200	0.30000	1.04303
3	0.400	0.70000	1.19373

8.4 Summary

The numerical procedure presented in this chapter, Runge-Kutta, can prove quite useful in finding the solution to differential equations. Remember that we have considered only first-order ordinary differential equations. The procedure can be revised to solve a series of differential equations (see Hildebrand, 1974). The Runge-Kutta method allows us to solve many problems occurring naturally in the physical environment.

References

Boyce, William E. and Richard C. DiPrima, *Elementary Differential Equations and Boundary Value Problems.* John Wiley and Sons, Inc., New York, N.Y., 1977.

Hildebrand, F.B., *Introduction to Numerical Analysis.* McGraw-Hill, Inc., New York, N.Y., 1974.

Exercises

1. At time $t=0$ there is Q_0 lbs. of a substance dissolved in 50 gallons of water in a barrel. Water is entering the barrel at the rate of 2 gallons/min. which contains $1/2$ lb. of the substance per gallon. The well-mixed solution leaves the barrel at the same rate. What is the amount of the substance in the barrel at $t=6$? Let $Q_0 = Q(t_0) = 5$ lbs.

    Hint: the differential equation for this system is

    $$Q'(t) = 1 - \frac{Q(t)}{25}$$

2. Solve the following first-order ordinary differential equation at $x=0, 1,$ and 2.

    $$y' = \frac{2x^2 + y}{6 - 9y - x} \qquad y(0) = 0$$

3. The following differential equation is not presented in standard form. Find the standard form and find g(x) at x=1, 1.5, and 2.

   $xy' + y = 1 - xy \qquad y(1) = 0$

4. Classify the order of the following differential equations. Which ones could you solve using the programs in this chapter?

   a. $x^2y'' + 3x(x-y) + 2xy' = 0$

   b. $y' + x^5 - 3x^4 + e^x - 3 = xy'$

   c. $y'' = e^{-5x}$

   d. $x^4y - 9x^3y' - 7 = x^6y' + \cos x$

   e. $(x^4 - 3x^2y)dx + x^2dy = xy$

5. The disintegration of a new radioactive isotope, Dittmannium, is proportional to the amount remaining. What is the expression of this phenomenon as a differential equation? If we had 300 milligrams of Dittmannium at t=0, and had 261 milligrams 6 days later, how much would we have in 24 days?

9

Linear Programming

In many cases of design and analysis, one needs to find the values of certain variables that result in the maximum or minimum value of a function. Further, these variables are often constrained in the values which they may take on. Such a problem is called a mathematical programming problem. In this chapter we examine this type of problem when all the functions are linear; it is therefore referred to as linear programming (LP). We will present the general structure of the LP model and briefly discuss the widely used solution procedure called simplex. A BASIC program is included to solve LP problems using the simplex procedure.

9.1 Linear Programming

Linear programming (LP) is one technique among many designated as mathematical programming. The use of the word "programming" implies nothing about the programming of a computer. In this context it refers to a specified, or programmed, solution procedure. For small problems solutions can be found through hand calculation. Larger problems—in fact, most problems—are solved much more efficiently on the computer.

Problem formulations which may be solved via LP, called linear programs, are distinguishable because all functions used are linear functions. If the functions are nonlinear the correct solution procedure is nonlinear programming, which we do not discuss in this book.

Linear programming has an extensive theoretical background. Time and space do not permit us to develop the theoretical foundations. Linear algebra concepts are extremely important to this development and are used extensively in the solution to problems, as you will see in Section 9.3. You may find it useful to supplement the material presented here with other sources. There are many fine books which discuss LP, from the easy and practical to the complex and theoretical. A few books which you may find helpful are Taha (1982), McMillan (1975), Hadley (1963), and Bazaraa and Jarvis (1977).

In which specific situations would you want to use linear programming? First of all, LP requires you to reduce a system or specific problem into a set of linear functions. The specification of these functions is discussed in greater detail in the next section. This conversion to a set of functions is itself a difficult task. It is, in fact, the most important part of the solution process. If you find the optimal solution to an LP formulation that is not accurate, you have nothing. We do not spend much time on this topic, but that is not indicative of its value. All the sources listed previously discuss the formulation of linear programs in greater detail.

The actual situations in which LP has been used are numerous. It is generally concerned with the optimal allocation of scarce resources. Since almost all resources are scarce for one reason or another, LP is quite applicable. It has been used in such varied situations as planning, scheduling, transportation, systems analysis, and systems design. It has applications in any type of industry or organization.

The general linear program is formulated in a very specific manner. It consists of a single objective function. This is a linear function of the variables of interest, called decision variables, that represents the desired objective of the problem. This objective can be maximized or minimized. For instance, a company may wish to minimize transportation cost, or maximize profit. A shop foreman may want to minimize machine down time. A designer of rocket engines may wish to maximize thrust. All of these would specify the objective function.

The remaining functions used are called constraints. These are functions which place restrictions on the system and the decision variables. The constraints limit the value of the objective function. The designer of the rocket engine may be required to use a particular kind of fuel. Thus, he wants to maximize thrust "subject to" the constraint of using a particular fuel. Restrictions, or constraints, on the shop foreman may be types of parts produced, number of machine set-ups required, training of employees, or types of machines. Thus, the entire LP formulation consists of a single objective function and a series of constraints.

In the following section, we present the general structure of a linear program including graphical solutions and an example problem. In Section 9.3 we discuss solution procedures for this type of problem, including the simplex procedure. Finally, a BASIC program for solving LP problems is presented in Section 9.4. This includes an example problem and complete explanation of the output provided.

9.2 Basic Formulation

In the previous section we discussed the general structure of a linear programming formulation; a single objective function and a series of constraints. All functions used in the formulation must be linear. There is a basic formulation that represents linear programs. The following basic formulation is offered:

$$\text{Maximize} = \sum_{j=1}^{n} c_j x_j \tag{1}$$

subject to

$$\sum_{j=1}^{n} a_{ij} x_j \leq b_i \qquad i=1,2,\ldots,m \tag{2}$$

$$x_j \geq 0 \qquad j=1,2,\ldots,n \tag{3}$$

where

x_j = decision variables; variables of interest
c_j = coefficients of the decision variables in the objective function
a_{ij} = coefficient of the jth decision variable in the ith constraint

b_i = maximum available amount of the ith resource
m = number of constraint functions
n = number of decision variables

This formulation represents a "maximization" problem with "less than or equal to" constraints. Notice from (3) that all the decision variables are required to be nonnegative.

We could also have a "minimization" problem or "greater than or equal to" constraints. If we had both the formulation may look like this:

$$\text{Minimize} = \sum_{j=1}^{n} c_j x_j \tag{4}$$

subject to

$$\sum_{j=1}^{n} a_{ij} x_j \geq b_i \quad i=1,2,\ldots,m \tag{5}$$

$$x_j \geq 0 \quad j=1,2,\ldots,n \tag{6}$$

As you know, working with inequalities can be significantly more troublesome than working with equalities. Thus, in order to solve LP problems analytically, we convert the inequality constraints to equality constraints. This procedure is explained thoroughly in Section 9.3.

Graphical Solution

For problems with two decision variables, the functions can all be plotted in two dimensions. We can use this to solve LP problems graphically. Consider the following problem:

Maximize $z = 2x_1 + 5x_2$

subject to

$$x_1 \leq 4$$
$$x_2 \leq 3$$
$$x_1 + 2x_2 \leq 8$$
$$x_1 \geq 0, \ x_2 \geq 0$$

We can graph the constraints to find a "feasible region." Only decision variables that lie within this region are possible solutions to the problem. Points lying outside the region constitute "infeasible" solutions. The feasible region for this problem is illustrated in Figure 9.1. The dashed line in Figure 9.1 is the objective function with a value of ten; that is $2x_1 + 5x_2 = 10$.

Any point (x_1, x_2) along $2x_1 + 5x_2 = 10$ of course yields an objective value of 10. From the graph you can see that many points in the feasible region satisfy this. We want to find the point (x_1, x_2) in the feasible region that yields the largest value of the objective function. This will be the "optimal solution."

As we move the objective function up and to the right in Figure 9.1, its value increases. We want to move the objective function to the point where it just touches the feasible region. If we do this for our problem, the result is presented in Figure 9.2.

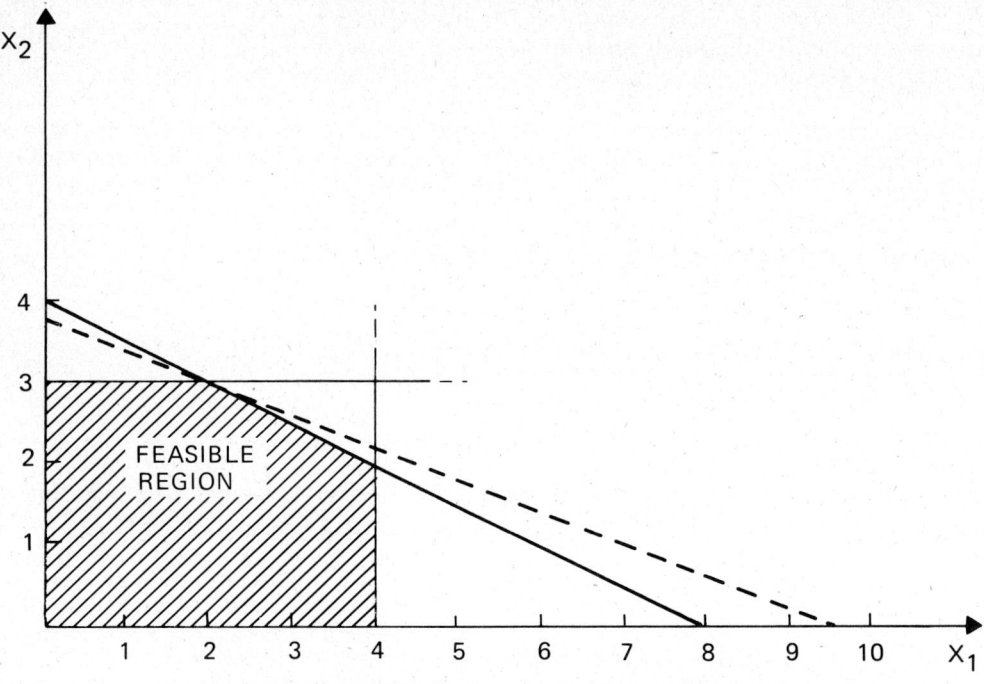

Figure 9.1—feasible region for the example problem

Notice that the optimal solution occurs at only one point: $x_1 = 2$, $x_2 = 3$. The resulting objective value is $z = 2(2) + 5(3) = 19$. The optimal point lies at one of the corners of the feasible region. With linear programming, this will always be the case. That is, the optimal solution will always be at a corner point of the feasible region. (In certain cases optimal solutions may also lie along one of the constraints, yielding an infinite number of optimal solutions. Consult the references for further discussion of this phenomenon.) This interesting fact allows us to solve LP problems much more simply. In our problem, since there are five corner points in the feasible region, we need only look at five points to find the optimal solution. For small problems, we could use total enumeration!

You can see that graphical solutions are only applicable for small problems (two decision variables). However, the same principles apply for any size problem. That is, the optimal solution will lie at an extreme point, or corner point, of the feasible region. This fact is used to help find the solution, employing techniques from linear algebra. We present this solution procedure later in the chapter.

If you can formulate a problem so that it fits one of the basic forms above (other forms are also possible; see any of the references listed previously), you can solve it using LP. Getting it into the proper form, that is, formulating the problem as a linear program, is not always an easy task. It takes extensive practice and experience to become confident and comfortable with the formulation process. Below we present a classic linear programming problem. It is used to illustrate the formulation process. Lack of space prevents us from examining more examples of this type. If you find LP useful, we encourage you to review the references at the end of the chapter and study additional formulation exercises.

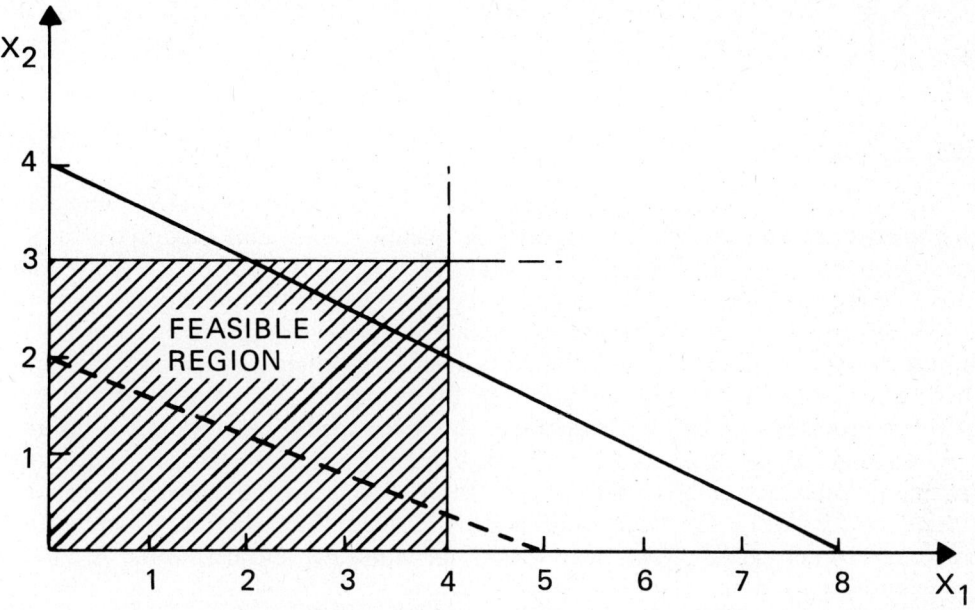

Figure 9.2—optimal solution to the example problem

Linear Programming Example

The following is the classic "peanut problem" adapted from Charnes, Cooper, and Henderson (1953): A manufacturer wishes to determine an optimal program for mixing three grades of nuts consisting of cashews, hazelnuts, and peanuts according to the specifications and prices listed in Table 1. Hazelnuts may be introduced into the mixture in any quantity, provided the specifications are met.

TABLE 1

Mixture	Specifications	Selling price: ¢/lb.
A	Not less than 50% cashews Not more than 25% peanuts	40
B	Not less than 25% cashews Not more than 50% peanuts	30
G	No specifications	25

If the symbols C for cashews, H for hazelnuts, and P for peanuts are adopted, the restrictions on marketable mixes may be stated in mathematical form:

$A_C \geq \tfrac{1}{2} A$

$A_P \leq \tfrac{1}{4} A$

$B_C \geq \frac{1}{4} B$

$B_P \leq \frac{1}{2} B$

where

$A_C + A_H + A_P = A$

$B_C + B_H + B_P = B$

The first four inequations state the restrictions on mixtures A and B. Whatever the quantity of A produced, the quantity of cashews going into this mixture, A_C, must account for at least half the total; peanuts going in, A_P, must not exceed one-fourth of the mixture. Similar restrictions apply to B, but no restrictions apply to G. Since mixture G enters only as a residual, it finds no explicit place in these statements of restrictions. Hazelnuts do not appear explicitly in the first group of inequations since no restrictions apply to these nuts. They may appear in any quantity provided that the mixtures meet specifications on cashews and peanuts.

The second group of equations represents definitional relations: The weight of the inputs must equal the weight of the outputs. Some of the quantities may, of course, be zero when the final solution is determined, but the total output of A, say, must be composed of the cashews, A_C, peanuts, A_P, and hazelnuts, A_H, that enter into its makeup.

Substituting these definitional relations into the right-hand side of the inequations, and multiplying by -1, where necessary, to point all inequality signs in the same direction, the set may be simplified to:

$-\frac{1}{2} A_C + \frac{1}{2} A_P + \frac{1}{2} A_H \leq 0$

$-\frac{1}{4} A_C + \frac{3}{4} A_P - \frac{1}{4} A_H \leq 0$

$-\frac{3}{4} B_C + \frac{1}{4} B_P + \frac{1}{4} B_H \leq 0$

$-\frac{1}{2} B_C + \frac{1}{2} B_P - \frac{1}{2} B_H \leq 0$

Now, suppose that the manufacturer has certain capacity limits on the amounts of inputs that can be employed. Let these limitations and the price of the inputs appear as in Table 2.

TABLE 2

Inputs	Capacity: lb./day	Price: ¢/lb.
C	100	65
P	100	25
H	60	35
Total	260	

These terms need to be interpreted with care before giving them mathematical expression. If "capacity" is interpreted as a maximal limitation on each type of nut, the conditions will be expressed as

$C = A_C + B_C + G_C \leq 100$

$P = A_P + B_P + G_P \leq 100$

$H = A_H + B_H + G_H \leq 60$

9.2 Basic Formulation

In other words, the total amount of cashews, C, going into A, B, and G cannot exceed 100 lb.; the same limitation applies to peanuts; and the total amounts of hazelnuts cannot exceed 60 lb.

There are then 16 unknowns to be determined from the 7 constraints and the nonnegativity constraints. Such a set of relations will, in general, yield an infinite number of solutions. From this infinite set of solutions it is desired to select the best variables according to some criterion or set of criteria. Such a criterion function is found in this problem by constructing the net receipts function. Referring to the set of prices listed in Tables 1 and 2, the following (linear) function may be constructed:

$$z = 40(A_C + A_P + A_H) + 30(B_C + B_P + B_H) + 25(G_C + G_P + G_H)$$
$$- 65(A_C + B_C + B_C) - 25(A_P + B_P + G_P) - 35(A_H + B_H + G_H)$$

or $z = -25A_C + 15A_P + 5A_H - 35B_C + 5B_P - 5B_H - 40G_C + 0G_P - 10G_H$

9.3 Solution Methods

We will discuss two solution procedures in this section. The first, solution via linear algebra, is not a practical method. However, it does assist us in establishing the simplex solution concepts which follow.

Solution Via Linear Algebra

In order to solve an LP problem using linear algebra, we must first put the problem in standard form. This standard form must contain an objective function subject to a set of *equality* constraints. That is, each constraint function is equal to the resource available. Our standard form will have a maximization objective. If the problem dictates a minimization objective, simply multiply the objective function by -1 to convert it to maximization. The standard form is:

$$\text{Maximize } Z = \sum_{j=1}^{n} c_j x_j \tag{7}$$

subject to

$$\sum_{j=1}^{n} a_{ij} x_j = b_i \qquad i = 1, 2, \ldots, m \tag{8}$$

$$x_j \geq 0 \qquad j = 1, 2, \ldots, n \tag{9}$$

The x_j now may not all be decision variables. Some variables are required to convert constraints from inequalities to equalities. These are called "slack" variables. We would modify our example problem to fit the standard form as follows:

$$\text{Maximize } Z = 2x_1 + 5x_2$$

subject to

$$x_1 + x_3 = 4$$
$$x_2 + x_4 = 3$$
$$x_1 + 2x_2 + x_5 = 8$$
$$x_1, x_2, x_3, x_4, x_5 \geq 0$$

Notice that variables x_3, x_4, and x_5 do not appear in the objective function (in other words, they have coefficients of zero). These variables merely take up any slack if the left hand side of an original inequality does not equal the right hand side.

The above procedure is used to convert less than or equal to constraints to equality constraints. Similarly, we can convert greater than or equal to constraints to equalities by subtracting "slack" variables. The references listed discuss this in detail. The constraints specifying nonnegativity will be implicit in the solution procedure.

With equality constraints, linear algebra can be used. Remember that the optimal solution will be at an extreme point of the feasible region. This is what is called a "basic" solution. A basic solution is found by assigning some variables the value of zero and solving for the remaining variables. For a problem with n variables and m equality constraints ($n \geq m$), a basic solution is found by setting n-m variables equal to zero (called nonbasic variables) and solving for the remaining m variables (called basic variables). Each basic solution corresponds to the intersection of m constraints in m-space. Some of these intersections will lie in the feasible region and will therefore be feasible solutions. Thus, each basic feasible solution represents a "corner" point of the feasible region. Intersections outside the feasible region yield infeasible solutions. The way we determine whether a solution is feasible or not is by examining the signs of the resulting basic variables. If all are nonnegative, the solution is feasible. If one or more variables are negative, the solution is infeasible. We are, of course, interested only in feasible solutions.

To solve our example problem, we could examine all basic solutions and choose that one which gives the highest value of the objective function. There are a total of eight basic solutions to our problem. We find two of these below. Finding a basic solution consists of solving a set of simultaneous equations. This procedure is not discussed here but is presented in detail in Chapter 5.

Basic solution #1

   Nonbasic variables: x_4, x_5

   Basic variables: x_1, x_2, x_3

   Constraint equations:
   $$x_1 + x_3 = 4$$
   $$x_2 = 3$$
   $$x_1 + 2x_2 = 8$$

   Solution: $x_1 = 2 \quad x_2 = 3 \quad x_3 = 2$

   Objective value: $Z = 19$

Basic solution #2

   Nonbasic variables: x_1, x_4

   Basic variables: x_2, x_3, x_5

Constraint equations:
$$x_3 = 4$$
$$x_2 = 3$$
$$2x_2 + x_5 = 8$$

Solution: $x_2 = 3 \quad x_3 = 4 \quad x_5 = 2$

Objective value: $Z = 15$

If we examined all basic solutions, we would find that the optimal solution is the first basic solution presented above (compare this with the graphical solution we found earlier).

As it turns out, we do not have to examine all the basic solutions to find the optimal solution to an LP problem. In the next section we briefly describe the procedure commonly used to solve LP problems. It utilizes the concepts developed so far in an efficient manner to reduce computational effort.

Simplex Procedure

The most commonly used procedure for solving linear programming problems is the simplex method. Developed by George Dantzig in the late 1940s, this is a simple technique which can solve linear programming problems in an efficient manner. The basic concept of the simplex method is easy to understand. We said earlier that the optimal solution to any linear programming problem always occurs at a corner point of the feasible region. All we have to do is find that particular corner point or points that yield the optimal value of the objective function. For large problems this can amount to many possible solution points that must be investigated. The simplex method provides an efficient pattern in which to investigate these points.

The simplex method proceeds from one corner point of the feasible region to another "adjacent" corner point. If the point reached yields the optimal solution, the procedure stops. If the point is not optimal, the solution moves to the next corner point. The process continues in this fashion until the optimal solution is found. An important fact to note is that the simplex method can detect the 'optimality' of a particular solution without examining any other solutions. This is important because once the optimal solution is found, it is not necessary to examine any additional points. We do not have to evaluate all corner points, but can stop when the optimal solution is found.

Let us refer back to the move from one corner point to an "adjacent" one. What does this mean? Recall the concept of basic variables from the previous section. A particular set of basic variables, a basic solution, corresponds to a particular corner point. To move from one corner point to another, the set of basic variables must change; at least one basic variable must be different. We move to an adjacent corner point by changing only one of the basic variables. Thus, the simplex method moves from one corner point to an adjacent one by making one of the basic variables a nonbasic variable, and making one of the nonbasic variables a basic variable.

The simplex method must also tell us which basic variable to make nonbasic, and which nonbasic variable should become basic. This decision is made according to its effect on the objective function and the feasibility of the solution. The nonbasic variable that "enters" the solution (becomes basic) is that nonbasic variable which will improve the objective function the most, per unit increase in that variable. This is the first decision that is made. The basic variable that "leaves" the solution (becomes nonbasic) is dependent on the variable that enters. The leaving variable is chosen so that the solution remains feasible. In this manner the simplex method guarantees that, starting with a basic solution in the feasible region, each successive basic solution will also be in the feasible region. There will be a leaving basic variable that meets this requirement. (Note: two important topics in this regard are "degeneracy" and "unboundedness." The references at the end of this chapter discuss these in detail.)

The simplex method stops when the current basic solution is optimal. The solution is optimal when none of the nonbasic variables, if they were entered into the solution, would improve the objective function.

A discussion of the actual operation of the simplex method is not possible in this book. You may want to consult the references to learn the mechanics of the procedure and the many special topics which could not be included here. The use of the simplex program is discussed in the following section.

9.4 Using the Simplex Program

Figure 9.3 presents a stand-alone program of the simplex method to solve linear programming problems. In this section we will discuss its use and evaluation of its output.

```
1000 REM THIS PROGRAM USES THE SIMPLEX PROCEDURE TO SOLVE LINEAR
1010 REM PROGRAMMING PROBLEMS. THE USER MUST FORMULATE THE PROBLEM
1020 REM AS A MAXIMIZATION PROBLEM WITH EQUALITY CONSTRAINTS. THE
1030 REM INITIAL BASIC VARIABLES MUST BE INCLUDED IN THE FORMULATION.
1040 REM INPUT MUST INCLUDE THE NUMBER OF CONSTRAINTS, THE NUMBER OF
1050 REM VARIABLES (INCLUDING THE INITIAL BASIC VARIABLES), THE
1060 REM COEFFICIENTS OF ALL VARIABLES IN THE OBJECTIVE FUNCTION,
1070 REM AND THE COEFFICIENTS OF ALL VARIABLES IN THE
1080 REM CONSTRAINTS WITH THE EXCEPTION OF INITIAL BASIC VARIABLES,
1090 REM WHOSE COEFFICIENTS ARE AUTOMATICALLY SUPPLIED BY THE
1100 REM PROGRAM.
1110 DEFDBL C,E,Z,V,T,R,P,A
1120 DEFINT I,J,N
1130 IT=0
1140 EP=.0000005#
1150 PRINT "ENTER THE FOLLOWING PARAMETERS"
1160 INPUT "NUMBER OF CONSTRAINT EQUATIONS, NC";NC
1170 PRINT
1180 INPUT "NUMBER OF VARIABLES, INCLUDING INITIAL BASIS";NV
1190 PRINT
1200 INPUT "ENTER 1--ALL TABLEAUS : 2--ONLY FIRST AND LAST
      TABLEAUS";IR
1210 INPUT "ENTER THE TITLE OF THE PROBLEM";A$
```

9.4 Using the Simplex Program

```
1220 PRINT:PRINT
1230 DIM A(NC,NV),P(NC),JV(NV),C(NV),IV(NV),CC(NV),CI(NV)
1240 DIM ZJ(NV),CM(NV)
1250 PRINT "ENTER THE VARIABLE IDENTIFICATION NUMBERS"
1260 PRINT "BEGINNING WITH THE INITIAL BASIC VARIABLES"
1270 FOR J=1 TO NV
1280 INPUT JV(J)
1290 NEXT J
1300 PRINT:PRINT
1310 PRINT "ENTER THE OBJECTIVE FUNCTION COEFFICIENTS OF THE
1320 PRINT "VARIABLES YOU HAVE JUST LISTED"
1330 FOR J=1 TO NV
1340 IF J>NC THEN GOTO 1390
1350 JJ=NV-NC+J
1360 PRINT "ENTER THE COST FOR VARIABLE";JJ;
1370 INPUT C(J)
1380 GOTO 1420
1390 J2=J-NC
1400 PRINT "ENTER THE COST FOR VARIABLE";J2;
1410 INPUT C(J)
1420 NEXT J
1430 ND=NC+1
1440 PRINT "FOR EACH CONSTRAINT EQUATION, ENTER FIRST THE"
1450 PRINT "VALUE OF THE RIGHT HAND SIDE AND THEN ENTER THE"
1460 PRINT "COEFFICIENTS OF ALL VARIABLES IN THAT CONSTRAINT."
1470 PRINT ("NOTICE THAT A VARIABLE NOT APPEARING IN A"
1480 PRINT "PARTICULAR CONSTRAINT HAS A COEFFICIENT OF ZERO.")
1490 PRINT "DO NOT ENTER COEFFICIENTS FOR ANY VARIABLES CLASSIFIED"
1500 PRINT "AS ORIGINAL BASIC VARIABLES."
1510 PRINT:PRINT:PRINT
1520 FOR I=1 TO NC
1530 PRINT:PRINT:PRINT:
1540 PRINT "THE FOLLOWING QUESTIONS PERTAIN TO CONSTRAINT ";I
1550 PRINT
1560 INPUT "ENTER THE VALUE OF THE RIGHT HAND SIDE";P(I)
1570 FOR J=ND TO NV
1580 PRINT "INPUT THE COEFFICIENT OF VARIABLE";J-NC;
1590 INPUT A(I,J)
1600 NEXT J
1610 NEXT I
1620 FOR I=1 TO NC
1630 FOR J=1 TO NC
1640 A(I,J)=0          'BEGIN SETTING COEFFICIENTS FOR BASIC VARIABLES
1650 NEXT J
1660 CI(I)=C(I)              'SAVE INITIAL COST COEFFICIENTS
1670 IV(I)=JV(I)
1680 A(I,I)=1        'SET IDENTIFY MATRIX
1690 NEXT I
1700 INPUT "WOULD YOU LIKE TO CHANGE INPUT DATA (Y/N)";C$
1710 IF C$="Y" THEN GOSUB 2960
1720 PRINT
1730 INPUT "OUTPUT TO CRT OR PRINTER (C/P)";Q$
```

```
1740 PRINT
1750 IF Q$="P" THEN INPUT "TURN ON PRINTER, CARRIAGE RETURN WHEN
     READY";S$
1760 IF Q$="P" THEN LPRINT "SOLUTION TO ";A$
1770 IF Q$="C" THEN PRINT "SOLUTION TO ";A$
1780 CZ=-8.999999E+19
1790 FOR J=1 TO NV
1800 Z=0
1810 FOR I=1 TO NC
1820 Z=Z+A(I,J)*CI(I)
1830 NEXT I
1840 TN=C(J)-Z
1850 ZJ(J)=Z
1860 CM(J)=TN
1870 IF CZ>TN THEN GOTO 1900
1880 CZ=TN
1890 JI=J
1900 NEXT J
1910 IT=IT+1
1920 IF IR<>2 THEN GOTO 1950
1930 IF IT=1 THEN GOTO 1950
1940 IF CZ>=EP THEN GOTO 2650
1950 VL=0
1960 FOR I=1 TO NC
1970 VL=VL+P(I)*CI(I)
1980 NEXT I
1990 IF Q$="P" THEN GOSUB 3920
2000 IF Q$="P" THEN GOTO 2630
2010 PRINT:PRINT:PRINT:PRINT
2020 PRINT "THE FOLLOWING RESULTS ARE FOR ITERATION NUMBER";IT
2030 PRINT "OBJ. COEF.";
2040 FOR J=1 TO NV
2050 JJ=J+NC
2060 IF JJ>NV THEN JJ=JJ-NV
2070 IF J<=8 THEN GOTO 2120
2080 ISP=INT((J-1)/8)
2990 IF J=(ISP*8)+1 THEN PRINT
2100 PRINT TAB(8+(J-(ISP*8))*8);
2110 GOTO 2130
2120 PRINT TAB(8+J*8);
2130 PRINT USING "####.##";C(JJ);
2140 NEXT J
2150 PRINT:
2160 PRINT "VARIABLES";
2170 FOR J=1 TO NV
2180 JJ=J+NC
2190 IF JJ>NV THEN JJ=JJ-NV
2200 IF J<=8 THEN GOTO 2250
2210 ISP=INT((J-1)/8)
2220 IF J=(ISP*8)+1 THEN PRINT
2230 PRINT TAB(12+(J-(ISP*8))*8);
2240 GOTO 2260
```

```
2250 PRINT TAB(12+J*8);
2260 PRINT USING "##";JV(JJ);
2270 NEXT J
2280 PRINT:
2290 FOR I=1 TO NC
2300 PRINT USING "##";IV(I);
2310 PRINT TAB(4);
2320 PRINT USING "#######.##";P(I);
2330 FOR J=1 TO NV
2340 JJ=J+NC
2350 IF JJ>NV THEN JJ=JJ-NV
2360 IF J<=8 THEN GOTO 2410
2370 ISP=INT((J-1)/8)
2380 IF J=(ISP*8)+1 THEN PRINT
2390 PRINT TAB(8+(J-(ISP*8))*8);
2400 GOTO 2420
2410 PRINT TAB(8+J*8);
2420 PRINT USING "####.##";A(I,JJ);
2430 NEXT J
2440 PRINT:
2450 NEXT I
2460 PRINT:
2470 PRINT:
2480 PRINT"RCST";
2490 PRINT TAB(5);
2500 PRINT USING "######.##";VL;
2510 FOR J=1 TO NV
2520 JJ=J+NC
2530 IF JJ>NV THEN JJ=JJ-NV
2540 IF J<=8 THEN GOTO 2590
2550 ISP=INT((J-1)/8)
2560 IF J=(ISP*8)+1 THEN PRINT
2570 PRINT TAB(8+(J-(ISP*8))*8);
2580 GOTO 2600
2590 PRINT TAB(8+J*8);
2600 PRINT USING "####.##";CM(JJ);
2610 NEXT J
2620 PRINT:
2630 IF CZ<EP THEN GOTO 2950
2640 INPUT "CARRIAGE RETURN TO CONTINUE SOLUTION";B$
2650 TT=9D+19
2660 FOR I=1 TO NC
2670 IF A(I,JI)<=EP THEN GOTO 2720
2680 RT=P(I)/A(I,JI)
2690 IF TT<RT THEN GOTO 2720
2700 TT=RT
2710 IL=I
2720 NEXT I
2730 IF TT<900000000000000# THEN GOTO 2760
2740 PRINT "THE OBJECTIVE FUNCTION IS UNBOUNDED"
2750 GOTO 2950
2760 PV=A(IL,JI)
```

```
2770 P(IL)=P(IL)/PV                  'Calculate the new right hand side
2780 FOR J=1 TO NV
2790 A(IL,J)=A(IL,J)/PV              'Calculate the new pivot row
2800 NEXT J
2810 FOR I=1 TO NC
2820 IF I=IL THEN GOTO 2890
2830 P(I)=P(I)-P(IL)*A(I,JI)         'New rhs for other basic variables
2840 FOR J=1 TO NV
2850 IF J=JI THEN GOTO 2870
2860 A(I,J)=A(I,J)-A(I,JI)*A(IL,J)   'Find the basic variable row
                                      elements
2870 NEXT J
2880 A(I,JI)=0                       'Create a unit vector column
2890 NEXT I
2900 EP=EP+.0000005
2910 CI(IL)=C(JI)                    'assign new costs to the basic c
                                      vector
2920 IV(IL)=JV(JI)                   'switch other variable assignments
2930 A(IL,JI)=1                      'To complete the unit vector column
2940 GOTO 1780
2950 END
2960 REM*****************************************************************
2970 REM Subroutine for editing input data
2980 REM
2990 PRINT "ENTER THE NUMBER FOR THE DESIRED CHANGE"
3000 PRINT "     1: RETURN TO DATA ENTRY"
3010 PRINT "     2: CHANGE OBJECTIVE COEFFICIENTS"
3020 PRINT "     3: CHANGE CONSTRAINT EQUATION COEFFICIENTS"
3030 PRINT "     4: CHANGE RIGHT HAND SIDE VALUES"
3040 PRINT "     5: EXIT EDIT ROUTINE"
3050 INPUT ED
3060 IF ED=5 THEN GOTO 3910
3070 IF ED=1 THEN GOTO 1150
3080 IF ED=2 THEN GOTO 3290
3090 IF ED=3 THEN GOTO 3550
3100 PRINT "CURRENT RIGHT HAND SIDE VALUES ARE:"
3110 FOR I=1 TO NC
3120 PRINT "FOR CONSTRAINT ";I;" RHS IS ";P(I)
3130 NEXT I
3140 INPUT "DISPLAY CONSTRAINT LIST AGAIN (Y/N)";C$
3150 IF C$="Y" THEN GOTO 3100
3160 INPUT "ENTER CONSTRAINT NUMBER FOR CHANGE";DCH
3170 INPUT "ENTER NEW RIGHT HAND SIDE VALUE";P(DCH)
3180 PRINT "ENTER:"
3190 PRINT "     M: MODIFY ANOTHER RIGHT HAND SIDE"
3200 PRINT "     R: RETURN TO EDIT MENU"
3210 PRINT "     E: EXIT EDIT ROUTINE"
3220 INPUT T$
3230 IF T$="M" THEN GOTO 3140
3240 IF T$="R" THEN GOTO 2990
3250 IF T$="E" THEN GOTO 3910
3260 REM Flag input error and re-enter
```

```
3270 PRINT "INPUT ERROR: RE-ENTER"
3280 GOTO 3180
3290 REM For modifying objective coefficients
3300 PRINT "CURRENT OBJECTIVE COEFFICIENTS"
3310 FOR I=1 TO NV
3320 J=I+NC
3330 IF I>NV-NC THEN J=I-NV+NC
3340 PRINT "FOR VARIABLE ";I;" COEFFICIENT IS";C(J)
3350 NEXT I
3360 INPUT "DISPLAY COEFFICIENT LIST AGAIN (Y/N)";C$
3370 IF C$="Y" THEN GOTO 3300
3380 INPUT "ENTER VARIABLE NUMBER FOR CHANGE";JJT
3390 IF JJT>NV-NC THEN GOTO 3420
3400 JJT=JJT+NC
3410 GOTO 3430
3420 JJT=JJT-NV+NC
3430 INPUT "ENTER NEW COEFFICIENT VALUE";C(JJT)
3440 PRINT "ENTER;"
3450 PRINT "    M: MODIFY ANOTHER OBJECTIVE COEFFICIENT"
3460 PRINT "    R: RETURN TO EDIT MENU"
3470 PRINT "    E: EXIT EDIT ROUTINE"
3480 INPUT T$
3490 IF T$="M" THEN GOTO 3360
3500 IF T$="R" THEN GOTO 2990
3510 IF T$="E" THEN GOTO 3910
3520 REM Flag input error and re-enter
3530 PRINT "INPUT ERROR: RE-ENTER"
3540 GOTO 3440
3550 REM For modifying constraint coefficients
3560 PRINT "BASIC VARIABLES WILL NOT BE CONSIDERED"
3570 INPUT "DISPLAY ALL CONSTRAINT COEFFICIENTS (Y/N)";C$
3580 IF C$="N" THEN GOTO 3670
3590 FOR I=1 TO NC
3600 PRINT
3610 PRINT "FOR CONSTRAINT ";I;" COEFFICIENTS ARE:"
3620 FOR J=NC+1 TO NV
3630 PRINT "VAR ";J-NC;" = ";A(I,J);
3640 NEXT J
3650 NEXT I
3660 PRINT
3670 INPUT "DISPLAY COEFFICIENTS FOR A PARTICULAR CONSTRAINT
     (Y/N)";C$
3680 IF C$="N" THEN GOTO 3760
3690 INPUT "CONSTRAINT NUMBER";B
3700 FOR J=NC+1 TO NV
3710 PRINT "VAR ";J-NC;" = ";A(B,J);
3720 NEXT J
3730 PRINT
3740 INPUT "DISPLAY ANOTHER CONSTRAINT (Y/N)";C$
3750 IF C$="Y" THEN GOTO 3690
3760 INPUT "ENTER CONSTRAINT NUMBER FOR CHANGE";CCH
3770 INPUT "ENTER VARIABLE NUMBER FOR CHANGE";DCH
```

```
3780 DCH=DCH+NC
3790 INPUT "ENTER NEW COEFFICIENT VALUE";A(CCH,DCH)
3800 PRINT "ENTER:"
3810 PRINT "    M: MODIFY ANOTHER CONSTRAINT COEFFICIENT"
3820 PRINT "    R: RETURN TO EDIT MENU"
3830 PRINT "    E: EXIT EDIT ROUTINE"
3840 INPUT T$
3850 IF T$="M" THEN GOTO 3560
3860 IF T$="R" THEN GOTO 2990
3870 IF T$="E" THEN GOTO 3910
3880 REM Flag input error and re-enter
3890 PRINT "INPUT ERROR: RE-ENTER"
3900 GOTO 3800
3910 RETURN
3920 REM*********************************************************************
3930 REM SUBROUTINE FOR OUTPUT TO THE PRINTER
3940 LPRINT:LPRINT:LPRINT:LPRINT
3950 LPRINT "THE FOLLOWING RESULTS ARE FOR ITERATION NUMBER";IT
3960 LPRINT "OBJ. COEF.";
3970 FOR J=1 TO NV
3980 JJ=J+NC
3990 IF JJ>NV THEN JJ=JJ-NV
4000 IF J<=8 THEN GOTO 4050
4010 ISP=INT((J-1)/8)
4020 IF J=(ISP*8)+1 THEN LPRINT
4030 LPRINT TAB(8+(J-(ISP*8))*8);
4040 GOTO 4060
4050 LPRINT TAB(8+J*8);
4060 LPRINT USING "####.##";C(JJ);
4070 NEXT J
4080 LPRINT:
4090 LPRINT "VARIABLES";
4100 FOR J=1 TO NV
4110 JJ=J+NC
4120 IF JJ>NV THEN JJ=JJ-NV
4130 IF J<=8 THEN GOTO 4180
4140 ISP=INT((J-1)/8)
4150 IF J=(ISP*8)+1 THEN LPRINT
4160 LPRINT TAB(12+(J-(ISP*8))*8);
4170 GOTO 4190
4180 LPRINT TAB(12+J*8);
4190 LPRINT USING "##";JV(JJ);
4200 NEXT J
4210 LPRINT:
4220 FOR I=1 TO NC
4230 LPRINT USING "##";IV(I);
4240 LPRINT TAB(4);
4250 LPRINT USING "#######.##";P(I);
4260 FOR J=1 TO NV
4270 JJ=J+NC
4280 IF JJ>NV THEN JJ=JJ-NV
4290 IF J<=8 THEN GOTO 4340
```

```
4300 ISP=INT((J-1)/8)
4310 IF J=(ISP*8)+1 THEN LPRINT
4320 LPRINT TAB(8+(J-(ISP*8))*8);
4330 GOTO 4350
4340 LPRINT TAB(8+J*8);
4350 LPRINT USING "####.##";A(I,JJ);
4360 NEXT J
4370 LPRINT:
4380 NEXT I
4390 LPRINT:
4400 LPRINT:
4410 LPRINT "RCST";
4420 LPRINT TAB(5);
4430 LPRINT USING "######.##";VL;
4440 FOR J=1 TO NV
4450 JJ=J+NC
4460 IF JJ>NV THEN JJ=JJ-NV
4470 IF J<=8 THEN GOTO 4520
4480 ISP=INT((J-1)/8)
4490 IF J=(ISP*8)+1 THEN LPRINT
4500 LPRINT TAB(8+(J-(ISP*8))*8);
4510 GOTO 4530
4520 LPRINT TAB(8+J*8);
4530 LPRINT USING "####.##";CM(JJ);
4540 NEXT J
4550 LPRINT:
4560 RETURN
```

Figure 9.3—simplex program

Program Input

Input to the program consists of the following:

1. Number of variables
2. Number of constraints
3. Identification numbers for variables
4. Coefficients of variables in the objective function
5. Coefficients of variables in the constraints
6. Right hand side value of each constraint

The program makes four major assumptions regarding your preparation of the problem for solution.

1. The objective function is to be maximized. If your objective function is to be minimized, simply multiply the function by -1 to convert it to a maximization problem.
2. All constraints have been converted to equality constraints.
3. All variables are restricted to non-negative values.
4. All right hand side values are non-negative.

Once the problem is in this format, you only need an initial solution to start the program. The stages of the solution are presented in a table called a "tableau." This will be discussed in detail in the output section.

The program requires an initial basic solution. This generally consists of slack variables. There will be a basic variable for each constraint. An initial basic variable must be one which appears in only one constraint and has a coefficient of +1 (if it is different from +1, you may be able to make it +1 by dividing the entire constraint equation by a constant). This makes slack variables a natural. Consider an earlier problem.

Maximize $Z = 2x_1 + 5x_2$

subject to

$$x_1 \leq 4$$
$$x_2 \leq 3$$
$$x_1 + 2x_2 \leq 8$$
$$x_1, x_2 \geq 0$$

With slack variables to create equality constraints:

Maximize $Z = 2x_1 + 5x_2$

subject to

$$x_1 + x_3 = 4$$
$$x_2 + x_4 = 3$$
$$x_1 + 2x_2 + x_5 = 8$$
$$x_1, x_2, x_3, x_4, x_5 \geq 0$$

In this problem, the slack variables x_3, x_4, and x_5 would be initial basic variables since they each appear in only one constraint and have coefficients of +1. However, any variable which meets these conditions can be an initial basic variable. Remember, there must be one from each constraint.

The problem above is in the form required to run the program. Let us look at the program operation as the inputs are requested and use this problem as an example. Below we present the input portion of the program and answer according to this problem.

```
ENTER THE FOLLOWING PARAMETERS
NUMBER OF CONSTRAINT EQUATIONS,NC? 3

NUMBER OF VARIABLES, INCLUDING INITIAL BASIS: 5

ENTER 1--ALL TABLEAUS : 2--ONLY FIRST AND LAST TABLEAUS? 1
ENTER THE TITLE OF THE PROBLEM? EXAMPLE PROBLEM NO 1
```

In the above questions we are defining the problem. We have three constraint equations and five variables (x_1 through x_5). The next question allows you the option of seeing all tableaus or just the first and last. Finally, you will want a title for your problem.

```
ENTER THE VARIABLE IDENTIFICATION NUMBERS
BEGINNING WITH THE INITIAL BASIC VARIABLES
?3
?4
?5
?1
?2
```

These are the last problem definition questions prior to inputting the actual problem numbers. We must be able to identify each variable with a number. The easiest to use are the subscript numbers as we have done here. Thus, 3 is x_3, 4 is x_4, etc. Notice that the initial basic variables are entered first. Their order must match the constraints. That is, x_3 is in constraint 1, x_4 is in constraint 2, and x_5 is in constraint 3. Order does not matter for the other variables.

```
ENTER THE OBJECTIVE FUNCTION COEFFICIENTS OF THE
VARIABLES YOU HAVE JUST LISTED
ENTER THE COST FOR VARIABLE 3 ? 0
ENTER THE COST FOR VARIABLE 4 ? 0
ENTER THE COST FOR VARIABLE 5 ? 0
ENTER THE COST FOR VARIABLE 1 ? 2
ENTER THE COST FOR VARIABLE 2 ? 5
```

Here we input the objective function coefficients of each variable. If a variable does not appear in the objective function, its coefficient is zero, as is the case for x_3, x_4, and x_5.

```
FOR EACH CONSTRAINT EQUATION, ENTER FIRST THE
VALUE OF THE RIGHT HAND SIDE AND THEN ENTER THE
COEFFICIENTS OF ALL VARIABLES IN THAT CONSTRAINT.
(NOTICE THAT A VARIABLE NOT APPEARING IN
A PARTICULAR CONSTRAINT HAS A COEFFICIENT OF ZERO).
DO NOT ENTER COEFFICIENTS FOR ANY VARIABLES CLASSIFIED
AS ORIGINAL BASIC VARIABLES.
```

This is printed as instruction for the next set of inputs; entering the constraint equations. The questions requesting the specific input provide additional guidance. The coefficients for the original basic variables are automatically provided by the program and should not be entered by the user.

```
THE FOLLOWING QUESTIONS PERTAIN TO CONSTRAINT 1

ENTER THE VALUE OF THE RIGHT HAND SIDE? 4
INPUT THE COEFFICIENT OF VARIABLE 1 ? 1
INPUT THE COEFFICIENT OF VARIABLE 2 ? 0

THE FOLLOWING QUESTIONS PERTAIN TO CONSTRAINT 2

ENTER THE VALUE OF THE RIGHT HAND SIDE? 3
INPUT THE COEFFICIENT OF VARIABLE 1 ? 0
INPUT THE COEFFICIENT OF VARIABLE 2 ? 1
```

```
THE FOLLOWING QUESTIONS PERTAIN TO CONSTRAINT 3
ENTER THE VALUE OF THE RIGHT HAND SIDE? 8
INPUT THE COEFFICIENT OF VARIABLE 1 ? 1
INPUT THE COEFFICIENT OF VARIABLE 2 ? 2
```

This is where you input the constraint equations. Recall that the order of the constraints and the order of the original basic variables must match. The program asks for specific right hand side values and coefficients for particular variables within each constraint.

The input is now complete and the computer can begin to use the simplex method to solve our problem. An editor is included in the event you wish to change your input prior to solving the problem. In the next section we will examine the output of the program.

There are many additional linear programming formulations and other issues which we have not been able to discuss. In order to fully utilize the program presented you should become familiar with these additional concepts. The references provide a wealth of knowledge which should prove very beneficial to you.

Program Output

The output of the program is somewhat controllable by you. That is, you can request an output at each iteration or only the first and last iterations. An iteration is a move from one corner point to another. At each iteration a "tableau" is printed. This tableau is full of information. Some of this information will be meaningless to you unless you delve into linear programming beyond our brief discussion here. It is provided for those who can put it to use. Since it follows the standard tableau format, it will be easy to understand.

The tableau reveals three very important conditions of the current solution: the basic variables, the solution value of each basic variable, and the nonbasic variables which would improve the objective function if they were made basic (this, in turn, tells us when we have reached the optimal solution). Figure 9.4 presents the initial tableau for our example problem. Let us analyze it.

```
THE FOLLOWING RESULTS ARE FOR ITERATION NUMBER 1

OBJ. COEF.              2     5     0     0     0
VARIABLES               1     2     3     4     5
   3          4.00   1.00  0.00  1.00  0.00  0.00
   4          3.00   0.00  1.00  0.00  1.00  0.00
   5          8.00   1.00  2.00  0.00  0.00  1.00

   RCST       0.00   2.00  5.00  0.00  0.00  0.00
```

Figure 9.4—tableau 1

Look first at the second row, labeled "VARIABLES." Each variable in the problem has its own column in the tableau. They are labeled according to the identification numbers you have provided. If there are more than eight variables in your problem, the variables after the eighth one will be shifted down one row. This will result in at least one column being used for at least two variables; one underneath the other. The variables assigned to each column will remain the same throughout the solution procedure.

The first row, labeled "OBJ.COEF.," lists the coefficients in the objective function for the respective variables listed in the second row. These also will not change during problem solution. You should check these to verify that your input was correct.

The left-most column lists the basic variables for the current solution. In Figure 9.4 the basic variables, going down the column, are variables 3(x_3), 4(x_4), and 5(x_5). These will change as the simplex method moves toward the optimal solution.

The column immediately to the right of the first column yields the solution values for the basic variables. In Figure 9.4, our solution is:

$x_3 = 4$
$x_4 = 3$
$x_5 = 8$

The last element in this column, the one not associated with a particular basic variable, represents the current value of the objective function. In Figure 9.4, this value is zero.

You may notice that we have not mentioned anything about variables x_1 and x_2. We do not need to. At this iteration they are nonbasic variables and therefore their value is zero.

In summary, Figure 9.4 presents this solution:

$x_1 = 0$ $x_2 = 0$ $x_3 = 4$ $x_4 = 3$ $x_5 = 8$ $Z = 0$

You should verify that this solution is correct by plugging these values into the problem formulation to see that all constraints are satisfied and the correct objective function value results.

The tableau also tells us whether or not the current solution is the optimal solution. This is found in the row labeled "RCST" (reduced cost). The first element of this row is the value of the objective function; the remaining elements reveal the condition of the solution. Notice that there is a value in this row for every variable in our problem. The values corresponding to basic variables will always be zero. We can see in Figure 9.4 that the values in the reduced cost row are zero in the columns of variables x_3, x_4, and x_5. The values in columns of nonbasic variables indicate the marginal desirability of allowing them to "enter" the solution. If the value is positive, then the entrance of the corresponding nonbasic variable into the solution will improve (increase) the value of the objective function, something we would want to do. If it is negative, then that variable should remain nonbasic. If it is zero, the objective function value will not change; we will leave the variable as nonbasic.

In Figure 9.4 the values in the reduced cost row for variables x_1 and x_2 are both positive, meaning they could both enter the solution. However, the simplex method chooses only one variable at a time to enter. The general procedure, and that followed by the program, is to choose that variable with the largest value in the reduced cost row. In this case this is x_2. In the next iteration, x_2 will become a basic variable and one of the current basic variables will become nonbasic.

The tableau resulting from this change is presented in Figure 9.5. You will notice that x_2 has replaced x_4 as a basic variable. The solution with these basic variables (x_3, x_2, and x_5) is found in Figure 9.5 as:

$x_1=0 \qquad x_2=3 \qquad x_3=4 \qquad x_4=0 \qquad x_5=2 \qquad Z=15$

```
THE FOLLOWING RESULTS ARE FOR ITERATION NUMBER 2
OBJ. COEF.             2     5     0     0     0
VARIABLES              1     2     3     4     5
     3      4.00     1.00  0.00  1.00  0.00  0.00
     2      3.00     0.00  1.00  0.00  1.00  0.00
     5      2.00     1.00  0.00  0.00 -2.00  1.00

   RCST    15.00     2.00  0.00  0.00 -5.00  0.00
```

Figure 9.5—tableau 2

If we examine the reduced cost row we find that the value in the x_1 column, 2, means that the objective function value can be increased if x_1 enters the solution. The value in the x_4 column, -5, means that x_4 should remain nonbasic.

If x_1 enters the solution, we obtain the tableau in Figure 9.6. The resulting solution is:

$x_1=2 \qquad x_2=3 \qquad x_3=2 \qquad x_4=0 \qquad x_5=0 \qquad Z=19$

```
THE FOLLOWING RESULTS ARE FOR ITERATION NUMBER 3
OBJ. COEF.             2     5     0     0     0
VARIABLES              1     2     3     4     5
     3      2.00     0.00  0.00  1.00  2.00 -1.00
     2      3.00     0.00  1.00  0.00  1.00  0.00
     1      2.00     1.00  0.00  0.00 -2.00  1.00

   RCST    19.00     0.00  0.00  0.00 -1.00 -2.00
```

Figure 9.6—tableau 3

This is the optimal solution to the problem. How do we know? Because there are no positive values in the reduced cost row (with the exception of the objective function value). This means that there are no nonbasic variables which, if they were to enter the solution, would increase the value of the objective function. Therefore, the solution in Figure 9.6 yields the largest value of the objective function for all choices of the variables which satisfy the stated constraints.

The option of seeing each iteration tableau is yours. The program will ask if you would like to see all tableaus or only the first and last. You will always want to see the first to verify your input, and the last for the optimal solution.

For formatting purposes, the program uses the PRINT USING statement. Objective coefficients are limited to 9999.99, constraint coefficients to 9999.99, and right hand side values to 9999999.99 for printing purposes. You will want to make sure that your problem is properly scaled. The current formatting will be sufficient for most problems. The remaining portions of the tableau are provided for those who spend additional time studying linear programming. For the novice, their purpose in unimportant. After further study, their meaning will become evident. Therefore, a discussion is omitted here.

Finally, if the problem is such that the objective function can increase without bound, the program will inform you of this situation. This occurrence usually implies that your model is not truly reflective of the modeled situation.

Program Coverage

Many special topics and considerations in linear programming were not explicitly discussed here, particularly with respect to the operation of the program. An attempt was made to provide a basic introduction to the concepts of the simplex method and the program application. The program can also be used in a more advanced manner, but the authors feel that a detailed description here is not feasible.

The program can be used in an imaginative way to address advanced problems but is dependent on the user's understanding of the simplex method. For example, the program can be used with artificial variables in a two-phase procedure. As the user becomes more knowledgeable in linear programming, a detailed description of the use of the program for more advanced problems will not be necessary.

9.5 Summary

In this chapter we have presented linear programming. This is a very powerful and popular tool which is applicable in many areas. In these few pages we were only able to touch the surface of the topic.

The "art" of linear programming is contained in the formulation of the model. We illustrated this with the "Peanut Problem" of Charnes, Cooper, and Henderson. We encourage you to work on the formulation examples. If the model is incorrect, the solution will be incorrect.

The "science" of linear programming is in finding the optimal solution. We briefly discussed graphical solution concepts and algebraic solution procedures. We also discussed the fundamental concepts behind the simplex method, which was a major breakthrough in the solution of linear programming problems. The computer program in this chapter used the simplex method.

In summary, this chapter provides a brief introduction to the topic of linear programming. The interested reader is encouraged to consult the references for more advanced topics.

References

Bazaraa, Mokhtar S. and John J. Jarvis, *Linear Programming and Network Flows*. John Wiley and Sons, Inc., New York, N.Y., 1977.

Charnes, A., W.W. Cooper, and A. Henderson, *An Introduction to Linear Programming*. John Wiley and Sons, Inc., New York, 1953.

Hadley, G., *Linear Programming*. Addison-Wesley Publishing Co., Reading, Mass., 1963.

McMillan, Claude, *Mathematical Programming*. John Wiley and Sons, Inc., New York, N.Y., 1975.

Taha, Hamdy A., *Operations Research*. Macmillan Publishing Company, Inc., New York, N.Y., 1982.

Exercises

1. Solve the following linear programming problem.

   Maximize $z = 8x_1 + 2x_2 + 5x_3$

   subject to

   $$2x_1 + 2x_2 + x_3 \leq 12$$
   $$3x_1 - x_2 + 2x_3 \leq 9$$
   $$x_2 + 3x_3 \leq 5$$
   $$x_1, x_2, x_3 \geq 0$$

2. An oil company requires two different types of crude oil to produce leaded, unleaded, and diesel fuels. A particular profit is made on each type of fuel, and each fuel consists of certain proportions of each crude type. Profit is limited because crude oil is limited. The table below presents the appropriate summary information.

   Percentage Required of each Crude Type for Fuels

Fuel Type	Crude Type 1	Crude Type 2	Profit Per bbl
leaded	0.4	0.6	0.12
unleaded	0.5	0.5	0.09
diesel	0.7	0.3	0.10
Available crude (bbl)	400	300	

   Formulate and solve a linear programming problem which maximizes profit while not exceeding crude oil availability.

3. Different crops are affected differently by various fertilizers. Each fertilizer/crop combination results in a particular crop yield. The table below illustrates these yields for a few crops of interest.

   Yields for Fertilizer/Crop Combinations
   (in bushels/lb. of Fertilizer)

Fertilizer	Corn	Wheat	Soybeans	Maximum Availability (lbs.)	Cost/lb.
A	15	31	20	100	$20
B	18	25	16	125	25
C	12	35	30	80	20
Selling Price Per Bushel	$2	$1.50	$1.80		

   Formulate and solve a linear programming problem to maximize profit from crop sales.

4. A manufacturing firm produces two types of power tools. Tool A requires 30 minutes of casting, 45 minutes of machining and 20 minutes of assembly. Tool B requires 40, 30, and 25 minutes, respectively, of these processes. The firm is limited to 40 hours of each process per week. If output of tool A is twice as valuable as output of tool B, how many units of each should be produced per week?

5. Solve the "Peanut Problem" presented in this chapter.

10

Forecasting with Exponential Smoothing

Forecasting is an important function because a good forecast reduces uncertainty and thereby aids in decision making. Unfortunately, much of forecasting is an art. Some techniques, however, have been developed and refined that assist us in making better forecasts. One of these, linear regression, was discussed in Chapter 4. Regression is almost always used when the forecast model contains a relationship between one or more variables and the dependent variable being forecast. For example, the projected number of new home starts for next month might be a function of the interest rate. In this case, the interest rate might be used to forecast the home starts.

Another forecasting technique called exponential smoothing will be introduced in this chapter. Exponential smoothing may be employed to forecast future values in a time series using only the historical values in the time series to make the projection. No other independent variable, such as interest rate, is involved. The monthly sales of microcomputers by firm ABC is a time series. Exponential smoothing could be used to forecast the future sales of microcomputers considering only the past sales.

Time-series forecasting techniques such as exponential smoothing are primarily used for short term forecasts (forecasts up to a year in the future). One reason is that these procedures extrapolate the historical observations into the future without considering changes to the underlying process, such as changes in economic conditions and demand.

10.1 Single Exponential Smoothing

An N-period moving average is a popular time-series forecasting technique. However, exponential smoothing can provide better results with less effort. An N-period moving average forecast is made by averaging the last N periods, then using this value to make the forecast. At the end of the next period, the oldest observation is removed from the series and the last observation is added, always keeping the N most recent observations.

$$X_{t+1} = \sum_{i=t-N+1}^{t} X_i/N \tag{1}$$

Each observation is equally weighted, $1/N$. Also, N elements of data must be maintained. This procedure works well if the process being modeled is level (horizontal) and not changing over time. Single exponential smoothing could also be used, and we will see that it has some advantages.

An expression for single exponential smoothing is

$$F_{t+1} = \alpha X_t + (1-\alpha)F_t \tag{2}$$

where F_t = forecast for period t,
X_t = actual observation for period t,
F_{t+1} = forecast for period t+1, and
α = smoothing constant, $0 \leq \alpha \leq 1$.

One advantage of exponential smoothing can be readily seen. Only three data elements are required: the most recent forecast, the most recent actual observation, and the smoothing constant. Because of these small data storage requirements, exponential smoothing is often used in computerized inventory control systems in which thousands of items must be forecast.

Some manipulation of expression (2) will reveal other interesting aspects of exponential smoothing. First we will expand this equation:

$$F_{t+1} = \alpha X_t + (1-\alpha)[\alpha X_{t-1} + (1-\alpha)F_{t-1}]$$
$$= \alpha X_t + \alpha(1-\alpha)X_{t-1} + (1-\alpha)^2 F_{t-1}$$

Additional expansions will yield

$$F_{t+1} = X_t + \alpha(1-\alpha)X_{t-1} + \alpha(1-\alpha)^2 X_{t-2} + \ldots \tag{3}$$

Looking at the coefficients of the time series, we can see that they decrease exponentially. For instance, if $\alpha = .5$, then equation (3) becomes

$$F_{t+1} = .5X_t + .25X_{t-1} + .125X_{t-2} + \ldots \tag{4}$$

Thus, the name exponential smoothing is applied. In control theory this expression is sometimes called a first order filter.

The coefficients in expression (4) can be interpreted as weights applied to each observation. The more recent observations receive the most weight. This is often desirable in time-series forecasting because the most recent observations may be more indicative of the future. Consequently, another advantage of exponential smoothing is that the more recent data receive the most weight.

If equation (2) is rearranged, we can see another interpretation:

$$F_{t+1} = F_t + \alpha(X_t - F_t)$$

The last term in parentheses is the error of the last forecast. Therefore,

$$F_{t+1} = F_t + \alpha E_t \tag{5}$$

This says that the new forecast is obtained by adjusting the old forecast by some portion of the last forecast error.

We can see that if α is close to 1, the forecast error has a substantial effect on the new forecast. However, if α is close to 0, the error has little impact. Therefore, if the underlying process is level but has substantial random variation, a small α will "filter" out most of this random variation. In fact, most computerized inventory control systems seldom use a smoothing constant greater than .3.

10.2 Exponential Smoothing with a Trend

If the underlying process is horizontal, single exponential smoothing is appropriate. However, if a trend is involved, forecasts using equation (5) will lag behind the actual observations. Consequently, an adjustment must be made for the trend.

There is more than one way to include trend considerations in an exponential smoothing model. Holt's [1] method will be used here:

$$F_t = \alpha X_t + (1-\alpha)(F_{t-1} + b_{t-1}) \tag{6}$$

$$b_t = \beta(F_t - F_{t-1}) + (1-\beta)b_{t-1} \tag{7}$$

$$F_{t+h} = F_t + b_t h \tag{8}$$

where b_t = estimate of the trend for period t,
β = trend smoothing constant, $0 \leq \beta \leq 1$, and
F_{t+h} = forecast for period t+h (a forecast lead time of h).

Equation (6) smooths the old forecast, F_{t-1}, adjusted by the trend, b_{t-1}, to obtain a forecast, F_t for period t. The estimate of the trend is smoothed in expression (7). This is done by assuming that the part of the difference in the last two forecasts $(F_t - F_{t-1})$ is due to changes in the trend. The last expression, (8), provides a forecast h periods ahead by adding to the forecast for period t the trend per period multiplied by the number of periods ahead to be forecast. The number of periods to be forecast ahead is called the *forecast lead time*. The number of periods forecast beyond the last observation in a time series is called the *forecast horizon*.

10.3 Seasonal Exponential Smoothing

A time series often exhibits seasonal patterns. Examples are the sales of toys, snow tires, and ice cream. The difference between a *seasonal* pattern and a *cyclical* pattern is that a seasonal pattern repeats itself in short intervals such as a week, month, or year. A cyclical pattern has a much larger duration. One approach to analyzing time-series data that contains seasonal patterns is to remove the seasonal component. Until this is done, it is difficult to distinguish between random effects and the seasonal effects. Several exponential smoothing methods have been developed which consider seasonal patterns; one of these is Winter's method [2].

Before presenting the equations for this method, we need to develop a procedure for calculating seasonal indices. These indices will be used to remove the seasonal component in a time series. Consider the following seasonal sales pattern of units sold: 100, 200, and 300, which is repeated every 3 periods. A seasonal index can be computed as follows:

(1) Compute an average sales/period:
$(100 + 200 + 300)/3 = 200$

(2) Compute a seasonal index, I_t, for each period by dividing the sales per period by the average sales per period.
$I_1 = 100/200 + .5$
$I_2 = 200/200 = 1.0$
$I_3 = 300/200 = 1.5$

The sum of the indices should equal the number of periods in a season. Using these indices we can remove the seasonal effect from the data. For instance, removing the seasonal component from the sales for period 1 results in:

$X'_1 = X_1/I_1,$

$= 100/.5 = 200$ units

This procedure will be used to remove the seasonal effects from a time series.

Now we can look at the equations for Winter's method, which considers trend and seasonality:

$$F_t = \alpha X_t/I_{t-k} + (1-\alpha)(F_{t-1} + b_{t-1}) \qquad (9)$$

$$b_t = \beta(F_t - F_{t-1}) + (1-\beta)b_{t-1} \qquad (10)$$

$$I_t = \gamma X_t/F_t + (1-\gamma)I_{t-k} \qquad (11)$$

$$F_{t+h} = (F_t + b_t h)I_{t-k+h} \qquad (12)$$

where K = the number of periods in a season
I_t = the seasonal index for period t, and
γ = the seasonal index smoothing constant, $0 \leq \gamma \leq 1$.

Equation (9) is the same as equation (6) used in Holt's method except that X_t is divided by the seasonal index to remove the seasonal fluctuations. The result of equation (9), F_t, is a smoothed estimate for period t. This is obtained by adding the trend to the forecast for the period $t-1$ and by removing the seasonality fluctuations.

Equation (10) is the same as Holt's equation (7). The next equation, (11), smooths the seasonal indices. As the seasonal indices are smoothed, they may no longer sum to the number of periods in a season. In the program that is presented later in the chapter, seasonal indices are normalized at the end of each season so that the sum of the indices equals the number of periods in a season. Equation (12) provides a forecast for period $t+1$. This forecast includes adjustments for trend and seasonality.

To implement Winter's method, it is desirable to have some procedures for estimating initial values for each of the components of the model (permanent, trend, and seasonal). Reference [2] provides procedures to estimate initial values for each component. These procedures will be presented here. It is assumed that the number of data elements used for initialization represents at least two seasons. The initial trend component may be estimated by:

10.3 Seasonal Exponential Smoothing

$$b_o = (\overline{X}_m - \overline{X}_1)/(m-1)K \tag{13}$$

where
- m = number of seasons
- $\overline{X}_m$ = average of the observations in season m
- K = length of a season
- b_o = initial estimate for the trend

This expression evaluates the differences in the average value between the first season and the last season, $(\overline{X}_m - \overline{X}_1)$. This value is then divided by the number of seasons between these averages to obtain the trend per season. It is assumed that $\overline{X}_i$ occurs in the middle of a season.

The initial permanent component may be estimated by

$$a_o = \overline{X}_1 - (K/2)b_o \tag{14}$$

Again, it is assumed that $\overline{X}_1$ occurs at the middle of the first season. Therefore, by subtracting $K/2 * b_o$ ($1/2$ of a season times the trend), we can obtain an estimate for the initial permanent component at the beginning of period 1.

Seasonal indices must be computed for each time period in the initialization time series. The procedure used is similar to the seasonal index calculations presented above. The following expression is used:

$$I_t = X_t / (\overline{X}_i - ((K+1)/2 - j)b_o, \ j=1,2,\ldots,K; \ t=1,2,\ldots,mK, \tag{15}$$

where j is the position of data element X_i within a season. Thus, if there are three periods in a season, X_3 would be in position 3 in the first season and X_4 would be in position 1 in the second season. The average seasonal value is adjusted by removing the estimated trend, leaving an estimate for the permanent component. Consequently equation (15) is the ratio of actual observations to the adjusted average seasonal observation for that season.

Equation (15) will result in K indices for each season. Using these estimates, an average seasonal index is computed:

$$\overline{I}_t = 1/m \sum_{i=0}^{m-1} I_{t+i+k} \quad , t=1,2,\ldots,K \tag{16}$$

The sum of the average indices should equal m. To insure this, the indices are normalized

$$I_t(0) = \overline{I}_t \ (m / \sum_{t=1}^{m} \overline{I}_t) \quad , t=1,2,\ldots,K \tag{17}$$

Having determined initial values for the model components (permanent, trend, and seasonal), we can then apply Winter's method to forecast values in a time series.

10.4 Exponential Smoothing Program

Figure 10.1 contains an exponential smoothing program incorporating Winter's method. The name of the program is EXSMOOTH. Some examples will be presented to illustrate how to use this program. A time series having as many as 100 elements and 24 periods in a season can be studied. A larger problem can be studied by changing the program dimension statements.

```
1000 REM TIME SERIES FORECASTING PROGRAM--WINTER'S METHOD
1010 REM This program will perform time series forecasting using
1020 REM the Winter's method of exponential smoothing. A time
1030 REM series having as many as 100 observations and 24 periods
1040 REM in a series can be studied. Larger problems can be
1050 REM studied by changing the program dimension statements.
1060 REM Additional instructions are provided during program
1070 REM execution. Definitions of major program variables:
1080 REM     LM=Maximum number periods in time series
1090 REM     LP=Maximum number periods in a season
1100 REM     LN=Maximum number of seasons in initialization
1110 REM     IT=Number of initialization periods
1120 REM     LT=Forecast lead time
1130 REM     IZN=Forecast horizon
1140 REM     X(I)=Time series data
1150 REM     A(I)=Permanent component
1160 REM     B(I)=Trend component
1170 REM     SF(I)=Season factor
1180 REM     FCS(I)=Forecast
1190 REM     ER(I)=Error in forecast
1200 DEFINT I,L
1210 LM=100        'Max no. periods in time series
1220 LP=24         'Max no. periods in season
1230 LN=24         'Max no. seasons in initialization
1240 DIM X(LM),A(LM),B(LM),SF(LP), AXS(LN), RSF(LN,LP)
1250 DIM ER(LM), FCS(LM), SS(LM), TEMP(LP)
1260 PRINT:PRINT
1270 PRINT "**TIME SERIES FORECASTING PROGRAM-WINTER'S METHOD**"
1280 PRINT:PRINT
1290 INPUT "WANT PROGRAM INSTRUCTIONS (Y/N)";Y$
1300 IF Y$<>"Y" THEN 1480
1310 PRINT:PRINT "PROGRAM INSTRUCTIONS:"
1320 PRINT "Time series data may be entered from keyboard or disk."
1330 PRINT "Time series data management options provided:"
1340 PRINT TAB(5) "-List data"
1350 PRINT TAB(5) "-Change data"
1360 PRINT TAB(5) "-Add to data"
1370 PRINT TAB(5) "-Delete data"
1380 PRINT TAB(5) "-Store data on disk"
1390 PRINT "User must specify number of data elements to be used"
1400 PRINT "for model initialization, number of periods in season,"
1410 PRINT "forecast lead time, and forecast horizon."
1420 PRINT "User has the option of inputting model components (permanent,"
1430 PRINT "trend, and seasonal factors) and smoothing constants (one"
1440 PRINT "constant for each type of model component) or specifying that"
1450 PRINT "the program determine these values. Program will optimize"
1460 PRINT "smoothing constants by performing a grid search using"
1470 PRINT "specified limits and increments within these limits."
```

10.4 Exponential Smoothing Program

```
1480 PRINT:INPUT "ENTER DATA FROM KEYBOARD OR DISK (K/D)";K$
1490 IF (K$<>"K") AND (K$<>"D") THEN PRINT "INVALID":GOTO 1480
1500 IF K$="K" THEN GOSUB 6630
1510 IF K$="D" THEN GOSUB 5590
1520 PRINT TAB(5)   "PROGRAM OPTIONS"
1530 PRINT TAB(10) "1-LIST TIME SERIES DATA"
1540 PRINT TAB(10) "2-CHANGE TIME SERIES DATA"
1550 PRINT TAB(10) "3-DELETE DATA FROM TIME SERIES"
1560 PRINT TAB(10) "4-ADD TO TIME SERIES"
1570 PRINT TAB(10) "5-STORE TIME SERIES DATA ON DISK"
1580 PRINT TAB(10) "6-PERFORM FORECAST COMPUTATIONS"
1590 PRINT TAB(10) "7-STUDY NEW MODEL"
1600 PRINT TAB(10) "8-QUIT"
1610 INPUT "OPTION'';IOP:PRINT
1620 IF(IOP<1) OR (IOP>8) THEN PRINT "INVALID":GOTO 1520
1630 IF IOP=1 THEN GOSUB 6180
1640 IF IOP=2 THEN GOSUB 6470
1650 IF IOP=3 THEN GOSUB 7000
1660 IF IOP=4 THEN GOSUB 6550
1670 IF IOP=5 THEN GOSUB 5560
1680 IF IOP=6 THEN 1720
1690 IF IOP=7 THEN CLEAR:GOTO 1210
1700 IF IOP=8 THEN PRINT TAB(25) "END OF PROGRAM":END
1710 GOTO 1520
1720 INPUT "NUMBER OF DATA ELEMENTS USED FOR INITIALIZATION";IT
1730 IF(IT<0) AND (IT>=LD)THEN PRINT"INVALID":GOTO 1720
1740 INPUT "NUMBER PERIODS IN A SEASON";LS
1750 IF(LS<0)OR(LS>IT)THEN PRINT"INVALID":GOTO 1740
1760 IF LS=0 THEN LS=1
1770 XL=LS
1780 IF(LS<=LP) THEN GOTO 1800
1790 PRINT "WILL NOT ACCEPT A SEASON OF THIS LENGTH":GOTO 1740
1800 INPUT "ESTIMATION OF MODEL PARAMETERS DESIRED (Y/N)";PY$
1810 IF(PY$<>"Y") AND (PY$<>"N") THEN PRINT "INVALID":GOTO 1800
1820 IF PY$="N" THEN 1880      'Ask for initial components
1830 IS=IT MOD LS
1840 IF(PY$="Y") AND (IS=0)THEN 1950
1850 INPUT "ESTIMATION OF MODEL PARAMETERS REQUIRES THAT NUMBER"
1860 INPUT "OF INITIALIZATION PERIODS BE A MULTIPLE OF SEASON LENGTH"
1870 GOTO 1720
1880 INPUT "ENTER INITIAL PERMANENT AND TREND COMPONENTS";A0,B0
1890 PRINT "ENTER INITIAL SEASONAL FACTORS"
1900 FOR I=1 TO LS
1910 PRINT "SEASONAL FACTOR";I;
1920 INPUT SF(I)
1930 TEMP(I)=SF(I)
1940 NEXT I
1950 INPUT "SMOOTHING CONSTANT OPTIMIZATION DESIRED (Y/N)";SY$
1960 IF(SY$<>"Y") AND (SY$<>"N") THEN PRINT "INVALID":GOTO 1950
1970 IF SY$="Y" THEN 2050
1980 INPUT "ENTER PERMANENT COMPONENT SMOOTHING FACTOR-ALPHA";AP
1990 IF(AP<0) OR (AP>1!) THEN PRINT "INVALID":GOTO 1980
```

```
2000 INPUT "ENTER TREND COMPONENT SMOOTHING FACTOR-BETA";BE
2010 IF(BE<0)OR(BE>1!) THEN PRINT "INVALID"; GOTO 2000
2020 INPUT "ENTER SEASONAL COMPONENT SMOOTHING FACTOR-GAMMA";GA
2030 IF(GA<0) OR (GA>1!) THEN PRINT "INVALID"; GOTO 2020
2040 GOTO 2270
2050 PRINT "ENTER LOWER LIMIT, UPPER LIMIT, STEP SIZE FOR ALPHA--"
2060 PRINT "PERMANENT COMPONENT SMOOTHING FACTOR"
2070 INPUT AL,AU,AI
2080 IF (AL<0) OR (AU<0) OR (AI<0) THEN PRINT "INVALID":GOTO 2050
2090 IF (AL>1) OR (AU>1) OR (AI>1) THEN PRINT "INVALID":GOTO 2050
2100 PRINT "ENTER LOWER LIMIT, UPPER LIMIT, STEP SIZE FOR BETA--"
2110 PRINT "TREND COMPONENT SMOOTHING FACTOR"
2120 INPUT BL,BU,BI
2130 IF (BL<0) OR (BU<0) OR (BI<0) THEN PRINT "INVALID":GOTO 2100
2140 IF (BL>1) OR (BU>1) OR (BI>1) THEN PRINT "INVALID":GOTO 2100
2150 PRINT "ENTER LOWER LIMIT, UPPER LIMIT, STEP SIZE FOR GAMMA--"
2160 PRINT "SEASONAL COMPONENT SMOOTHING FACTOR"
2170 INPUT GL,GU,GI
2180 IF (GL<0) OR (GU<0) OR (GI<0) THEN PRINT "INVALID":GOTO 2150
2190 IF (GL>1) OR (GU>1) OR (GI>1) THEN PRINT "INVALID":GOTO 2150
2200 P1=(AU-AL)/AI+1
2210 P2=(BU-BL)/BI+1
2220 P3=(GU-GL)/GI+1
2230 T=P1*P2*P3
2240 PRINT T;" TRIALS ARE REQUIRED TO FIND OPTIMUM VALUES FOR"
2250 INPUT "SMOOTHING CONSTANTS. OKAY TO CONTINUE (Y/N)";Y$
2260 IF Y$="N" THEN 2050
2270 INPUT "ENTER FORECAST LEAD TIME";LT
2280 IF LT<0 THEN PRINT "INVALID";GOTO 2270
2290 INPUT "NUMBER PERIODS IN FORECAST HORIZON";IZN
2300 IF IZN<0 THEN PRINT "INVALID":GOTO 2290
2310 XLT=LT:IE=IT
2320 IF IT>(LD-LT) THEN IE=LD-LT
2330 INPUT "WOULD YOU LIKE TO SEE INPUT DATA DISPLAYED (Y/N)";Y$
2340 IF Y$<>"Y" THEN 2360
2350 GOSUB 5790
2360 PRINT:INPUT "ARE DATA CORRECT (Y/N)"; Y$
2370 IF Y$= "Y" THEN 2410
2380 INPUT "DO YOU WANT TO REENTER DATA (Y/N)"; Y$
2390 IF Y$= "Y" THEN 1720
2400 GOTO 1520
2410 INPUT "RESULTS TO APPEAR AT CRT OR PRINTER (C/P)";C$
2420 IF (C$<>"C") AND (C$<>"P") THEN PRINT "INVALID";GOTO 2410
2430 IF C$="C" THEN 2480
2440 X$=STRING$(75,45)
2450 LPRINT:LPRINT:LPRINT X$:LPRINT:LPRINT
2460 LPRINT TAB(10) "TIME SERIES FORECASTING-WINTER'S METHOD"
2470 LPRINT:LPRINT
2480 IF PY$="N" THEN 3120
2490 PRINT:PRINT "INITIAL VALUES FOR PERMANENT, TREND, AND SEASONAL"
2500 PRINT "COMPONENTS TO BE ESTIMATED FROM DATA"
2510 IK=IT/LS                    'No. of seasons
```

10.4 Exponential Smoothing Program

```
2520 I1=1
2530 I2=LS
2540 FOR I=1 TO IK
2550 AXS(I)=0
2560 FOR J=I1 TO I2
2570 AXS(I)=AXS(I)+X(J)
2580 NEXT J
2590 AXS(I)=AXS(I)/XL          'Average X per season I
2600 I1=I2+1
2610 I2=I1+LS-1
2620 NEXT I
2630 R=IT-LS
2640 B0=(AXS(IK)-AXS(1))/R     'Estimate trend component
2650 A0=AXS(1)-(XL-1)/2!*B0          'Estimate permanent component
2660 I0=0
2670 FOR I1=1 TO IK
2680 FOR I2=1 TO LS
2690 IJ=I2+I0*LS
2700 R=I2
2710 REM Calculate seasonal values for each season
2720 RSF(I1,I2)=X(IJ)/(AXS(I1)-(((XL+1!)/2!)-R)*B0)
2730 NEXT I2
2740 I0=I0+1
2750 NEXT I1
2760 GTOT=0!
2770 R=IK
2780 FOR J=1 TO LS
2790 TOT=0!
2800 FOR I=1 TO IK
2810 TOT=TOT+RSF(I,J)
2820 NEXT I
2830 REM Average seasonal factor for period J in a season.
2840 SF(J)=TOT/R
2850 REM Total of all seasonal factors
2860 GTOT=GTOT+SF(J)
2870 NEXT J
2880 IF C$="C" THEN 2980
2890 LPRINT "ESTIMATED PERMANENT COMPONENT ";A0
2900 LPRINT "ESTIMATED TREND COMPONENT ";B0
2910 LPRINT:LPRINT "PERIOD","ESTIMATED SEASONAL FACTOR"
2920 FOR J=1 TO LS
2930 SF(J)=SF(J)*(XL/GTOT)
2940 TEMP(J)=SF(J)
2950 LPRINT J,,SF(J)
2960 NEXT J
2970 LPRINT:GOTO 3120
2980 PRINT:PRINT "ESTIMATED PERMANENT COMPONENT ";A0
2990 PRINT "ESTIMATED TREND COMPONENT "; B0
3000 PRINT:PRINT "PERIOD", "ESTIMATED SEASONAL FACTOR"
3010 FOR J=1 TO LS
3020 REM Normalize seasonal factors
3030 SF(J)=SF(J)*(XL/GTOT)
```

```
3040 TEMP(J)=SF(J)
3050 IC=1
3060 IF J<>(IC*15) THEN 3090
3070 IC=IC+1
3080 PRINT:INPUT "PRESS ENTER TO CONTINUE";Y$
3090 PRINT J, , SF(J)
3100 NEXT J
3110 PRINT:INPUT "PRESS ENTER TO CONTINUE";Y$
3120 IF SY$="N" THEN 3740
3130 PRINT:PRINT "SMOOTHING CONSTANT OPTIMIZATION--PROGRAM EXECUTING!"
3140 IA=P1:IB=P2:IG=P3
3150 MER=1E+30
3160 AP=AL:BE=BL:GA=GL
3170 A1=AL
3180 FOR II=1 TO IA
3190 AM=1-A1
3200 B1=BL
3210 FOR IJ=1 TO IB
3220 BM=1-B1
3230 G1=GL
3240 FOR IK=1 TO IG
3250 FOR IL=1 TO LS
3260 SF(IL)=TEMP(IL)
3270 NEXT IL
3280 GM=1-G1
3290 A(1)=A1*(X(1)/SF(1))+AM*A0
3300 B(1)=B1*(A(1)-(A0-B0))+BM*B0
3310 SF(1)=G1*(X(1)/A(1))+GM*SF(1)
3320 IH=1+LT
3330 IF IH<=LS THEN 3350
3340 IH=IH-LS:GOTO 3330
3350 EST=(A(1)+XLT*B(1))*SF(IH)
3360 DIF=(X(1+LT)-EST)^2
3370 FOR I=2 TO IE
3380 IP=I MOD LS
3390 IF IP=0 THEN IP=LS
3400 A(I)=A1*(X(I)/SF(IP))+AM*(A(I-1)+B(I-1))
3410 B(I)=B1*(A(I)-A(I-1))+BM*B(I-1)
3420 SF(IP)=G1*(X(I)/A(I))+GM*SF(IP)
3430 IF IP<>LS THEN 3510
3440 GTOT=0
3450 FOR J=1 TO LS
3460 GTOT=GTOT+SF(J)
3470 NEXT J
3480 FOR J=1 TO LS
3490 SF(J)=SF(J)*(XL/GTOT)
3500 NEXT J
3510 IH=IP+LT
3520 IF IH≤LS THEN 3540
3530 IH=IH-LS:GOTO 3520
3540 EST=(A(I)+XLT*B(I))*SF(IH)
3550 DIF=DIF+(X(I+LT)-EST)^2
```

10.4 Exponential Smoothing Program

```
3560 NEXT I
3570 IF DIF>MER THEN 3600
3580 AP=A1:BE=B1:GA=G1
3590 MER=DIF
3600 G1=G1+GI
3610 NEXT IK
3620 B1=B1+BI
3630 NEXT IJ
3640 A1=A1+AI
3650 NEXT II
3660 IF C$="C" THEN 3710
3670 LPRINT:LPRINT "OPTIMUM SMOOTHING CONSTANTS"
3680 LPRINT "ALPHA=";AP,"BETA=";BE,"GAMMA=";GA
3690 LPRINT "SUM OF ERRORS SQUARED=";MER
3700 LPRINT:GOTO 3740
3710 PRINT:PRINT "OPTIMUM SMOOTHING CONSTANTS"
3720 PRINT "ALPHA="; AP, "BETA="; BE, "GAMMA="; GA
3730 PRINT "SUM OF ERRORS SQUARED="; MER
3740 IF LT=0 THEN LT=1      'Start model initialization
3750 FOR I=1 TO LT
3760 ER(I)=0:FCS(I)=0
3770 NEXT I
3780 REM Perform smoothing and calculate forecasts
3790 FOR I=1 TO LS
3800 SF(I)=TEMP(I)
3810 NEXT I
3820 AM=1!-AP
3830 BM=1!-BE
3840 GM=1!-GA
3850 A(1)=AP*(X(1)/SF(1))+AM*A0
3860 B(1)=BE*(A(1)-(A0-B0))+BM*B0
3870 SF(1)=GA*(X(1)/A(1))+GM*SF(1)
3880 IH=1+LT
3890 IF IH<=LS THEN 3910
3900 IH=IH-LS:GOTO 3890
3910 FCS(1+LT)=(A(1)+XLT*B(1))*SF(IH)
3920 SS(1)=SF(1)
3930 ER(1+LT)=X(1+LT)-FCS(1+LT)
3940 SUMSQ=ER(1)^2
3950 XMAD=ABS(ER(1))
3960 FOR I=2 TO IT
3970 IL=I MOD LS
3980 IF IL=0 THEN IL=LS
3990 A(I)=AP*(X(I)/SF(IL))+AM*(A(I-1)+B(I-1))
4000 B(I)=BE*(A(I)-A(I-1))+BM*B(I-1)
4010 SF(IL)=GA*(X(I)/A(I))+GM*SF(IL)
4020 SS(I)=SF(IL)
4030 IF IL<>LS THEN 4120
4040 REM Normalize seasonal factors
4050 GTOT=0
4060 FOR J=1 TO LS
4070 GTOT=GTOT+SF(J)
```

```
4080 NEXT J
4090 FOR J=1 TO LS
4100 SF(J)=SF(J)*(XL/GTOT)
4110 NEXT J
4120 IH=IL+LT
4130 IF IH<=LS THEN 4150
4140 IH=IH-LS:GOTO 4130
4150 FCS(I+LT)=(A(I)+XLT*B(I))*SF(IH)
4160 IF (I+LT)>LD THEN 4200
4170 ER(I+LT)=X(I+LT)-FCS(I+LT)
4180 SUMSQ=SUMSQ+ER(I)^2
4190 XMAD=XMAD+ABS(ER(I))
4200 NEXT I
4210 XMAD=XMAD/(IT-LT):MSE=SUMSQ/(IT-LT)
4220 IF C$="C" THEN 4370
4230 LPRINT:LPRINT TAB(5) "RESULTS OF INITIALIZATION PHASE"
4240 LPRINT:LPRINT "SEASON LENGTH=";LS,"FORECAST LEAD TIME=";LT
4250 LPRINT "PERIOD","ACTUAL DATA","FCS. VALUE","ERROR"
4260 FOR I=1 TO IT
4270 LPRINT I,X(I),FCS(I),ER(I)
4280 NEXT I
4290 LPRINT:LPRINT:LPRINT TAB(20) "MODEL COMPONENTS"
4300 LPRINT "PERIOD","PERMANENT","TREND","SEASONAL"
4310 FOR I=1 TO IT
4320 LPRINT I,A(I),B(I),SS(I)
4330 NEXT I
4340 LPRINT:LPRINT "MEAN SQUARED ERROR="; MSE
4350 LPRINT "MEAN ABSOLUTE DEVIATION=";XMAD
4360 GOTO 4590
4370 PRINT:PRINT TAB(5) "RESULTS OF INITIALIZATION PHASE"
4380 PRINT:PRINT "SEASON LENGTH="; LS, "FORECAST LEAD TIME=";LT
4390 PRINT "PERIOD", "ACTUAL DATA", "FCS. VALUE", "ERROR"
4400 IC=1
4410 FOR I=1 TO IT
4420 IF I<>(IC*15) THEN 4450
4430 IC=IC+1
4440 PRINT:INPUT "PRESS ENTER TO CONTINUE"; Y$:PRINT
4450 PRINT I, X(I), FCS(I), ER(I)
4460 NEXT I
4470 PRINT:INPUT "PRESS ENTER TO CONTINUE"; Y$
4480 PRINT:PRINT TAB(20) "MODEL COMPONENTS"
4490 PRINT"PERIOD", "PERMANENT", "TREND", "SEASONAL"
4500 IC=1
4510 FOR I=1 TO IT
4520 IF I<>(IC*18) THEN 4550
4530 PRINT:INPUT "PRESS ENTER TO CONTINUE"; Y$
4540 IC=IC+1
4550 PRINT I, A(I), B(I), SS(I)
4560 NEXT I
4570 PRINT:PRINT "MEAN SQUARED ERROR="; MSE
4580 PRINT "MEAN ABSOLUTE DEVIATION="; XMAD
4590 I1=IT+1            'Start forecast phase
```

```
4600 IF (IT=LD) AND (IZN=0) THEN 5320
4610 IF IT=LD THEN 4870
4620 SUMSQ=0
4630 XMAD=0
4640 REM Perform smoothing and calculate forecasts
4650 FOR I=I1 TO LD
4660 IL=I MOD LS
4670 IF IL=0 THEN IL=LS
4680 FCS(I)=(A(I-LT)+XLT*B(I-LT))*SF(IL)
4690 A(I)=AP*(X(I)/SF(IL))+AM*(A(I-1)+B(I-1))
4700 B(I)=BE*(A(I)-A(I-1))+BM*B(I-1)
4710 SF(IL)=GA*(X(I)/A(I))+GM*SF(IL)
4720 SS(I)=SF(IL)
4730 IF IL<>LS THEN 4820
4740 REM Normalize seasonal factors
4750 GTOT=0
4760 FOR J=1 TO LS
4770 GTOT=GTOT+SF(J)
4780 NEXT J
4790 FOR J=1 TO LS
4800 SF(J)=SF(J)*(XL/GTOT)
4810 NEXT J
4820 ER(I)=X(I)-FCS(I)
4830 SUMSQ=SUMSQ+ER(I)∧2
4840 XMAD=XMAD+ABS(ER(I))
4850 NEXT I
4860 XMAD=XMAD/(LD-IT): MSE = SUMSQ/(LD-IT)
4870 IF C$="P" THEN 4990
4880 PRINT:INPUT "PRESS ENTER TO CONTINUE"; Y$
4890 PRINT:PRINT TAB(5) "RESULTS OF FORECASTING PHASE"
4900 PRINT "PERIOD","ACTUAL DATA", "FCS. VALUE", "ERROR"
4910 IF IT=LD THEN 4990
4920 IC=1
4930 FOR I=I1 TO LD
4940 IF(I-I1+1)<>(IC*20) THEN 4970
4950 IC=IC+1
4960 PRINT:INPUT "PRESS ENTER TO CONTINUE";Y$:PRINT
4970 PRINT I, X(I), FCS(I), ER(I)
4980 NEXT I
4990 IF IZN=0 THEN 5080
5000 FOR I=LD+1 TO LD+IZN
5010 IL= I MOD LS
5020 IF IL=0 THEN IL=LS
5030 IF(I-LT)>LD THEN 5060
4040 FCS(I)=(A(I-LT)+XLT*B(I-LT))*SF(IL)
4050 GOTO 5070
5060 FCS(I)=(A(LD)+(I-LD)*B(LD))*SF(IL)
5070 NEXT I
5080 IF C$="P" THEN 5350
5090 IF IZN=0 THEN 5190
5100 PRINT TAB(7) "FORECAST HORIZON"
5110 IL1=LD-I1+1
```

206   Chapter 10—Forecasting with Exponential Smoothing

```
5120 FOR I=LD+1 TO LD+IZN
5130 IL1=IL1+1
5140 IF IL1<>(IC*20) THEN 5170
5150 IC=IC+1
5160 PRINT:INPUT "PRESS ENTER TO CONTINUE";Y$:PRINT
5170 PRINT I,,FCS(I)
5180 NEXT I
5190 IF IT=LD THEN 5320
5200 PRINT:INPUT "PRESS ENTER TO CONTINUE";Y$
5210 PRINT:PRINT TAB(28) "MODEL COMPONENTS"
5220 PRINT "PERIOD", "PERMANENT", "TREND", "SEASONAL"
5230 IC=1
5240 FOR I=I1 TO LD
5250 IF (I-I1+1)<>(IC*20) THEN 5280
5260 IC=IC+1
5270 PRINT:INPUT "PRESS ENTER TO CONTINUE";Y$
5280 PRINT I,A(I),B(I),SS(I)
5290 NEXT I
5300 PRINT:PRINT "FORECAST PHASE MEAN SQUARED ERROR=";MSE
5310 PRINT "MEAN ABSOLUTE DEVIATION IN FORECAST PHASE="; XMAD
5320 PRINT:PRINT TAB(10) "END OF RUN"
5330 PRINT:INPUT "PRESS ENTER TO CONTINUE";Y$:PRINT
5340 GOTO 1520
5350 LPRINT:LPRINT TAB(5) "RESULTS OF FORECASTING PHASE"
5360 LPRINT "PERIOD","ACTUAL DATA","FCS. VALUE","ERROR"
5370 IF IT=LD THEN 5410
5380 FOR I=I1 TO LD
5390 LPRINT I,X(I),FCS(I),ER(I)
5400 NEXT I
5410 IF IZN=0 THEN 5460
5420 LPRINT TAB(7) "FORECAST HORIZON"
5430 FOR I=LD+1 TO LD+IZN
5440 LPRINT I,,FCS(I)
5450 NEXT I
5460 IF IT=LD THEN 5540
5470 LPRINT:LPRINT TAB(28) "MODEL COMPONENTS"
5480 LPRINT "PERIOD","PERMANENT","TREND","SEASONAL"
5490 FOR I= I1 TO LD
5500 LPRINT I,A(I),B(I),SS(I)
5510 NEXT I
5520 LPRINT:LPRINT "FORECAST PHASE MEAN SQUARED ERROR=";MSE
5530 LPRINT "MEAN ABSOLUTE DEVIATION=";XMAD
5540 PRINT:PRINT TAB(10) "END OF RUN":PRINT
5550 GOTO 1520
5560 REM SUBROUTINE: STORE DATA ON DISK
5570 PRINT:INPUT "ENTER NAME OF DISK:FILE";NAM$
5580 OPEN NAM$ FOR OUTPUT AS #3
5590 WRITE#3,LD
5600 FOR I=1 TO LD
5610 WRITE#3, X(I)
5620 NEXT I
5630 CLOSE#3
```

10.4 Exponential Smoothing Program

```
5640 PRINT:RETURN
5650 REM SUBROUTINE: READ DATA FROM DISK
5660 PRINT:INPUT "ENTER NAME OF DISK:FILE";NAM$
5670 OPEN NAM$ FOR INPUT AS #3
5680 INPUT#3, LD
5690 FOR I=1 TO LD
5700 INPUT#3, X(I)
5710 NEXT I
5720 CLOSE#3
5730 PRINT "TIME SERIES CONTAINS ";LD;" DATA ELEMENTS":PRINT
5740 RETURN
5750 REM SUBROUTINE: LIST MODEL DATA
5760 CLS:PRINT TAB(10) "LISTING OF INPUT DATA":PRINT
5770 PRINT "NUMBER OF DATA POINTS IN TIME SERIES ";LD
5780 PRINT "NUMBER OF DATA POINTS USED FOR INITIALIZATION "; IT
5790 PRINT "NUMBER OF PERIODS IN A SEASON "; LS
5800 PRINT "FORECAST LEAD TIME "; LT
5810 PRINT "PERIODS IN FORECAST HORIZON ";IZN
5820 IF SY$="Y" THEN 5860
5830 PRINT:PRINT TAB(10) "SMOOTHING CONSTANTS"
5840 PRINT "ALPHA="; AP, "BETA="; BE, "GAMMA="; GA:PRINT
5850 GOTO 5910
5860 PRINT:PRINT TAB(10) "SMOOTHING CONSTANT RANGES"
5870 PRINT "TYPE", "LOWER LIMIT", "UPPER LIMIT", "INCREMENT"
5880 PRINT "ALPHA", AL, AU, AI
5890 PRINT "BETA", BL, BU, BI
5900 PRINT "GAMMA", GL, GU, GI:PRINT
5910 IF PY$="Y" THEN RETURN
5920 PRINT "PERMANENT AND TREND COMPONENTS "; A0, B0
5930 IF LS=0 THEN RETURN
5940 PRINT:PRINT TAB(5) "PERIODS" TAB(25) "**SEASONAL FACTORS**"
5950 IR=LS MOD 4
5960 IP=INT (LS/4)
5970 IF (IR<>0) THEN IP=IP+1
5980 FOR I=1 TO IP
5990 I1=I:I2=I+3
6000 IF(I<>11) THEN 6030
6010 INPUT "PRESS ENTER TO CONTINUE"; Y$
6020 PRINT:PRINT TAB(5)"PERIODS" TAB(25)"
6030 IF I<IP THEN 6150
6040 IF(I=IP) AND (IR=0) THEN 6150
6050 ON IR GOTO 6060, 6090, 6120
6060 I2=I1
6070 PRINT I1; " THRU "; I2,,SF(I)
6080 GOTO 6160
6090 I2=I1+1
6100 PRINT I1; " THRU "; I2,,SF(I), SF(I+1)
6110 GOTO 6160
6120 I2=I1+2
6130 PRINT I1; " THRU "; I2, SF(I), SF(I+1), SF(I+2)
6140 GOTO 6160
6150 PRINT I1; " THRU "; I2, SF(I), SF(I+1), SF(I+2), SF(I+3)
```

Chapter 10—Forecasting with Exponential Smoothing

```
6160 NEXT I
6170 RETURN
6180 REM SUBROUTINE: LIST TIME SERIES DATA
6190 PRINT:PRINT TAB(10) "LISTING OF TIME SERIES DATA ELEMENTS":PRINT
6200 PRINT TAB(5) "ELEMENT NOS." TAB(25) "**DATA ELEMENT VALUES**"
6210 IC=1
6220 IR=LD MOD 4
6230 IP=INT(LD/4)
6240 IF(IR<>0) THEN IP=IP+1
6250 FOR I=1 TO IP
6260 I1=4*(I-1)+1:I2=I1+3
6270 IF I<>(IC*20) THEN 6310
6280 IC=IC+1
6290 INPUT "PRESS ENTER TO CONTINUE"; Y$
6300 PRINT:PRINT "ELEMENT NOS. ", "**DATA ELEMENT"
6310 IF I<IP THEN 6430
6320 IF (I=IP) AND (IR=0) THEN 6430
6330 ON IR GOTO 6340, 6370, 6400
6340 I2=I1
6350 PRINT TAB(8)"ELEMENT ";I1,X(I1)
6360 GOTO 6440
6370 I2=I1+1
6380 PRINT "ELEMENTS";I1; " THRU "; I2,X(I1), X(I1+1)
6390 GOTO 6440
6400 I2=I1+2
6410 PRINT "ELEMENTS";I1; " THRU "; I2,X(I1), X(I1+1), X(I1+2)
6420 GOTO 6440
6430 PRINT "ELEMENTS";I1; " THRU "; I2,X(I1), X(I1+1), X(I1+2), X(I1+3)
6440 NEXT I
6450 PRINT:INPUT "PRESS ENTER TO CONTINUE";Y$:PRINT
6460 RETURN
6470 REM SUBROUTINE: CHANGE DATA
6480 PRINT
6490 INPUT "ENTER POSITION OF DATA ELEMENT TO BE CHANGED";I
6500 IF(I<0) OR (I>LD) THEN PRINT "IN VALID"': GOTO 6490
6510 INPUT "ENTER DATA ELEMENT";X(I)
6520 INPUT "WANT TO CHANGE OTHER DATA ELEMENTS (Y/N)";Y$
6530 IF Y$="Y" THEN 6490
6540 RETURN
6550 REM SUBROUTINE: ADD TO TIME SERIES
6560 PRINT
6570 PRINT "TIME SERIES NOW CONTAINS "; LD;"DATA ELEMENTS"
6580 INPUT "HOW MANY DATA ELEMENTS ARE TO BE ADDED";IADD
6590 IF(IADD<0) THEN PRINT "MUST BE >=0";GOTO 6580
6600 IF IADD=0 THEN RETURN
6610 IF ((IADD+LD)>LM) THEN PRINT "TOTAL MUST BE <=";LM:GOTO 6580
6620 PRINT "ENTER DATA"
6630 FOR I=LD+1 TO LD+IADD
6640 PRINT "VALUE FOR DATA ELEMENT ";I;
6650 INPUT X(I)
6660 NEXT I
6670 LD=LD+IADD
```

```
6680 PRINT:RETURN
6690 REM SUBROUTINE: ENTER TIME SERIES DATA FROM KEYBOARD
6700 INPUT "NUMBER OF DATA POINTS IN TIME SERIES"; LD
6710 IF (LD<0)OR(LD>LM) THEN PRINT "INVALID":GOTO 6700
6720 PRINT "ENTER TIME SERIES DATA ELEMENTS (4 ELEMENTS PER LINE)"
6730 IR=LD MOD 4
6740 IP=INT(LD/4)
6750 IF (IR<>0) THEN IP=IP+1
6760 FOR I=1 TO IP
6770 I1=4*(I-1)+1:12=I1+3
6780 IF I<IP THEN 6930
6790 IF (I=IP) AND (IR=0) THEN 6930
6800 ON IR GOTO 6810, 6850, 6890
6810 I2=I1
6820 PRINT "ELEMENT "; 12;
6830 INPUT X(I1)
6840 GOTO 6950
6850 I2=I1+1
6860 PRINT "ELEMENTS "; I1; " THRU "; I2;
6870 INPUT X(I1), X(I1+1)
6880 GOTO 6950
6890 I2=I1+2
6900 PRINT "ELEMENTS "; I1; " THRU "; I2;
6910 INPUT X(I1), X(I1+1), X(I1+2)
6920 GOTO 6950
6930 PRINT "ELEMENTS "; I1; " THRU "; I2;
6940 INPUT X(I1), X(I1+1), X(I1+2), X(I1+3)
6950 NEXT I
6960 INPUT "WOULD YOU LIKE DATA ELEMENTS LISTED (Y/N)";Y$
6970 IF Y$="N" THEN RETURN
6980 GOSUB 6180
6990 RETURN
7000 REM SUBROUTINE: DELETE DATA ELEMENTS
7010 PRINT
7020 PRINT "ENTER BEGINNING AND ENDING POSITIONS OF DATA ELEMENTS"
7030 INPUT "TO BE DELETED FROM TIME SERIES";IBEG,IEND
7040 IF (IBEG<0) OR (IEND<0) THEN PRINT "INVALID":GOTO 7020
7050 IF (IBEG>LD) OR (IEND>LD) THEN PRINT "INVALID":GOTO 7020
7060 IF (IEND-IBEG+1)<0 THEN PRINT "INVALID":GOTO 7020
7070 IF IEND=LD THEN 7110
7080 FOR I=IEND+1 TO LD
7090 X(I-(IEND-IBEG+1))=X(I)
7100 NEXT I
7110 LD=LD-(IEND-IBEG+1)
7120 PRINT "DATA ELEMENTS HAVE BEEN DELETED"
7130 PRINT:RETURN
```

Figure 10.1—exponential smoothing forecasting program

Data can be entered in two ways, from the keyboard or from a previously saved disk file. The following data management options are provided by the program:

1. List time series data.
2. Change time series data.
3. Delete data from time series.
4. Add to time series.
5. Store time series data on disk.

Having entered the data, the exponential smoothing computations will be performed when the PERFORM FORECAST COMPUTATIONS option is specified. These computations are organized into two major parts: initialization phase and forecast phase.

The initialization phase may be used to develop estimates for the model components (permanent, trend, and seasonal) and to optimize the values for the smoothing constants associated with each of the component types. Note that at least two seasons must be available for the program to develop these estimates. Also, the number of periods used in initialization must be a multiple of the season length.

These initial estimates are used in the initialization phase. During this phase, the values are smoothed so the best values will be used to start the forecasting phase. At the end of this phase, the program displays a comparison of forecast versus actual data for each period used in the initialization phase. The user may input initial values for the model components and the smoothing constants. In this case the initialization phase is used only to smooth these input values to obtain the best values to be used in the forecast phase.

During the forecast phase, forecasts are made for all periods in the time series beyond the last period used in the initialization phase. If a forecast horizon is specified (periods beyond the last period in the time series), forecasts are also made for these periods. The results of the initialization and forecast phases can be displayed on your CRT or your printer. If they are displayed on your CRT, program execution is periodically interrupted to give you an opportunity to read the results before they are scrolled off the screen. Details of using this program will now be illustrated with some examples. In these examples, user input is denoted by underlined characters.

EXAMPLE 1. A local firm manufactures turbo chargers for trucks. This firm would like some help in forecasting impeller wheel failures for the next two quarters. They have six years (24 quarters) of historical data representing failures by quarter.

Year 1	202	242	293	195
Year 2	257	281	343	244
Year 3	307	359	438	318
Year 4	397	437	528	379
Year 5	481	529	626	445
Year 6	564	601	725	507

Figure 10.2 is a plot of these data. Note that trend and seasonal effects are exhibited.

SOLUTION. The Time Series Forecasting Program is stored in the file EX-SMOOTH.

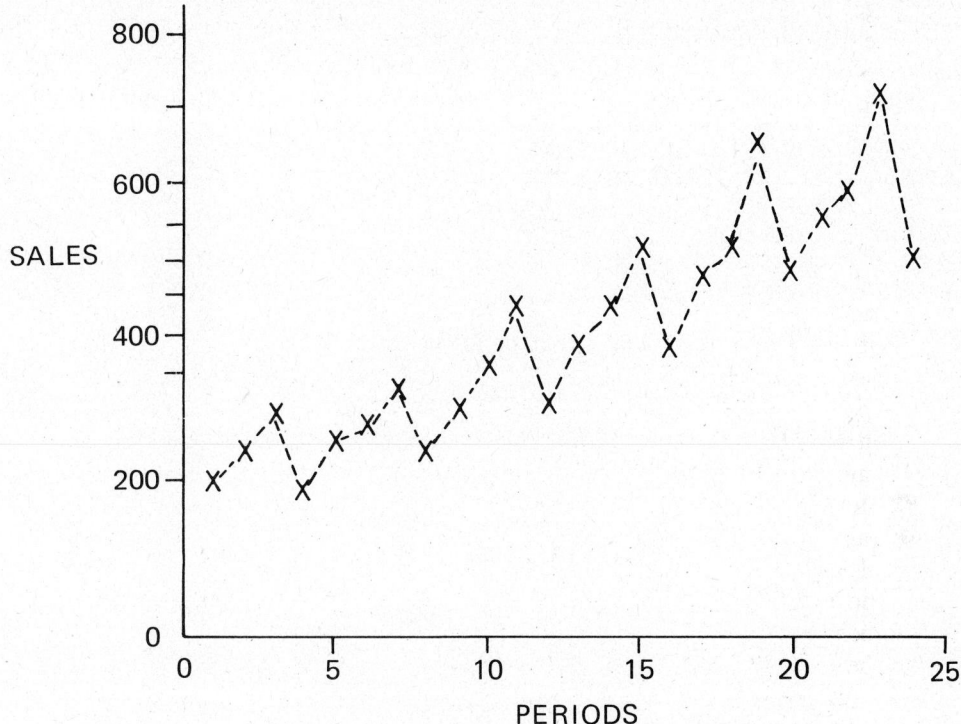

Figure 10.2—impeller wheel failures example

```
LOAD "B:EXSMOOTH"
RUN

     **TIME SERIES FORECASTING PROGRAM-WINTER'S METHOD**

WANT PROGRAM INSTRUCTIONS (Y/N)? Y

PROGRAM INSTRUCTIONS:

Time series data may be entered from keyboard or disk.
Time series data management options provided:
     -List data
     -Change data
     -Add to data
     -Delete data
     -Store data on disk
User must specify number of data elements to be used
for model initialization, number of periods in season,
forecast lead time, and forecast horizon.
User has the option of inputting model components (permanent,
trend, and seasonal factors) and smoothing constants (one
constant for each type of model component) or specifying that
the program determine these values. Program will optimize
```

smoothing constants by performing a grid search using specified limits and increments within these limits.

```
ENTER DATA FROM KEYBOARD OR DISK (K/D)? K

NUMBER OF DATA POINTS IN TIME SERIES? 24
ENTER TIME SERIES DATA ELEMENTS (4 ELEMENTS PER LINE)
ELEMENTS 1 THRU 4? 202,242,293,195
ELEMENTS 5 THRU 8? 257,281,343,244
ELEMENTS 9 THRU 12? 307,359,438,318
ELEMENTS 13 THRU 16? 397,437,528,379
ELEMENTS 17 THRU 20? 481,529,626,445
ELEMENTS 21 THRU 24? 564,601,725,507

WOULD YOU LIKE DATA ELEMENTS LISTED (Y/N)? Y

     LISTING OF TIME SERIES DATA ELEMENTS

       ELEMENT NOS.         **DATA VALUES**
     ELEMENTS 1 THRU 4      202   242   293   195
     ELEMENTS 5 THRU 8      257   281   343   244
     ELEMENTS 9 THRU 12     307   359   438   318
     ELEMENTS 13 THRU 16    397   437   528   379
     ELEMENTS 17 THRU 20    481   529   626   445
     ELEMENTS 21 THRU 24    564   601   725   507

PRESS ENTER TO CONTINUE?

          PROGRAM OPTIONS
              1-LIST TIME SERIES DATA
              2-CHANGE TIME SERIES DATA
              3-DELETE DATA FROM TIME SERIES
              4-ADD TO TIME SERIES
              5-STORE TIME SERIES DATA ON DISK
              6-PERFORM FORECAST COMPUTATIONS
              7-STUDY NEW MODEL
              8-QUIT
```

Having entered the data and verified that they are correct, we will store them in a disk file named WINEXAMP. Then the forecast computations will be performed using the first 16 periods to initialize the model. The remaining 8 periods can be used to evaluate the model. Since little is known about the time series being studied, we will let the program estimate the model parameters and determine the best smoothing constants.

```
OPTION? 5

ENTER NAME OF DISK:FILE? B:WINEXAMP

          PROGRAM OPTIONS
              1-LIST TIME SERIES DATA
              2-CHANGE TIME SERIES DATA
              3-DELETE DATA FROM TIME SERIES
              4-ADD TO TIME SERIES
              5-STORE TIME SERIES DATA ON DISK
```

10.4 Exponential Smoothing Program

```
            6-PERFORM FORECAST COMPUTATIONS
            7-STUDY NEW MODEL
            8-QUIT
OPTION? 6

NUMBER OF DATA ELEMENTS USED FOR INITIALIZATION? 16
NUMBER PERIODS IN A SEASON? 4
ESTIMATION OF MODEL PARAMETERS DESIRED (Y/N)? Y
SMOOTHING CONSTANT OPTIMIZATION DESIRED (Y/N)? Y
```

To determine the optimum smoothing constants, the program performs a grid search using the lower and upper limits and step size specified for each constant. Remember that the smoothing constants must have a value in the range of 0 to 1.

In this example, lower and upper limits of .1 and .3 and a step size of .1 were specified for each of the smoothing constants. Consequently, 27 different combinations are possible (.1, .2, .3 for each of the constants, giving $3 \times 3 \times 3 = 27$). Note that you should be careful not to make the number of trials too large. For instance, five minutes of computer time may be required to evaluate 127 combinations (trials). Each combination of smoothing constants is evaluated by using each one as initial values in the initialization phase. The combination resulting in the smallest value for the sum of errors squared is chosen as the optimum.

```
ENTER LOWER LIMIT, UPPER LIMIT, STEP SIZE FOR ALPHA--
PERMANENT COMPONENT SMOOTHING FACTOR
? .1,.3..1
ENTER LOWER LIMIT, UPPER LIMIT, STEP SIZE FOR BETA--
PERMANENT COMPONENT SMOOTHING FACTOR
? .1,.3..1
ENTER LOWER LIMIT, UPPER LIMIT, STEP SIZE FOR GAMMA--
PERMANENT COMPONENT SMOOTHING FACTOR
? .1,.3..1
27 TRIALS ARE REQUIRED TO FIND OPTIMUM VALUES FOR
SMOOTHING CONSTANTS. OKAY TO CONTINUE (Y/N)? Y
```

We will assume a forecast lead time of 1 period (1 quarter) and a forecast horizon of 2 periods (each period represents 1 quarter).

```
ENTER FORECAST LEAD TIME? 1
NUMBER PERIODS IN FORECAST HORIZON? 2
WOULD YOU LIKE TO SEE INPUT DATA DISPLAYED (Y/N)? Y

LISTING OF INPUT DATA

NUMBER OF DATA POINTS IN TIME SERIES 24
NUMBER OF DATA POINTS USED FOR INITIALIZATION 16
NUMBER OF PERIODS IN A SEASON 4
FORECAST LEAD TIME 1
PERIODS IN FORECAST HORIZON 2
```

```
            SMOOTHING CONSTANT RANGES
TYPE         LOWER LIMIT      UPPER LIMIT     INCREMENT
ALPHA            .1               .3              .1
BETA             .1               .3              .1
GAMMA            .1               .3              .1
```

ARE DATA CORRECT (Y/N)? Y

RESULTS TO APPEAR AT CRT OR PRINTER (C/P)? C

INITIAL VALUES FOR PERMANENT, TREND, AND SEASONAL
COMPONENTS TO BE ESTIMATED FROM DATA

ESTIMATED PERMANENT COMPONENT 207.7188
ESTIMATED TREND COMPONENT 16.85417

```
PERIOD          ESTIMATED SEASONAL FACTOR
  1                     .9661799
  2                    1.038803
  3                    1.194782
  4                     .8002349
```

PRESS ENTER TO CONTINUE

OPTIMUM SMOOTHING CONSTANTS
ALPHA= .3   BETA= .3   GAMMA= .1
SUM OF ERRORS SQUARED= 2298.504

Note that the smoothing constants for the permanent and trend components are at their specified upper limits and the seasonal smoothing constant is at the lower limit. Look at the values for the model components displayed below; you will note that the permanent and trend components are not very stable. The model is trying to adjust to this instability by using larger values for the associated smoothing constants. Remember that a small smoothing constant is used for a stable process.

```
         RESULTS OF INITIALIZATION PHASE
SEASON LENGTH=  4              FORECAST LEAD TIME=  1
PERIOD      ACTUAL DATA      FCS VALUE         ERROR
  1             202              0                0
  2             242           233.8348         8.165238
  3             293           292.8908          .1092224
  4             195           210.3501       -15.35013
  5             257           264.0346        -7.034577
  6             281           298.1124       -17.11237
  7             343           352.6972        -9.697204
  8             244           243.4822          .517807
  9             307           308.6226        -1.622559
 10             359           344.7361        14.26395
 11             438           418.3526        19.6474
 12             318           295.7983        22.20169
 13             397           382.9904        14.00964
 14             437           437.8469         -.8468322
```

PRESS ENTER TO CONTINUE?

15	528	526.6478	1.352234
16	379	368.206	10.79398

PRESS ENTER TO CONTINUE?

MODEL COMPONENTS

PERIOD	PERMANENT	TREND	SEASONAL
1	208.1244	16.97.585	.9666192
2	227.4583	17.68327	1.041316
3	245.169	17.6915	1.194814
4	257.1058	15.96511	.7960556
5	270.8884	15.31033	.9650904
6	281.2702	13.83177	1.037369
7	292.6678	13.10153	1.192852
8	305.9644	13.16005	.7964122
9	318.6211	13.00906	.9667351
10	335.7467	14.244	1.042493
11	354.9218	15.72333	1.1992
12	378.9909	18.22707	.8021641
13	401.577	19.53479	.9666238
14	420.8675	19.46148	1.039599
15	440.6681	19.56324	1.196247
16	464.2789	20.77748	.8016729

SUM OF ERRORS SQUARED= 142.6191
MEAN ABSOLUTE DEVIATION = 9.514988

PRESS ENTER TO CONTINUE?

RESULTS OF FORECASTING PHASE

PERIOD	ACTUAL DATA	FCS VALUE	ERROR
17	481	468.3819	12.61816
18	529	530.613	−1.612915
19	626	636.0072	−10.07721
20	445	441.1069	3.893128
21	564	554.9628	9.037231
22	601	621.7125	−20.71246
23	725	732.1151	−7.115051
24	507	506.0732	.926894

FORECAST HORIZON

25		631.7901	
26		696.5431	

Looking at the forecast phase results, we see that 632 and 697 impeller wheel failures are predicted to occur during the next two quarters (25th and 26th periods). Since we do not know of any changes in the process being modeled, these numbers will be used as our forecast of failures for these periods.

MODEL COMPONENTS

PERIOD	PERMANENT	TREND	SEASONAL
17	488.9766	21.95355	.9674299
18	510.4642	21.81377	1.038302
19	529.7481	21.05482	1.193678
20	552.2613	21.49233	.8013368

21	576.5566	22.33322	.9683464
22	592.9041	20.53751	1.035663
23	611.6531	20.00097	1.192641
24	632.0011	20.10508	.8012898

```
FORECAST PHASE SUM OF ERRORS SQUARED= 105.0859
MEAN ABSOLUTE DEVIATION IN FORECAST PHASE= 8.249126

        END OF RUN

    PROGRAM OPTIONS
        1-LIST TIME SERIES DATA
        2-CHANGE TIME SERIES DATA
        3-DELETE TIME SERIES DATA FROM TIME
        4-ADD TO TIME SERIES
        5-STORE DATA ON DISK
        6-PERFORM FORECAST COMPUTATIONS
        7-STUDY NEW MODEL
        8-QUIT

OPTION? 8
        END OF PROGRAM
```

EXAMPLE 2. Two quarters have passed and the company that manufactures turbo chargers was very pleased with our projections of impeller wheel failures for this time period. As a result, this firm would like our assistance in developing a forecast for the next year (4 quarters). In reviewing the future data, errors were found in the figures for periods 23 and 24. Also, the failure figures for periods 25 and 26 are now available:

Period	Failures
23	730 (revised data)
24	510 (revised data)
25	640
26	700

SOLUTIONS. Since the original time series containing 24 quarters of failure data was saved in a disk file named WINEXAMP, we can enter this data using the enter data from disk option. However, the data associated with periods 23 and 24 must be changed and data for periods 25 and 26 must be added to the time series.

```
RUN

    **TIME SERIES FORECASTING PROGRAM-WINTER'S METHOD**

WANT PROGRAM INSTRUCTIONS (Y/N)? N

ENTER DATA FROM KEYBOARD OR DISK (K/D)? D

ENTER NAME OF DISK:FILE? B:WINEXAMP
TIME SERIES CONTAINS 24 DATA ELEMENTS
```

10.4 Exponential Smoothing Program

```
PROGRAM OPTIONS
    1-LIST TIME SERIES DATA
    2-CHANGE TIME SERIES DATA
    3-DELETE DATA FROM TIME SERIES
    4-ADD TO TIME SERIES
    5-STORE TIME SERIES DATA ON DISK
    6-PERFORM FORECAST COMPUTATIONS
    7-STUDY NEW MODEL
    8-QUIT
```

There are two ways we could revise the data for periods 23 and 24. One way is to use option 2, CHANGE TIME SERIES DATA to change the incorrect data elements. Another way is to use option 3, DELETE DATA FROM TIME SERIES, in conjunction with option 4, ADD TO TIME SERIES. The latter approach was used here so that option 3 could be illustrated.

```
OPTION? 3

ENTER BEGINNING AND ENDING POSITIONS OF DATA ELEMENTS
TO BE DELETED FROM TIME SERIES? 23,24
DATA ELEMENTS HAVE BEEN DELETED

    PROGRAM OPTIONS
        1-LIST TIME SERIES DATA
        2-CHANGE TIME SERIES DATA
        3-DELETE DATA FROM TIME SERIES
        4-ADD TO TIME SERIES
        5-STORE DATA ON DISK
        6-PERFORM FORECAST COMPUTATIONS
        7-STUDY NEW MODEL
        8-QUIT

OPTION? 4

TIME SERIES NOW CONTAINS 22 DATA ELEMENTS
HOW MANY DATA ELEMENTS ARE TO BE ADDED? 4
ENTER DATA
VALUE FOR DATA ELEMENT 23
? 730
VALUE FOR DATA ELEMENT 24
? 510
VALUE FOR DATA ELEMENT 25
? 640
VALUE FOR DATA ELEMENT 26
? 700


    PROGRAM OPTIONS
        1-LIST TIME SERIES DATA
        2-CHANGE TIME SERIES DATA
        3-DELETE DATA FROM TIME SERIES
        4-ADD TO TIME
        5-STORE TIME SERIES DATA ON DISK
```

218 Chapter 10—Forecasting with Exponential Smoothing

```
            6-PERFORM FORECAST COMPUTATIONS
            7-STUDY NEW MODEL
            8-QUIT
OPTION? 1

            LISTING OF TIME SERIES DATA ELEMENTS

      ELEMENT NOS.      **DATA ELEMENT VALUES**
      ELEMENTS  1 THRU  4    202   242   293   195
      ELEMENTS  5 THRU  8    257   281   343   244
      ELEMENTS  9 THRU 12    307   359   438   318
      ELEMENTS 13 THRU 16    397   437   528   379
      ELEMENTS 17 THRU 20    481   529   626   445
      ELEMENTS 21 THRU 24    564   601   730   510
      ELEMENTS 25 THRU 26    640   700

PRESS ENTER TO CONTINUE?
```

Looking at the listing of the time series data, we see that the data are correct. Therefore, we are ready to perform the forecast computations. Our model will be initialized using the first 20 time periods. Remember that the number of time periods used for initialization must be a multiple of the number of periods in a season (the season length is 4). Also, initial estimates for the model components can be obtained from example 1; these estimates will be used here. Likewise, the optimum values from example 1 will be input for the smoothing constants.

```
            PROGRAM OPTIONS
                  1-LIST TIME SERIES DATA
                  2-CHANGE TIME SERIES DATA
                  3-DELETE DATA FROM TIME SERIES
                  4-ADD TO TIME SERIES
                  5-STORE TIME SERIES DATA ON DISK
                  6-PERFORM FORECAST COMPUTATIONS
                  7-STUDY NEW MODEL
                  8-QUIT

OPTION? 6

NUMBER OF DATA ELEMENTS USED FOR INITIALIZATION? 20
NUMBER PERIODS IN A SEASON? 4
ESTIMATION OF MODEL PARAMETERS DESIRED (Y/N)? N
ENTER INITIAL PERMANENT AND TREND COMPONENTS? 208,17
ENTER INITIAL SEASONAL FACTORS
SEASONAL FACTOR 1?  .97
SEASONAL FACTOR 2? 1.04
SEASONAL FACTOR 3? 1.19
SEASONAL FACTOR 4?  .80
SMOOTHING CONSTANT OPTIMIZATION DESIRED (Y/N)? N
ENTER PERMANENT COMPONENT SMOOTHING FACTOR-ALPHA? .3
ENTER TREND COMPONENT SMOOTHING FACTOR-BETA? .3
ENTER SEASONAL COMPONENT SMOOTHING FACTOR-GAMMA? .1
```

10.4 Exponential Smoothing Program

ENTER FORECAST LEAD TIME? <u>1</u>
NUMBER PERIODS IN FORECAST HORIZON? <u>4</u>
WOULD YOU LIKE TO SEE INPUT DATA DISPLAYED (Y/N)? <u>N</u>

ARE DATA CORRECT (Y/N)? <u>Y</u>
RESULTS TO APPEAR AT CRT OR PRINTER (C/P)? <u>P</u>

We would like our results displayed at the printer.

```
              TIME SERIES FORECASTING-WINTER'S METHOD
       RESULTS OF INITIALIZATION PHASE
```

SEASON LENGTH= 4   FORECAST LEAD TIME= 1

PERIOD	ACTUAL DATA	FCS VALUE	ERROR
1	202	0	0
2	242	234.1003	2.899674
3	293	291.6466	−1.353455
4	195	210.5845	−15.58446
5	257	265.3491	−8.349121
6	281	298.3824	−17.38242
7	343	351.187	−8.187012
8	244	243.6003	.3997193
9	307	309.8833	−2.883301
10	359	344.8895	14.1105
11	438	416.5466	21.45346
12	318	295.8472	22.15283
13	397	384.3578	12.64221
14	437	437.9949	−.9949036
15	528	524.5345	3.465576
16	379	368.2131	10.7869
17	481	469.8964	11.10367
18	529	530.8119	−1.81189
19	626	633.7522	−7.752197
20	445	441.0946	3.905426

MODEL COMPONENTS

PERIOD	PERMANENT	TREND	SEASONAL
1	208.0742	17.02227	.9700808
2	227.3753	17.7059	1.042432
3	245.4224	17.80826	1.190386
4	257.3864	16.055	.7957615
5	270.8603	15.28066	.9682481
6	281.1402	13.78044	1.038453
7	292.8486	13.16166	1.188828
8	306.1703	13.20684	.7961195
9	318.4857	12.93939	.9696401
10	335.4929	14.15975	1.04357
11	355.0552	15.78051	1.195545
12	379.166	18.27961	.8018749
13	401.3674	19.45614	.969275
14	420.5368	19.37011	1.040639
15	440.7788	19.63169	1.192926
16	464.4569	20.84559	.8013753

17	488.7428	21.87768	.9698449
18	510.0976	21.72082	1.039294
19	529.8668	21.13534	1.190646
20	552.4656	21.5744	.801264

SUM OF ERRORS SQUARED= 126.0884
MEAN ABSOLUTE DEVIATION= 9.064143

RESULTS OF FORECASTING PHASE

PERIOD	ACTUAL DATA	FCS VALUE	ERROR
21	564	556.6168	7.383179
22	601	621.9786	−20.97864
23	730	729.6818	.3182373
24	510	507.3605	2.639557
25	640	636.167	3.833008
26	700	702.5945	−2.594483

FORECAST HORIZON

27		830.8483	
28		575.8969	
29		718.2465	
30		788.2722	

MODEL COMPONENTS

PERIOD	PERMANENT	TREND	SEASONAL
21	576.3243	22.25967	.9705449
22	592.527	20.4426	1.036605
23	613.0498	20.46665	1.190441
24	634.4375	20.76329	.801155
25	656.453	21.11863	.971258
26	676.821	20.89343	1.036662

FORECAST PHASE MEAN SQUARED ERROR = 87.18439
MEAN ABSOLUTE DEVIATION = 6.291184

Looking at the results, we see that the following failures are predicted:

Period	Failures
27	831
28	576
29	718
30	788

Again, since we know of no changes in the process being modeled and because our previous predictions were very good, we will use the above figures for our failures forecast.

```
END OF RUN

PROGRAM OPTIONS
    1-LIST TIME SERIES DATA
    2-CHANGE TIME SERIES DATA
    3-DELETE DATA FROM TIME SERIES
    4-ADD TO TIME SERIES
    5-STORE DATA ON DISK
```

```
            6-PERFORM FORECAST COMPUTATIONS
            7-STUDY NEW MODEL
            8-QUIT

OPTION? 8

            END OF PROGRAM
```

10.5 Summary

In this chapter exponential smoothing, a popular time series forecasting technique, was presented. Winter's method for incorporating trend and seasonal considerations into a model was implemented in a computer program. Several data management options were also incorporated into the computer program to enhance its usefullness. When using time series forecasting techniques you should remember that historical observations are being projected into the future. Therefore, these techniques are normally used for short term forecasting. A plot of the historical data can also be a valuable aid in developing a forecast. The plot subroutine from Chapter 2 can make this task easier.

References

Makridakis, Spyros, and Steven C. Wheelwright, *Forecasting Methods and Applications*. John Wiley & Sons, New York, 1978.

Montgomery, Douglas C., and Lynwood A. Johnson, *Forecasting and Time Series Analysis*. McGraw-Hill, New York, 1976.

Exercises

1. The following data depict the monthly average stock price of a microcomputer firm.

Month	Average Stock Price
1	$3.5
2	3.6
3	3.2
4	3.7
5	4.2
6	4.8
7	5.1
8	5.4
9	5.8

    a. Use a 3-period moving average to estimate the stock price for the 10th month.
    b. Use single exponential smoothing with $\alpha = .3$ to estimate the stock price for the 10th month.

2. The unit sales of 1200 baud modems for the Speciality Modems Corporation have been increasing. The sales data for the last 18 months are given below. You are to develop a unit sales forecast for the next 6 months (months 19-24) using Winter's exponential smoothing method. Also, you are to plot this data using the plot subroutine from Chapter 2.

Month	Sales (1000's of Units)
1	11
2	9
3	11
4	13
5	15
6	16
7	14
8	17
9	17
10	18
11	20
12	22
13	21
14	24
15	27
16	29
17	30
18	33

3. For the time series of 1, 3, 5, 7, 9, 11, 13, 15, 17, compute a forecast for period 10 using:
   a. a 3-period moving average
   b. single exponential smoothing with $\alpha = .5$.
   c. Holt's method of exponential smoothing with a trend.
      Let α and $\beta = .1$
   d. Winter's method (use program provided in this chapter).
   Which is the better method? Why?
4. For the time series of 1, 4, 7, 10, 1, 4, 7, 10, 1, 4, 7, 10, 1, 4, 7, 10, compute a forecast for period 17 using:
   a. Single exponential smoothing with $\alpha = .5$.
   b. Winter's method (use program provided in this chapter).
5. The time series representing the number of units of microcomputers selling for less than $500 has a seasonal pattern. The unit sales for the last 24 months are given below. You are to develop a sales forecast for the next 6 months using Winter's method. Also, you are to plot this data.

Month	Sales (1000's of Units)
1	18
2	19
3	22
4	19
5	20
6	24
7	22
8	24
9	26
10	28
11	32
12	34
13	24
14	24
15	26
16	26
17	29
18	32
19	32

20 36
21 40
22 44
23 46
24 50

11

Project Planning & Scheduling with CPM

The critical path method (CPM) is a popular project planning and scheduling technique. Working together, Remington Rand and DuPont developed CPM around 1960. At about the same time, a similar method, called program evaluation and review technique (PERT), was developed by the Navy and Lockheed Aircraft Corporation. CPM is the most popular; consequently, it will be used in this chapter. The interested reader can refer to references [1] and [2] for an in-depth discussion of both techniques.

11.1 The CPM Technique

One objective of CPM is to develop a schedule such that a project will be completed on time. This technique requires a list of all activities that make up a project, an estimated duration time for each activity, and a list of any activities that must be completed before a particular activity can be started (immediate predecessors). Table 11.1 contains an example of the required information. Note that activities B and C cannot start until activity A has been completed. Therefore, activity A is an immediate predecessor to activities B and C. We can also say that B and C are immediate successors to activity A.

TABLE 11.1
ACQUIRE A COMPUTER
PROJECT ACTIVITIES

Activity	Description	Duration (weeks)	Immediate Predecessors
A	Contact potential vendors	1	—
B	Review bid proposals	2	A
C	Locate financing	2	A
D	Select vendor	4	B
E	Negotiate contract	2	D
F	Prepare facilities	5	C
G	Install computer	2	E, F

H	Train operations manager	2	E, F
I	Implement application software	6	G, H
J	Train users	3	H
K	Validate installation	1	I, J

The data from Table 11.1 can be used to draw a CPM network such as the one depicted in Figure 11.1. The circles in the network are called nodes (or events) and each is numbered. The arrow between two nodes represents an activity. Thus, we can say activity A begins at node 1 and ends at node 2. The letter on each arrow in Figure 11.1 denotes a particular activity and the number denotes the activity duration. In addition to representing a particular activity, the arrows show which activities must precede other activities. In Figure 11.1 activity A must precede activities B and C.

A dashed line in CPM network represents a "dummy" activity. Such an activity has no duration time; it is only used to denote precedence relationships. In Figure 11.1, the activity between nodes 7 and 8 is a dummy activity showing that activity I cannot start until activity H has been completed. Activity H cannot be represented by an arrow between nodes 6 and 8 because activity G is represented there. Only one activity arrow can be drawn between any two nodes.

One of the greatest benefits of using CPM is the resulting detailed planning. In order to draw the network, detailed planning is required. Drawing the network also brings out interactions between the activities, because, once the network is drawn, the entire project can be better visualized. In addition, because CPM is a formal technique understood by many people, communications are improved.

Having completed the CPM network, the next step is to calculate the minimum project duration. The first step in calculating this value is to determine the earliest time all activities ending at a node can be completed. This is done for all nodes making sure that precedence relationships are satisfied. Looking at Figure 11.2, we see that the earliest finish time for each node has been calculated; the times appear in the box by each node. The minimum time in which a computer can be acquired and implemented is 18 weeks. This number is obtained from the earliest finish time of the last node in the network.

The next step is to calculate the latest time each activity can be started without delaying the completion time (18 weeks) of the project. To do this we begin at the end of the network and work backward.

The critical path through a CPM network is the sequence of activities that have no slack. We define slack for an activity to be:

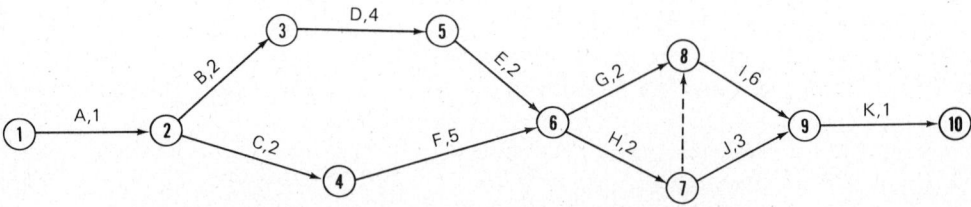

Figure 11.1—acquire a computer project network

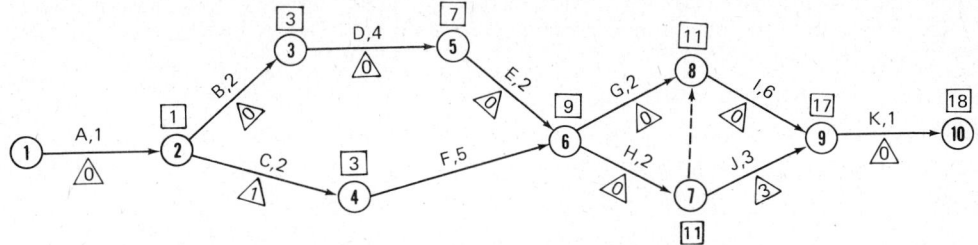

Figure 11.2—acquire a computer project with CPM computations

Activity slack = latest start—earliest start,
where

latest start = the latest time that an activity can start and not affect the project finish date

earliest start = the earliest time that an activity can start

One interpretation of slack is that an activity having slack can be delayed by this amount of time without delaying the completion time of the project. The activity slack appears in a triangle by each activity in Figure 11.2. Using these figures, we can list the activities on the critical path:

A—B—D—E—G—H—I—K

These activities all have a slack = 0.

11.2 CPM Program

Figure 11.3 contains a program that will determine the expected project duration and the critical path through a project network. As many as 100 activities can be in the network. This number can be changed by changing the value of ITOT at the beginning of the program.

```
1000 REM PROJECT PLANNING AND SCHEDULING PROGRAM: CPM
1010 REM This program uses the critical path method, CPM, to
1020 REM determine the estimated completion time of a project and
1030 REM the associated critical path(s). Each activity in the
1040 REM project network is represented by a beginning node and
1050 REM an ending node. The beginning node number must be less
1060 REM than the ending node number. The network can have only
1070 REM one initial node and one terminal node. Each activity is
1080 REM identified by an activity code. A network may have as many
1090 REM as 100 activities. Larger networks may be analyzed if the
1100 REM program dimension statements are changed. Data management
1110 REM options are provided that facilitate
1120 REM    -Adding new activities
1130 REM    -Changing activity data
1140 REM    -Deleting activities
1150 REM    -Listing activity data
1160 REM    -Storing network data in a disk file
```

```
1170 DEFINT I,L
1180 ITOT=100                       'Max. no. activities in network
1190 DIM D$(ITOT),C$(ITOT),ST(ITOT),ED(ITOT),D(ITOT),AS(ITOT)
1200 DIM TT(2),TD$(2),EF(ITOT),IR(ITOT),ES(ITOT),XS(ITOT),XF(ITOT)
1210 REM Definition of output formats
1220 D1$="ENTER ACTIVITY CODE, DESCRIPTION, BEG. NODE, END NODE,
     DURATION"
1230 D2$="    ACTIVITY             BEGINNING   ENDING   ACTIVITY"
1240 D3$="CODE    DESCRIPTION        NODE NO   NODE NO  DURATION"
1250 D4$="\           \ "
1260 D5$="\                    \ "
1270 D7$="    ACTIVITY          DUR-    EARLY     EARLY     LATE"
1280 D8$="      LATE       SLACK"
1290 D9$="CODE  DESCRIPTION     ATION   START    FINISH    START"
1300 D0$="     FINISH      TIME"
1310 F7$="#######.#"
1320 INPUT "DATA TO BE INPUT FROM KEYBOARD OR DISK (K/D)"; Y$
1330 IF (Y$<>"K") AND (Y$<>"D") THEN PRINT "INVALID ENTRY":
     GOTO 1320
1340 IF Y$="K" THEN 1470
1350 INPUT "ENTER NAME OF DISK:FILE";N$
1360 OPEN N$ FOR INPUT AS #3
1370 INPUT #3,T$, DAT$, NAM$, INA
1380 FOR I=1 TO INA
1390 INPUT #3,C$(I), D$(I), ST(I), ED(I), D(I)
1400 NEXT I
1410 CLOSE #3
1420 PRINT "DATE FROM DISK FILE IS: "; DAT$
1430 INPUT "DO YOU WANT TO CHANGE DATE (Y/N)"; Y$
1440 IF Y$<>"Y" THEN 1540
1450 INPUT "ENTER NEW DATE"; DAT$
1460 GOTO 1540
1470 INPUT "ENTER TITLE OF PROJECT"; T$
1480 INPUT "ENTER YOUR NAME";NAM$
1490 INPUT "ENTER DATE";DAT$
1500 INPUT "TOTAL NUMBER OF ACTIVITIES"; INA
1510 IF INA>ITOT THEN PRINT "MUST<= ";ITOT:GOTO 1500
1520 II=1
1530 GOSUB 3570                      'Network data input
1540 GOSUB 3750                      'Sort network act.
1550 PRINT:PRINT TAB(5) "PROGRAM OPTIONS"
1560 PRINT TAB(10) "1-ADD NEW ACTIVITIES"
1570 PRINT TAB(10) "2-CHANGE EXISTING ACTIVITY DATA"
1580 PRINT TAB(10) "3-DELETE ACTIVITIES FROM NETWORK"
1590 PRINT TAB(10) "4-LIST DATA AT CRT"
1600 PRINT TAB(10) "5-LIST DATA AT PRINTER"
1610 PRINT TAB(10) "6-STORE DATA ON DISK"
1620 PRINT TAB(10) "7-PERFORM CPM CALCULATIONS"
1630 PRINT TAB(10) "8-QUIT"
1640 INPUT "ENTER OPTION";IOP:PRINT
1650 IF (IOP<0) OR (IOP>8) THEN PRINT "INVALID": GOTO 1550
1660 IF IOP=1 THEN IFL=1:GOTO 1940
```

11.2 CPM Program

```
1670 IF IOP=2 THEN IFL=1:GOTO 2410
1680 IF IOP=3 THEN IFL=1:GOTO 2040
1690 IF IOP=4 THEN 1770
1700 IF IOP=5 THEN 2630
1710 IF IOP=6 THEN 3480
1720 IF IOP=7 THEN 2770
1730 IF IOP=8 THEN PRINT TAB(15) "END OF PROGRAM";END
1740 GOTO 1550
1750 REM*******************************************************************
1760 REM LIST DATA AT CRT
1770 PRINT:PRINT TAB(10) "PROJECT: ";T$
1780 PRINT:PRINT "DATE: ";DAT$; TAB(20) "NETWORK BY: ";NAM$
1790 PRINT D2$
1800 PRINT D3$
1810 IC=1
1820 FOR I=1 TO INA
1830 IF I<> (15*IC)THEN 1870
1840 IC-IC+1
1850 PRINT:INPUT "PRESS ENTER TO CONTINUE";Y$:PRINT
1860 PRINT D2$:PRINT D3$
1870 PRINT USING D4$; C$(I);
1880 PRINT USING D5$; D$(I);
1890 PRINT USING F7$; ST(I); ED(I); D(I)
1900 NEXT I
1910 PRINT:INPUT "PRESS ENTER TO CONTINUE";Y$:PRINT
1920 GOTO 1550
1930 REM*******************************************************************
1940 REM ADD ACTIVITIES
1950 INPUT "ENTER NUMBER OF ACTIVITIES TO BE ADDED"; LA
1960 IF LA<0 THEN PRINT "INVALID":GOTO 1950
1970 IF LA=0 THEN 1890    1550
1980 IF (LA+INA)>ITOT THEN PRINT"MUST BE <= "; ITOT:GOTO 1950
1990 II=INA+1
2000 INA=INA+LA
2010 GOSUB 3570
2020 GOTO 1550
2030 REM*******************************************************************
2040 REM DELETE ACTIVITIES
2050 INPUT "ENTER NUMBER OF ACTIVITIES TO BE DELETED"; LA
2060 IF LA<0 THEN PRINT "INVALID":GOTO 2050
2070 IF LA=0 THEN 1550
2080 IF LA>INA THEN PRINT"MUST BE <= "; INA:GOTO 2050
2090 II=0
2100 FOR I=1 TO LA
2110 INPUT "ENTER ACTIVITY CODE OF ACTIVITY TO BE DELETED"; AC$
2120 FOR L=1 TO INA
2130 IF C$(L)<> AC$ THEN 2180
2140 C$(L)=" ":D$(L)=" ":ST(L)=0:ED(L)=0:D(L)=0
2150 II=II+1
2160 IR (II)=L
2170 GOTO 2200
2180 NEXT L
```

```
2190 PRINT "NO MATCH FOUND FOR ACTIVITY CODE ";AC$
2200 NEXT I
2210 IF II=0 THEN GOTO 1550
2220 REM Move activity data into arrays where data was deleted.
2230 LFT=INA-II
2240 FOR IL=LFT+1 TO INA
2250 IF C$(IL)=" " THEN 2370
2260 IF IR(I)=INA THEN 2370
2270 FOR L=1 TO II
2280 IF IR(L)>LFT THEN 2350
2290 LL=IR(L)
2300 C$(LL)=C$(IL):D$(LL)=D$(IL):ST(LL)=ST(IL)
2310 ED(LL)=ED(IL):D(LL)=D(IL)
2320 C$(IL)=" ":D$(IL)=" ":ST(IL)=0:ED(IL)=0:D(IL)=0
2330 IR(L)=INA
2340 GOTO 2370
2350 NEXT L
2360 PRINT "COULD NOT FIND ANY SPACE IN DATA ARRAYS"
2370 NEXT IL
2380 INA=LFT
2390 GOTO 1550
2400 REM*****************************************************************
2410 REM CHANGE ACTIVITY DATA
2420 INPUT "ENTER NUMBER OF ACTIVITIES TO BE CHANGED";LA
2430 IF LA<0 THEN PRINT "INVALID":GOTO 2420
2440 IF LA=0 THEN 1550
2450 IF LA>INA THEN PRINT "MUST BE <=";INA:GOTO 2420
2460 FOR I=1 TO LA
2470 PRINT:INPUT "ENTER ACTIVITY CODE OF ACTIVITY TO BE CHANGED";AC$
2480 FOR L=1 TO INA
2490 IF AC$<> C$(L) THEN 2570
2500 PRINT "CHANGES ARE FOR ACTIVITY ";L;" HAVING ACTIVITY CODE ";C$(L)
2510 PRINT D1$
2520 PRINT "ACTIVITY NO ";L;
2530 INPUT C$(L),D$(L),ST(L),ED(L),D(L)
2540 IF ST(L)<ED(L) THEN 2590
2550 PRINT "BEGINNING NODE NO. MUST BE LESS THAN ENDING NODE NO."
2560 PRINT "REENTER DATA FOR THIS NODE":GOTO 2520
2570 NEXT L
2580 PRINT "NO MATCH FOUND FOR ACTIVITY CODE ";AC$
2590 NEXT I
2600 GOTO 1550
2610 REM*****************************************************************
2620 REM LIST DATA AT PRINTER
2630 LPRINT:LPRINT
2640 LPRINT TAB(10) "PROJECT:";T$
2650 LPRINT:LPRINT "DATE: ";DAT$;TAB(40)"NETWORK BY: ";NAM$
2660 LPRINT D2$
2670 LPRINT D3$
2680 FOR I=1 TO INA
2690 LPRINT USING D4$;C$(I);
```

```
2700 LPRINT USING D5$;D$(I);
2710 LPRINT USING F7$;ST(I);ED(I);D(I)
2720 NEXT I
2730 LPRINT:LPRINT
2740 GOTO 1550
2750 REM*****************************************************************
2760 REM CPM CALCULATIONS
2770 REM Perform forward pass
2780 IF IFL=1 THEN GOSUB 3750
2790 ES(1)=0:EF(1)=D(1)
2800 FOR I=2 TO INA
2810 IF ST(I)=ST(1) THEN ES(I)=0:EF(I)=D(I):GOTO 2900
2820 MAX=0
2830 FOR L=1 TO INA
2850 IF ED(L)<> ST(I) THEN 2880
2860 IF EF(L)>MAX THEN MAX=EF(L)
2870 ES(I)=MAX
2880 NEXT L
2890 EF(I)=ES(I)+D(I)
2900 NEXT I
2910 REM Perform backward pass
2920 LN=ED(INA)
2930 DX=0
2940 FOR I=INA TO 1 STEP -1
2950 IF ED(I)<>LN THEN 2970
2960 IF EF(I)>DX THEN DX=EF(I)
2970 NEXT I
2980 XF(INA)=DX
2990 FOR I=INA TO 1 STEP -1
3000 IF ED(I)=ED(INA) THEN XF(I)=DX:XS(I)=XF(I)-D(I):GOTO 3090
3010 MIN=999999!
3020 FOR L=INA TO 1 STEP -1
3030 IF ST(L)<ED(I) THEN 3080
3040 IF ED(I)<>ST(L) THEN 3070
3050 IF XS(L)<MIN THEN MIN=XS(L)
3060 XF(I)=MIN
3070 NEXT L
3080 XS(I)=XF(I)-D(I)
3090 NEXT I
3100 REM Perform slack calculations
3110 FOR I=1 TO INA
3120 AS(I)=XF(I)-EF(I)
3130 NEXT I
3140 INPUT "DISPLAY RESULTS ON CRT OR PRINTER (C/P)";Y$
3150 IF Y$<>"C" THEN 3330
3160 PRINT:PRINT TAB(25) "CPM RESULTS"
3170 PRINT TAB(10) "PROJECT:";T$
3180 PRINT "DATE: ";DAT$; TAB(30) "NETWORK BY: ";NAM$
3190 IC=1
3200 PRINT D7$; D8$
3210 PRINT D9$; DO$
3220 FOR I=1 TO INA
```

```
3230 IF I<>(15*IC) THEN 3260
3240 IC=IC+1
3250 PRINT:INPUT "PRESS ENTER TO CONTINUE";Y$
3260 PRINT USING D4$;C$(I);
3270 PRINT USING D5$;D$(I);
3280 PRINT USING F7$;D(I);ES(I);EF(I);XS(I);XF(I);AS(I)
3290 NEXT I
3300 PRINT:PRINT "PROJECT DURATION:";DX:PRINT
3310 INPUT "LIST RESULTS ON THE PRINTER (Y/N)";Y$
3320 IF Y$<> "Y" THEN 3440
3330 LPRINT TAB(25)"CPM RESULTS":LPRINT TAB(10) "PROJECT: ";T$
3340 LPRINT "DATE: ";DAT$; TAB(30) "NETWORK BY: ";NAM$
3350 LPRINT
3360 LPRINT D7$;D8$
3370 LPRINT D9$;DO$
3380 FOR I=1 TO INA
3390 LPRINT USING D4$;C$(I);
3400 LPRINT USING D5$;D$(I);
3410 LPRINT USING F7$;D(I);ES(I);EF(I);XS(I);XF(I);AS(I)
3420 NEXT I
3430 LPRINT:LPRINT "PROJECT DURATION:";DX
3440 PRINT:PRINT TAB(20) "END OF CPM ANALYSIS"
3450 GOTO 1550
3460 REM*********************************************************************
3470 REM STORE NETWORK DATA ON DISK
3480 INPUT "ENTER NAME OF DISK:FILE";N$
3490 OPEN N$ FOR OUTPUT AS #3
3500 WRITE #3, T$,DAT$,NAM$,INA
3510 FOR I=1 TO INA
3520 WRITE #3,C$(I),D$(I),ST(I),ED(I),D(I)
3530 NEXT I
3540 CLOSE #3
3550 GOTO 1550
3560 REM*********************************************************************
3570 REM SUBROUTINE: NETWORK DATA INPUT
3580 IC=1
3590 INE=0
3600 PRINT D1$
3610 FOR I=II TO INA
3620 INE=INE+1
3630 IF INE<>(20*IC) THEN 3670
3640 IC=IC+1
3650 INPUT "PRESS ENTER TO CONTINUE"; Y$
3660 PRINT D1$
3670 PRINT "ACTIVITY NO ";I;
3680 INPUT C$(I),D$(I),ST(I),ED(I),D(I)
3690 IF ST(I)<ED(I) THEN 3720
3700 PRINT "BEGINNING NODE NO. MUST BE LESS THAN ENDING NODE NO"
3710 PRINT "REENTER DATA FOR THIS ACTIVITY":GOTO 3670
3720 NEXT I
3730 PRINT:RETURN
```

```
3740 REM************************************************************
3750 REM SUBROUTINE: SORT ACTIVITIES USING BEGINNING ACTIVITY NO.
3760 PRINT:PRINT "ACTIVITIES ARE BEING SORTED!":PRINT
3770 I1=1
3780 FOR I=I1 TO INA-1
3790 IE=0
3800 FOR L=I+1 TO INA
3810 IF ST(I)<ST(L) THEN 3930
3820 IF ST(I)>ST(L) THEN 3850
3830 IF ED(I)<ED(L) THEN 3930
3840 IF ED(I)=ED(L) THEN 3950
3850 IM=L
3860 IE=1
3870 TT(1)=ST(IM):TT(2)=ED(IM)
3880 TD$(1)=C$(IM):TD$(2)=D$(IM):TD=D(IM)
3890 ST(IM)=ST(I):ED(IM)=ED(I)
3900 C$(IM)=C$(I):D$(IM)=D$(I):D(IM)=D(I)
3910 ST(I)=TT(1):ED(I)=TT(2)
3920 C$(I)=TD$(1):D$(I)=TD$(2):D(I)=TD
3930 NEXT L
3940 GOTO 3970
3950 PRINT "SAME STARTING AND ENDING NODE NUMBERS-INVALID"
3960 PRINT "CHECK DATA FOR ACTIVITY CODES "; C$(I), C$(L)
3970 NEXT I
3980 IF IE=0 THEN 4010
3990 I1=I1+1
4000 GOTO 3780
4010 IFL=0
4020 RETURN
```

Figure 11.3—CPM program

Network data may be input from the keyboard or from a disk file. One restriction is that an activity beginning-node number must be less than the ending node number. Another restriction is that there can only be one beginning (initial) node and one terminal node for a project network. This latter restriction may seem severe; however, by using dummy activities we can easily satisfy this restriction. Figure 11.4(a) contains the ending nodes of some project network. Only the ending two activities are shown. Figure 11.4(b) shows how we can use dummy activities (having no duration) to satisfy the requirement that the project can have only one ending node. In a similar manner the restriction of only one beginning node can be satisfied.

When data are input from the keyboard, the program prompts for the following information for each activity:

1. CODE
2. DESCRIPTION
3. BEGINNING NODE NUMBER
4. ENDING NODE NUMBER
5. DURATION

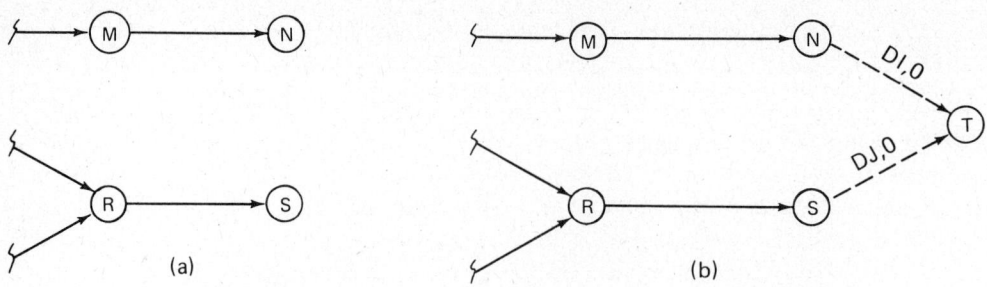

Figure 11.4—satisfying the one beginning node restriction

The activity code can be an alphanumeric value with five characters; the activity code should be unique (no other activity in the same network should have the same code). An activity description can have as many as sixteen alphanumeric characters. Node numbers and activity durations are limited by the size of the largest single-precision number that BASIC will permit (must be less than or equal to 9999999.).

If data are input from a disk file, the user is prompted for the file name. After the data are input from the keyboard or disk, the user must select an option from the following list:

1. ADD NEW ACTIVITIES
2. CHANGE EXISTING ACTIVITY DATA
3. DELETE ACTIVITIES FROM NETWORK
4. LIST DATA AT CRT
5. LIST DATA AT PRINTER
6. STORE DATA ON DISK
7. PERFORM CPM CALCULATIONS
8. QUIT

Several data management options are included in this list.

Option 1, ADD NEW ACTIVITIES, permits you to add additional activities to a network that is already in the computer. The program prompts you for the required information. Option 2, CHANGE EXISTING ACTIVITY DATA, lets you modify the data associated with an activity that is already in the computer. The program identifies all activities by the activity code; the activity description is only used to make the reports more meaningful. Option 3, DELETE ACTIVITIES FROM NETWORK, needs no explanation. Likewise, options 4 and 5, LIST DATA AT CRT and LIST DATA AT PRINTER, need no explanation. To avoid having to reenter your network from the keyboard, you can store your network in a disk file using option 6, STORE DATA ON DISK.

Once you are sure that your network is represented correctly in the computer, option 7, PERFORM CPM CALCULATIONS, can be executed. A report is output to the monitor containing the early start, early finish, late start, late finish, and slack for each activity. If you desire, a copy of this report will be listed at the printer.

11.3 Example CPM Problem

The data in Table 11.1 will be used to illustrate using the CPM program. User input appears in underlined characters.

EXAMPLE 1. Use the CPM program to determine the critical path through the network appearing in Figure 11.1.

SOLUTION. This program is stored in a file named CPM. Data can be input from the keyboard or from a disk file. Each time you run the program you are given an opportunity to specify the date the CPM analysis was performed. We will enter the activity information from the keyboard.

```
LOAD "B:CPM"
RUN
DATA TO BE INPUT FROM KEYBOARD OR DISK (K/D)? K
ENTER TITLE OF PROJECT? ACQUIRE A COMPUTER
ENTER YOUR NAME? PHILIP WOLFE
ENTER DATE? 6-5-83
TOTAL NUMBER OF ACTIVITIES? 12
ENTER ACTIVITY CODE, DESCRIPTION, BEG. NODE, END NODE, DURATION
ACTIVITY NO  1 ? A,CONTACT VENDORS,1,2,1
ACTIVITY NO  2 ? B,REVIEW PROPOSALS,2,3,2
ACTIVITY NO  3 ? C,LOCATE FINANCING,2,4,2
ACTIVITY NO  4 ? D,SELECT VENDOR,3,5,4
ACTIVITY NO  5 ? E,NEGOTIATE CONTRACT,5,6,2
ACTIVITY NO  6 ? F,PREPARE FACILITIES,4,6,5
ACTIVITY NO  7 ? G,INSTALL COMPUTER,6,8,2
ACTIVITY NO  8 ? H,TRAIN OP. MANAGER,6,7,2
ACTIVITY NO  9 ? I,IMPLEMENT APPLICATIONS,8,9,6
ACTIVITY NO 10 ? J,TRAIN USERS,7,9,3
ACTIVITY NO 11 ? DUM1,DUMMY ACTIVITY,7,8,0
ACTIVITY NO 12 ? K,VALIDATE INSTALLATION,9,10,1
ACTIVITIES BEING SORTED
```

After the activity data have been entered, the activities are sorted using the beginning node numbers. Because larger networks may take some time to sort, the program displays a message to let the user know what is occurring. Next we will store our network data in a disk file to avoid having to reenter the data if this network is studied at a later date. The data will be stored in a file named COMPUTER.

```
PROGRAM OPTIONS
  1-ADD NEW ACTIVITIES
  2-CHANGE EXISTING ACTIVITY DATA
  3-DELETE ACTIVITIES FROM NETWORK
  4-LIST DATA AT CRT
  5-LIST DATA AT PRINTER
```

```
         6-STORE DATA ON DISK
         7-PERFORM CPM CALCULATIONS
         8-QUIT
   ENTER OPTION? 6

   ENTER NAME OF DISK:FILE? B:COMPUTER

      PROGRAM OPTIONS
         1-ADD NEW ACTIVITIES
         2-CHANGE EXISTING ACTIVITY DATA
         3-DELETE ACTIVITIES FROM NETWORK
         4-LIST DATA AT CRT
         5-LIST DATA AT PRINTER
         6-STORE DATA ON DISK
         7-PERFORM CPM CALCULATIONS
         8-QUIT
```

Before proceeding with the CPM analysis, we will display the network data at our CRT to see if the data are correct. Then the CPM calculations will be performed.

```
   ENTER OPTION? 4
               PROJECT:   ACQUIRE A COMPUTER
   DATE:  6-5-83         NETWORK BY: PHILIP WOLFE
           ACTIVITY           BEGINNING    ENDING      ACTIVITY
   CODE    DESCRIPTION        NODE NO      NODE NO     DURATION
   A       CONTACT VENDORS       1.0         2.0         1.0
   B       REVIEW PROPOSALS      2.0         3.0         2.0
   C       LOCATE FINANCING      2.0         4.0         2.0
   D       SELECT VENDOR         3.0         5.0         4.0
   F       PREPARE FACILITY      4.0         6.0         5.0
   E       NEGOTIATE CONTRA      5.0         6.0         2.0
   H       TRAIN OP. MANAGE      6.0         7.0         2.0
   G       INSTALL COMPUTER      6.0         8.0         2.0
   DUM1    DUMMY ACTIVITY        7.0         8.0         0.0
   J       TRAIN USERS           7.0         9.0         3.0
   I       IMPLEMENT APPLIC      8.0         9.0         6.0
   K       VALIDATE INSTALL      9.0        10.0         1.0

   PRESS ENTER TO CONTINUE

      PROGRAM OPTIONS
         1-ADD NEW ACTIVITIES
         2-CHANGE EXISTING ACTIVITY DATA
         3-DELETE ACTIVITIES FROM NETWORK
         4-LIST DATA AT CRT
         5-LIST DATA AT PRINTER
         6-STORE DATA ON DISK
         7-PERFORM CPM CALCULATIONS
         8-QUIT
   ENTER OPTION? 7
```

11.3 Example CPM Problem

DISPLAY RESULTS ON CRT OR PRINTER (C/P)? <u>C</u>

```
        CPM RESULTS
           PROJECT:ACQUIRE A COMPUTER
DATE:  6-5-83            NETWORK BY: PHILIP WOLFE
        ACTIVITY        DUR-    EARLY   EARLY   LATE    LATE    SLACK
CODE    DESCRIPTION     ATION   START   FINISH  START   FINISH  TIME
A       CONTACT VENDORS  1.0    0.0     1.0     0.0     1.0     0.0
B       REVIEW PROPOSALS 2.0    1.0     3.0     1.0     3.0     0.0
C       LOCATE FINANCING 2.0    1.0     3.0     2.0     4.0     1.0
D       SELECT VENDOR    4.0    3.0     7.0     3.0     7.0     0.0
E       NEGOTIATE CONTRA 2.0    7.0     9.0     7.0     9.0     0.0
F       PREPARE FACILITY 5.0    3.0     8.0     4.0     9.0     1.0
H       TRAIN OP. MANAGE 2.0    9.0     11.0    9.0     11.0    0.0
G       INSTALL COMPUTER 2.0    9.0     11.0    9.0     11.0    0.0
DUM1    DUMMY ACTIVITY   0.0    11.0    11.0    11.0    11.0    0.0
J       TRAIN USERS      3.0    11.0    14.0    14.0    17.0    3.0
I       IMPLEMENT APPLIC 6.0    11.0    17.0    11.0    17.0    0.0
K       VALIDATE INSTALL 1.0    17.0    18.0    17.0    18.0    0.0
```

PROJECT DURATION: 18

LIST RESULTS ON THE PRINTER (Y/N)? <u>N</u>
   END OF CPM ANALYSIS

The results correspond to the project duration (18 weeks) and critical path depicted in Figure 14.2 It appears that we have a valid solution.

PRESS ENTER TO CONTINUE:

```
   PROGRAM OPTIONS
      1-ADD NEW ACTIVITIES
      2-CHANGE EXISTING ACTIVITY DATA
      3-DELETE ACTIVITIES FROM NETWORK
      4-LIST DATA AT CRT
      5-LIST DATA AT PRINTER
      6-STORE DATA ON DISK
      7-PERFORM CPM CALCULATIONS
      8-QUIT
ENTER OPTION? 8
```

   END OF PROGRAM

EXAMPLE 2. After some time had elapsed, management decided that another activity should be added to the Acquire A Computer project network. It was decided that activity K, VALIDATE INSTALLATION, would be deleted and replaced by the following activities:

Code	Description	Beginning Node	Ending Node	Duration
L	Audit System	9	10	2
M	Validate Installation	10	11	1

Our task is to incorporate this new information into the project network and then perform the CPM analysis.

SOLUTION. The CPM program will be used to perform the required analysis. The task is relatively simple because we had stored the original network in a disk file named COMPUTER.

```
RUN
DATA TO BE INPUT FROM KEYBOARD OR DISK (K/D)? D
ENTER NAME OF DISK:FILE? B:COMPUTER
DATE FROM DISK FILE IS: 6-5-83
DO YOU WANT TO CHANGE DATE (Y/N)? Y
ENTER NEW DATE: 7-1-83

ACTIVITIES ARE BEING SORTED!

   PROGRAM OPTIONS
     1-ADD NEW ACTIVITIES
     2-CHANGE EXISTING ACTIVITY DATA
     3-DELETE ACTIVITIES FROM NETWORK
     4-LIST DATA AT CRT
     5-LIST DATA AT PRINTER
     6-STORE DATA ON DISK
     7-PERFORM CPM CALCULATIONS
     8-QUIT
```

First we will delete activity K from the network, and then add activities L and M. After entering activities L and M, we note that the durations are incorrect. Therefore, these activities must be changed. After these changes are made we will display the data at the CRT to make sure that the data are correct. Next, the CPM calculations will be performed.

```
ENTER OPTION? 3

ENTER NUMBER OF ACTIVITIES TO BE DELETED: 1
ENTER ACTIVITY CODE OF ACTIVITY TO BE DELETED? K

   PROGRAM OPTIONS
     1-ADD NEW ACTIVITIES
     2-CHANGE EXISTING ACTIVITY DATA
     3-DELETE ACTIVITIES FROM NETWORK
     4-LIST DATA AT CRT
     5-LIST DATA AT PRINTER
     6-STORE DATA ON DISK
     7-PERFORM CPM CALCULATIONS
     8-QUIT
ENTER OPTION? 1

ENTER NUMBER OF ACTIVITIES TO BE ADDED? 2
ENTER ACTIVITY CODE, DESCRIPTION, BEG. NODE, END NODE, DURATION
ACTIVITY NO 12? L,AUDIT SYSTEM,9,10,1
ACTIVITY NO 13 ? M,VALIDATE INSTALLATION,10,11,2
```

11.3 Example CPM Problem

```
   PROGRAM OPTIONS
     1-ADD NEW ACTIVITIES
     2-CHANGE EXISTING ACTIVITY DATA
     3-DELETE ACTIVITIES FROM NETWORK
     4-LIST DATA AT CRT
     5-LIST DATA AT PRINTER
     6-STORE DATA ON DISK
     7-PERFORM CPM CALCULATIONS
     8-QUIT
ENTER OPTION? 2

ENTER NUMBER OF ACTIVITIES TO BE CHANGED? 2

ENTER ACTIVITY CODE OF ACTIVITY TO BE CHANGED? L
CHANGES ARE FOR ACTIVITY 12 HAVING ACTIVITY CODE L
ENTER ACTIVITY CODE, DESCRIPTION, BEG. NODE, END NODE, DURATION
ACTIVITY NO 12 ? L,AUDIT SYSTEM,9,10,2

ENTER ACTIVITY CODE OF ACTIVITY TO BE CHANGED? M
CHANGES ARE FOR ACTIVITY 13 HAVING ACTIVITY CODE M
ENTER ACTIVITY CODE, DESCRIPTION, BEG. NODE, END NODE, DURATION
ACTIVITY NO 13 ? M,VALIDATE INSTALLATION,10,11,1

   PROGRAM OPTIONS
     1-ADD NEW ACTIVITIES
     2-CHANGE EXISTING ACTIVITY DATA
     3-DELETE ACTIVITIES FROM NETWORK
     4-LIST DATA AT CRT
     5-LIST DATA AT PRINTER
     6-STORE DATA ON DISK
     7-PERFORM CPM CALCULATIONS
     8-QUIT
ENTER OPTION? 4
              PROJECT: ACQUIRE A COMPUTER
DATE: 7-1-83         NETWORK BY: PHILIP WOLFE
```

ACTIVITY CODE	ACTIVITY DESCRIPTION	BEGINNING NODE NO	ACTIVITY NODE NO	DURATION
A	CONTACT VENDORS	1.0	2.0	1.0
B	REVIEW PROPOSALS	2.0	3.0	2.0
C	LOCATE FINANCING	2.0	4.0	2.0
D	SELECT VENDOR	3.0	5.0	4.0
F	PREPARE FACILITY	4.0	6.0	5.0
E	NEGOTIATE CONTRA	5.0	6.0	2.0
H	TRAIN OP. MANAGE	6.0	7.0	2.0
G	INSTALL COMPUTER	6.0	8.0	2.0
DUM1	DUMMY ACTIVITY	7.0	8.0	0.0
J	TRAIN USERS	7.0	9.0	3.0
I	IMPLEMENT APPLIC	8.0	9.0	6.0
L	AUDIT SYSTEM	9.0	10.0	2.0
M	VALIDATE INSTALL	10.0	11.0	1.0

```
PRESS ENTER TO CONTINUE
```

```
     PROGRAM OPTIONS
        1-ADD NEW ACTIVITIES
        2-CHANGE EXISTING ACTIVITY DATA
        3-DELETE ACTIVITIES FROM NETWORK
        4-LIST DATA AT CRT
        5-LIST DATA AT PRINTER
        6-STORE DATA ON DISK
        7-PERFORM CPM CALCULATIONS
        8-QUIT
     ENTER OPTION? 7

     ACTIVITIES ARE BEING SORTED!

     DISPLAY RESULTS ON CRT OR PRINTER (C/P)? C

                    CPM RESULTS
             PROJECT: ACQUIRE A COMPUTER
     DATE: 7-1-83           NETWORK BY: PHILIP WOLFE
          ACTIVITY          DUR-    EARLY   EARLY   LATE    LATE    SLACK
     CODE DESCRIPTION       ATION   START   FINISH  START   FINISH  TIME
     A    CONTACT VENDORS   1.0     0.0     1.0     0.0     1.0     0.0
     B    REVIEW PROPOSALS  2.0     1.0     3.0     1.0     3.0     0.0
     C    LOCATE FINANCING  2.0     1.0     3.0     2.0     4.0     1.0
     D    SELECT VENDOR     4.0     3.0     7.0     3.0     7.0     0.0
     F    PREPARE FACILITY  5.0     3.0     8.0     4.0     9.0     1.0
     E    NEGOTIATE CONTRA  2.0     7.0     9.0     7.0     9.0     0.0
     H    TRAIN OP. MANAGE  2.0     9.0     11.0    9.0     11.0    0.0
     G    INSTALL COMPUTER  2.0     9.0     11.0    9.0     11.0    0.0
     DUM1 DUMMY ACTIVITY    0.0     11.0    11.0    11.0    11.0    0.0
     J    TRAIN USERS       3.0     11.0    14.0    14.0    17.0    3.0
     I    IMPLEMENT APPLIC  6.0     11.0    17.0    11.0    17.0    0.0
     L    AUDIT SYSTEM      2.0     17.0    19.0    17.0    19.0    0.0
     M    VALIDATE INSTALL  1.0     19.0    20.0    19.0    20.0    0.0

     PROJECT DURATION: 20
```

The expected time to complete the revised project, twenty weeks, seems reasonable.

```
     LIST RESULTS ON THE PRINTER (Y/N)? N

          END OF CPM ANALYSIS

     PROGRAM OPTIONS
        1-ADD NEW ACTIVITIES
        2-CHANGE EXISTING ACTIVITY DATA
        3-DELETE ACTIVITIES FROM NETWORK
        4-LIST DATA AT CRT
        5-LIST DATA AT PRINTER
        6-STORE DATA ON DISK
        7-PERFORM CPM CALCULATIONS
        8-QUIT
     ENTER OPTION? 8

          END OF PROGRAM
```

11.4 Summary

At some time most engineers and scientists will find themselves involved in project planning. As a result, a program was included in this book that implements the CPM project planning and scheduling technique. Space does not permit in-depth coverage of this topic. Books have been written on this subject covering such topics as cost/time trade-offs and resource allocations. Reference [2] contains a very good bibliography on this subject.

References

Moder, J. J., and C. R. Phillips, *Project Management with CPM and PERT.* Reinhold, New York, 1964.

Weist, Jerome D., and Ferdinand K. Levy, *A Management Guide to PERT/CPM*, 2nd Edition. Prentice-Hall, Englewood Cliffs, New Jersey, 1977.

Exercises

1. A project manager has asked for some assistance in determining the critical activities in a project. From the following data you are to:
   (a) draw a CPM network, and
   (b) determine the critical path and the expected project completion date.

Activity	Description	Duration Days	Immediate Predecessor
A	Activity 1	3	—
B	Activity 2	2	A
C	Activity 3	3	B
D	Activity 4	1	C
E	Activity 5	3	B
F	Activity 6	3	B
G	Activity 7	2	F,H
H	Activity 8	3	A
I	Activity 9	5	A
J	Activity 10	1	H
K	Activity 11	2	D,E,G,J
L	Activity 12	2	I
M	Activity 13	7	I
N	Activity 14	4	K,L,M

2. You are to find the critical path in the following network where the arrows represent activities and the numbers, by each arrow, represent duration in days.

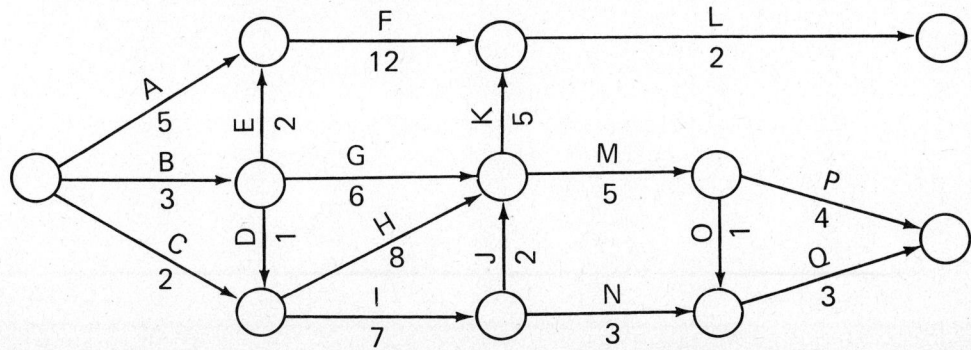

3. A small construction project consists of 13 jobs. The precedence relationships are identified below in terms of node numbers. You are to:
   a. Draw the network.
   b. Determine the early and late start times for each job.
   c. Identify the critical path.

Job	Initial Node, Final Node	Duration (Days)
J1	1, 2	3
J2	1, 3	1
J3	2, 4	5
J4	3, 4	2
J5	4, 5	4
J6	4, 6	3
J7	3, 6	2
J8	5, 9	2
J9	6, 7	1
J10	6, 8	3
J11	7, 8	1
J12	8, 10	2
J13	9, 10	1

4. A new research and develop project is being planned in a leading medical products manufacturing firm. The project network has been given to you in tabular form. You are to determine the expected duration of the project and the critical path.

Activity	Duration (Weeks)	Immediate Predecessor
A	2	—
B	4	A
C	3	A
D	2	A
E	2	C
F	1	B, E
G	2	C
H	4	C
I	5	C, D
J	2	F
K	6	G, N
L	4	F, K
M	1	G, N
N	2	H, I
O	7	H, I
P	3	O
Q	2	L, M
R	4	J, Q
S	2	J, Q
T	1	L, M, P
U	2	O
V	3	U
W	5	T
X	3	V
Y	2	S
Z	1	R, Y
AB2	2	W, X
MN4	2	AB2
DC5	2	Z
END	5	AB2, DC5

12

Sorting

One common task is the sorting of data according to some specified order. Arranging a list of numbers from smallest to largest and placing alphabetic information in alphabetical order are examples of sorting. This procedure can be a very time consuming task even when using a computer, especially if thousands of data elements are involved. As a result, several sorting procedures have been developed with the objective of making this task as efficient as possible. In this section, two sorting procedures will be presented.

12.1 Bubble Sort

The bubble sort is not very efficient, but it is easy to understand. Therefore, it is a good procedure to use as an introduction to sorting. Assume that you have a list of numbers that you want to sort into increasing order:

8 2 6 1 4

Using the bubble sort, the first two numbers are compared. If out of order, the positions of the numbers are swapped:

2 8 6 1 4

Then the numbers in positions two and three are compared, and if out of order, they are swapped:

2 6 8 1 4

This process continues until the numbers in the last two positions are compared and swapped if they are out of order:

2 6 1 8 4
2 6 1 4 8

If no swaps were made in the above process, the list is in sorted order. Note what has happened in this example; the largest number (8) has "bubbled" to the bottom of the list. At this point, the largest number is in the correct position; therefore, this number need not be considered in any future comparisons.

Another pass is made through the list. At the end of this pass, the list would be in the following order:

2 1 4 6 8

Now the bottom two numbers are in the correct positions.
The third pass results in:

1 2 4 6 8

The fourth pass results in no swaps; therefore, the list is in increasing order.

Bubble Sort Computer Program

A subroutine that will sort a list of numbers into ascending order using the bubble sort procedure appears in Figure 12.1. The list to be sorted must be in the array X and the number of entries in the list must be provided in the variable N.

```
23000 REM SUBROUTINE: BUBBLE SORT
23010 REM Program sorts into increasing order a one dimensional
23020 REM array, X, having N numbers. The array X and its size
23030 REM N are supplied by the main program. Changing the >=
23040 REM in line 23090 to <= will result in the array being
23050 REM sorted in decreasing order.
23060 FOR I=1 TO N-1
23070 IFL=0
23080 FOR J=1 TO N-I
23090 IF X(J+1)>=X(J) GOTO 23120
23100 SWAP X(J+1), X(J)
23110 IFL=1
23120 NEXT J
23130 IF IFL=0 GOTO 23150
23140 NEXT I
23150 RETURN
```

Figure 12.1—bubble sort program

EXAMPLE 1. Use the bubble sort procedure to sort the following list of numbers into ascending order:

38 21 7 32 2 45 13 60 18 55

SOLUTION. Figure 12.2 contains a program which was written to prompt the user for the list of numbers to be sorted. After the list is entered, the bubble sort subroutine is utilized to sort the numbers. The sorted list is then printed out.

```
100 DIM X(100)
110 INPUT "HOW MANY NUMBERS ARE TO BE SORTED";N
120 PRINT "ENTER";N;" DATA ELEMENTS"
130 FOR I=1 TO N
140 PRINT "ENTER ELEMENT";I;
150 INPUT X(I)
160 NEXT I
170 GOSUB 23000
```

```
180 FOR I=1 TO N
190 PRINT X(I)
200 NEXT I
210 END
```

Figure 12.2—bubble sort application

12.2 Quicksort

The quicksort procedure will usually require less time than the bubble sort procedure to sort a randomly ordered list. However, if the list is already sorted or nearly sorted, the bubble sort can be faster.

The quicksort algorithm partitions the initial list into two lists with all the numbers in one subset being smaller than all of the numbers in the other subset. Then this procedure is repeated on each of the subsets, partitioning each of them into subsets. This process continues until each subset contains just one element. At this point the array is sorted.

This procedure is best explained with an example. Assume the list of numbers

7 2 14 10 6 4 9

is to be sorted into increasing order using the quicksort procedure. Initially, some number has to be chosen as a "pivot." This procedure will rearrange the above list into two subsets such that all of the numbers to the left are smaller than the pivot. All of the numbers to the right will be larger than the pivot.

For this example, the first number in the list will be chosen as the pivot.

P
7 2 14 10 6 4 9

The pivot element is then compared with the last number. If the last number is smaller than the pivot number, the numbers are swapped. Looking at our list, we see that 7 and 9 are in the proper order. The pivot is then compared to the next to last number, 4. Here the numbers are out of order, so the 7 and 4 are swapped:

         P
4 2 14 10 5 7 9
L   R

We will use an L and R to indicate which numbers at the left and right ends of the list have been compared with the pivot. Everything to the left of the L will have been compared with the pivot, and everything to the right of the R will have been compared with the pivot. Attention is now directed to the left end of the list, because we know that everything to the right of the number 7 is larger than 7 (at this point, only the number 9 is to the right of the pivot, 7).

Looking at the left end, we know that the last number exchanged with the pivot is less than the pivot because a swap was just made. Therefore, we start with the next number to the right, the number 2. Since 2 is less than the pivot, no swap occurs. However, 7 and 14 are out of order, so a swap occurs:

```
          P
4  2  7  10  6  14  9
      L      R
```

At this point, everything to the left of the pivot is less than 7 and everything from the number 14 to the right end is greater than 7. Therefore, the comparisons start with the first number to the left of the number 14. Another swap is made here because 6 is less than 7:

```
           P
4  2  6  10  7  14  9
      L  R
```

Only the number 10 has not been considered. Again a swap is made:
```
         P
4  2  6  7  10  14  9
         L
         R
```

The list has now been partitioned into two subsets. Everything to the left of the pivot is less than the pivot, and everything to the right is greater than the pivot.

This same procedure is now applied to these subsets. First, consider the subset:

```
P
4  2  6
L     R
```

Select the first element (4) as the pivot. Since 6 and 4 are in the proper order, only one swap would be made:

```
   P
2  4  6
   L
   R
```

Everything to the left of the pivot (4) is less than the pivot and everything to the right is greater than the pivot. The result is two subsets containing one element each. Consequently, these numbers are in the proper order.

We have one remaining subset to sort:

```
P
10  14  9
L    R
```

Applying this procedure once again, we will select 10 as the pivot. Comparing the pivot with the rightmost element, we see that 10 is greater than 9, so these numbers must be swapped.

```
        P
9  14  10
   L   R
```

Then 10 and 14 are swapped

```
     P
9   10   14
     L
```

Again we have partitioned a list into subsets of one number each. Our original list is now in increasing order

2 4 6 7 9 10 14

Consequently, the quicksort procedure has successfully sorted our original list.

The next step is to develop a computer program to implement the quicksort procedure. Comparing the various numbers is not difficult. However, developing a method for storing the various subset of numbers to be partitioned requires some thought. We will start by establishing an arbitrary rule: when a set is partitioned, the right subset will be stored and the left subset will be partitioned. However, the right subset will only be saved to be partitioned later if it contains more than one element. Likewise, the left subset will only be partitioned if it contains more than one element.

A subset can be stored by saving the positions in the list of beginning and ending elements. These positions can be saved in two, one-dimensional arrays, one array containing the beginning position and one array containing the ending position. As the quicksort procedure is executed and subsets are created, the subsets on the right will be added to this list of subsets to be partitioned. When there are no more subsets on the left to be partitioned, the stored subsets will be removed from the one dimensional arrays in a last-in, first-out order (LIFO).

To illustrate this process, we will consider the following list:

7 2 14 10 6 4 9

The array LEF will be used to store the position of the first element in a subset and the array RIT will be used to store the position of the last element in a subset. Initially, we will have

LEF(1) = 1

RIT(1) = 7,

because the subset to be processed is the entire list, containing seven elements.

After partitioning the first subset, we obtain:

```
         P
4  2  6  7  10  14  9
```

Saving the right subset gives us

LEF(1) = 5
RIT(1) = 7

Remember, that LEF(1) and RIT(1) were not being used after the first subset, beginning in position 1 and ending in position 7, was removed to be partitioned.

Next, the left subset is partitioned, giving

```
    P
2   4   6
```

Neither of the resulting subsets would be saved for additional processing since they contain only one element. Therefore, another subset must be removed from our subset array, that is the subset beginning in position 5 and ending in position 7. Thus, we get

```
10  14  9
```

Upon partitioning this subset, we get

```
        P
9   10  14
```

Each of the resulting subsets contains only one element and our arrays LEF and RIT are empty. Consequently, the list is sorted in an increasing order. The procedure stops.

Refinements to the Quicksort Procedure

Before a program is introduced which uses the quicksort procedure, we will present some ideas which can improve the basic process. First, if the pivot element chosen is the largest or smallest element in the list, the result is a subset of one element and another subset of $n-1$ elements. Consequently, a better procedure for specifying the pivot element is to choose the median of the elements $\{X(1), X[(1+n)/2], X(n)\}$, where X is an array holding the list to be sorted.

The amount of memory required to store the end positions of subsets remaining to be processed can be reduced. This can be accomplished by always storing the larger subset and immediately processing the smaller subset. Another improvement involves how the numbers being swapped are handled. Assume that a median value (as described above) is chosen as the pivot. This element can be removed from the array. If the first element is not chosen as the pivot element, the first element is moved to the empty position. The empty first position is then filled by comparing the elements in the list with the pivot element, starting from the right end of the list. The first element to be found less than the pivot element is moved into the empty position.

Consider our initial array:

```
7   2   14  10  6   4   9
```

The pivot element is selected from

$$X(1) = 7,$$
$$X[(1+7)/2] = X(4) = 10, \text{ and}$$
$$X(7) = 9.$$

The median is $X(7) = 9$.

The pivot, 9, is removed from this list and the first element is moved to the empty position:

_ 2 14 10 6 4 7

Starting with the right end, elements are compared with the pivot element, 9, until one is found to be less than the pivot. In this case 7<9; therefore, the 7 is moved into the empty position:

7 2 14 10 6 4 _

This leaves the end position open; it can be filled by working from the left end until a larger element is found.

7 2 _ 10 6 4 14

Continuing this process results in:

7 2 4 10 6 _ 14

7 2 4 _ 6 10 14

7 2 4 6 _ 10 14

At this point, the list is partitioned into two subsets, and the pivot element is placed in the remaining empty position.

7 2 4 6 9 10 14

We are left with two subsets to be partitioned, one beginning in position 1 and ending in position 4, and another beginning in position 6 and ending in position 7. The largest subset will be stored to be partitioned later.

Since the small subset contains only two elements, we will choose the left element as the pivot. Comparing the elements, we see that no interchange is required; therefore, this subset is in the correct order.

The subset beginning in position 1 and ending in position 4 remains to be partitioned. We choose the pivot as follows:

$X(1)=7$
$X(4)=6$
$X[(4+1)/2)]=X(3)=4$

Therefore, the pivot is 4. After removing the pivot and moving the element in position 1 to the empty position, we have:

_ 2 7 6

The remaining steps required to sort this example list will be left to the reader.

Quicksort Computer Program

Figure 12.3 contains a subroutine that will sort a list of numbers into ascending order using the quicksort procedure. The refinements to this procedure, discussed above, have been incorporated into this subroutine. The list to be sorted must be in the array X and the number of entries in the list must be provided in the variable N.

```
23200 REM SUBROUTINE: QUICKSORT
23210 REM This subroutine uses the QUICKSORT procedure to sort a list
```

```
23220 REM of numbers into ascending order. The list must be
      provided to
23230 REM this subroutine in the array X. The number of elements
      in the
23240 REM list must be supplied in the variable N. The arrays LEF
      and RIT
23250 REM are used to store subsets to be partitioned. LEF and RIT
23260 REM contain the position of the first element and the last
      element
23270 REM of a subset to be partitioned. LS and RS maintain the
      left and
23280 REM right end positions of the subset being partitioned. L and R
23290 REM indicate the elements that have been compared with the
      pivot,
23300 REM PV. PT indicates the position in the arrays LEF and RIT
      of the
23310 REM last subset stored to be partitioned. If all the REM
      statements
23320 REM are removed from within the sort procedure, the subroutine
      will
23330 REM sort a list in less time.
23340 REM Changing the > in line 23710 to < and changing the < in
23350 REM line 23750 to >will result in the array being sorted in
23360 REM decreasing order.
23370 SIZ%=50
23380 DIM LEF(SIZ%),RIT(SIZ%)
23390 LEF%(1)=1                     'Left end of subset
23400 RIT%(1)=N                     'Right end of subset
23410 REM Position in LEF and RIT of last subset stored
23420 PT%=1
23430 REM Set left side pointer
23440 L%=LEF%(PT%)
23450 REM Save starting value of left side pointer
23460 LS%=L%
23470 REM Set right side pointer
23480 R%=RIT%(PT%)
23490 REM Save starting value of right side pointer
23500 RS%=R%
23510 PT%=PT%-1
23520 REM If only 1 element in subset, do not partition
23530 IF(R%-L%)<=0 THEN 24210
23540 REM Temporarily set pivot to left element
23550 PV=X(L%)
23560 REM If 5 or less elements in subset leave pivot as is
23570 IF(R%-L%)<5 THEN 23700
23580 REM Set pivot to median
23590 MED=(L%+R%)/2
23600 IF(PV<X(R%)) AND (PV>X(MED)) THEN 23700
23610 IF(PV < X(MED)) AND (PV > X(R%)) THEN 23700
23620 IF (X(R%) < PV) AND (X(R%) > X(MED)) THEN 23680
23630 IF (X(R%) < X(MED)) AND (X(R%) > PV) THEN 23680
```

12.2 Quicksort

```
23640 REM Set pivot=median and move left pointer to hole
23650 PV=X(MED)
23660 X(MED)=X(L%)
23670 GOTO 23700
23680 PV=X(R%)
23690 X(R%)=X(L%)
23700 IF(L% >= R%) THEN 23840      'If L=R, subset partitioned
23710 IF (PV > X(R%)) THEN 23780   'If true, swap
23720 R% = R%-1                    'Move pointer
23730 GOTO 23700
23740 IF (L%> = R%) THEN 23840     'If L=R, subset partitioned
23750 IF (PV < X(L%)) THEN 23810   'If true, swap
23760 L%=L%+1                      'Move pointer
23770 GOTO 23740
23780 X(L%)=X(R%)                  'Perform swap
23790 L%=L%+1                      'Move pointer
23800 GOTO 23740
23810 X(R%)=X(L%)                  'Perform swap
23820 R%=R%-1                      'Move pointer
23830 GOTO 23700
23840 X(L%)=PV                     'Place pivot in hole
23850 REM Determine which subset to store; largest is to be stored
23860 IF(L%-LS%) <=(RS%-L%) THEN 24040
23870 REM Right subset is smallest; if it contains more than 1
23880 REM element, store left subset
23890 IF(RS%-L%)<=1 THEN 23990
23900 LEF%(PT%+1)=LS%
23910 RIT%(PT%+1)=L%-1
23920 L%=L%+1
23930 LS%=L%
23940 R%=RS%
23950 PT%=PT%+1
23960 GOTO 23530
23970 REM Right subset contained only 1 element; partition left
23980 REM subset if it contains more than 1 element
23990 IF(L%-LS%)<=1 THEN 24210
24000 R%=L%-1
24010 RS%=L%
24020 L%=LS%
24030 GOTO 23530
24040 IF(L%-LS%)<=1 THEN 24140
24050 REM Left subset if smallest; if it contains more than 1
24060 REM element, store right subset
24070 LEF%(PT%+1)=L%+1
24080 RIT%(PT%+1)=RS%
24090 R%=L%-1
24100 RS%=R%
24110 L%=LS%
24120 PT%=PT%+1
24130 GOTO 23530
24140 IF(RS%-L%)<=1 THEN 24210
24150 REM Right subset contained more than 1 element; partition it
```

```
24160 L%=L%+1
24170 LS%=L%
24180 R%=RS%
24190 GOTO 23530
24200 REM Any more subsets to be partitioned?
24210 IF(PT%<1) THEN 24230
24220 GOTO 23440
24230 RETURN
```

Figure 12.3—quicksort program

EXAMPLE 2. Use the quicksort procedure to sort the list of numbers used in Example 1.

SOLUTION. The only difference between the solution to this example and the solution to Example 1 is that line 170 would be changed to

```
170 GOSUB 23200
```

specifying that the quicksort subroutine is to be used. References [1] and [2] provide additional material on sorting procedures.

12.3 Summary

Many computer applications require data to be sorted. Consequently, to make this text as useful as possible, two sort subroutines (bubble sort and quicksort) were presented. These subroutines can be used as provided.

References

Baase, Sara, *Computer Algorithms: Introduction to Design and Analysis*. Addison-Wesley, Massachusetts, 1978.
Scheid, Francis, *Computer and Programming*. McGraw-Hill, New York, 1982.

Exercises

1. Modify the bubble sort subroutine so that the array X is sorted into decreasing order. Test the resulting program using the data from example 1.
2. Modify the quicksort subroutine so that the array X is sorted into decreasing order. Test the resulting program using the data from example 1.
3. The following statements will generate 100 random integers in the range of $0 \leq x \leq 100$ (Chapter 15 provides background for this program):

```
10  DIM X(100)
20  FOR I=1 TO 100
30  X(I)=INT(100*RND)+1
40  NEXT I
```

Also the statement

```
50  TIME$="0:0:0"
```

will set a clock inside your IBM Personal Computer to 0 (see your *BASIC Reference Manual* for details regarding the TIME$ statement). Using these statements generate 100 random integers in the range of $0 \leq x \leq 100$. Then set your internal clock to 0. Next, sort these 100 numbers in increasing order using the bubble sort subroutine. After the numbers are sorted, print out the time using the statement

PRINT TIME$

4. Repeat exercise 3 using the quicksort subroutine.
5. Repeat exercises 3 and 4 using a list of 100 numbers that is already sorted before you attempt to sort them. Can you explain the results?

13

Disk Data Files

Your IBM personal Computer has the capability to support several types of memory storage devices. These devices can be grouped into two major types: primary memory and secondary memory. ROM (Read Only Memory) and RAM (Random Access Memory) are examples of primary storage devices. Cassette BASIC resides in ROM; RAM is used to store your programs and data as the programs are executed. A cassette tape and a floppy diskette are examples of secondary storage devices.

When RAM devices are used to store information for long periods of time, they have a characteristic that can cause a problem. These devices are volatile; when power to a RAM device is turned off, the contents of the memory device are lost. One solution is to type in your program every time you turn on the computer; obviously, this is not an attractive one. A better alternative is to store the programs on a secondary storage medium, e.g., floppy disk, and load the program into RAM whenever you want to use it. Data can be stored on a secondary storage medium like a floppy disk in a similar manner; however, there are some simple concepts that must first be understood. These concepts will be discussed in this chapter. Note that devices other than diskettes can be used for secondary storage. This text, however, assumes that a disk will be used. Refer to the *BASIC Reference Manual* for details on using other secondary storage devices.

13.1 Recording Information on a Diskette

In order for a computer to read and write information to a diskette, some procedures must be established. Preferably, these procedures should be performed quickly and without error by the computer. They should also be easy for the user to understand and use.

Figure 13.1 illustrates how information is organized on a diskette. Each diskette contains 40 concentric tracks on one side. If your computer uses double sided disk drives, there are 40 tracks on each side for a total of 80 tracks per diskette. Each of these tracks are divided into smaller pie-shaped segments, called sectors. There are 8 sectors per track.

Theoretically, your IBM Personal Computer can store 163,840 bytes on one side of a diskette. This value is obtained as follows: 40 tracks $\times$ 8 sectors/track $\times$ 512 bytes/sector=163,840 bytes. However, not all of this space is available for data storage, because the disk operating system (DOS) must use some of the space to manage the information stored on a diskette. The information maintained by DOS on each diskette includes a file allocation table and a directory. One primary function of the file allocation table is to identify unused sectors. Thus, if you are creating a new file, the file allocation table is used by DOS to determine where on the diskette the file would be stored.

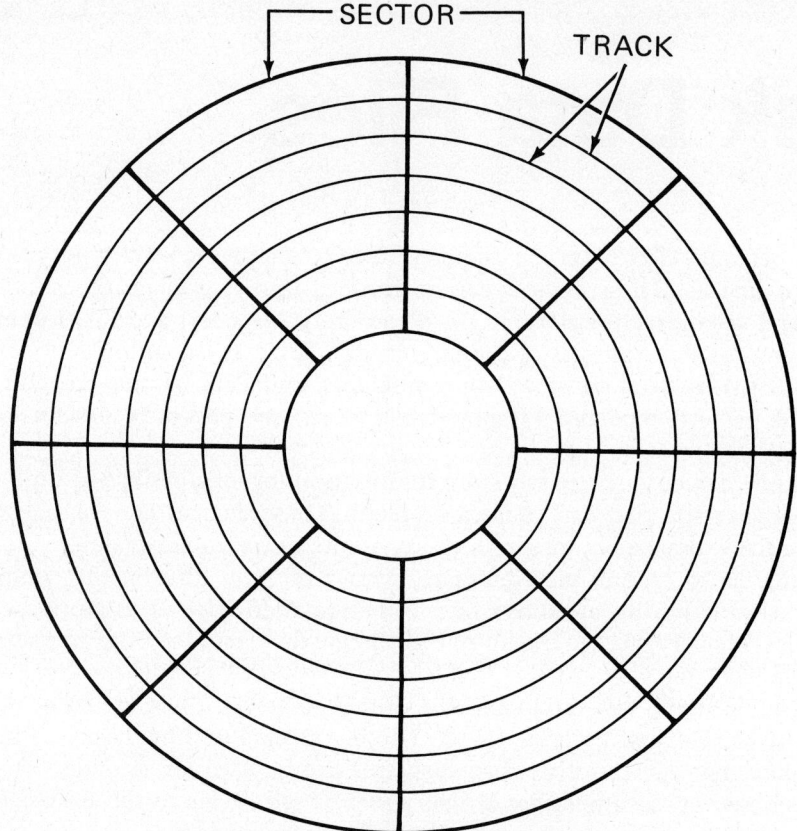

Figure 13.1—floppy disk organization

The primary function of the DOS diskette directory is to maintain a table containing the names of files that have been saved on the diskette. This table also contains other information such as starting sector of the file and date the file was created or last updated. There is room in the directory for 64 file names. As a result, 64 is the maximum number of files that can be saved on a diskette even though not all of the sectors have been used. If you try to save more than 64 files, you will receive an error message. If you are running a BASIC program the error message you will receive is:

```
Too many files
```

The DOS error message is:

```
No room in disk directory
```

13.1 Recording Information on a Diskette

For additional information regarding the file allocation table and the diskette directory, refer to your *Disk Operating System Manual*.

After space is allocated for DOS usage, there are 313 sectors (160,256 bytes) available for data on one side of a diskette, assuming that the diskette does not contain a copy of DOS. If the diskette also contains DOS, there are 286 sectors available for data. If your Personal Computer has two sided disk drives, you have an additional 315 sectors (161,280 bytes) available for data. The number of sectors required to store DOS may vary with the version of DOS. Version 1.10 was used to write this text.

As long as your programs and data are stored within the RAM memory of your computer, you will have little trouble identifying this information. However, when you store information on a diskette, it must be managed so that you can easily retrieve the appropriate information whether it is a program or data. Information is managed on secondary storage devices, such as a disk or tape, using the concept of a file. A *file* is a collection of related information that is kept somewhere other than in the primary memory of your Personal Computer.

For example, you might have a program that is used for curve fitting. This program can be saved on a diskette so that you could easily retrieve the program and use it again. In order to identify the program on the diskette you must assign a name to the file. Remember, one diskette can have as many as 64 files; therefore, a name provides an easy way to identify the various files.

Data files and program files can exist on the same diskette. However, each file on the diskette must have a unique name. The names of files on a particular diskette are maintained in the directory of that diskette. A file name must conform to the following rules:

1. The *name* may be one to eight characters long.
2. The name may be followed by a three character extension provided the extension is separated from the name by a period (.): *name.extension*.
3. The following are valid characters for the name and extension:

    The letters A through Z, and the numbers 0 through 9.

Refer to your DOS manual for a complete specification of characters. We will use only letters and numbers to specify file names.

Here are some valid names for a diskette file:

ANALYSIS.BAS
DESIGN
DESIGN.JAN
TESTA 82.JAN
82TEST.123
RESULT1

If your system has more than one disk drive, specifying a file name may not be enough for the computer to locate the file. In addition, you may need to specify the disk drive containing the diskette on which the specified file is stored. Therefore, the complete file specification is

device:file name

Some examples are:

B:DESIGN
B:DESIGN.JAN

In these examples, the B identifies the disk drive. If the file resides on the default disk drive, the device specification is not needed. A default disk drive is the drive that DOS assumes will contain the diskette or file you specify in a command unless another drive is specified.

13.2 Sequential Access Files

There are two types of diskette data files that can be used in BASIC: sequential and random access. Sequential access methods are easier to learn; however, you will find that this type of I/O (input/output) is limited in flexibility and speed. These limitations will be discussed after random access is explained. The sequential access method assumes that data are written to a file one entry after another. If you want to retrieve the tenth entry, you must retrieve the first nine entries before the tenth entry can be retrieved. You might visualize a sequential file as residing on a tape similar to a tape on your home stereo. Before you can play a song, (a specific file) you must position the tape at the beginning at the desired song (file). If you want to play the third song, you must first pass the first two songs on the tape. This is an example of sequential access.

Disk BASIC uses a temporary storage area (buffer) to manage data that are written to and read from a disk. Sequential access requires a buffer area for writing to the disk and for reading from the disk. Figure 13.2 illustrates these buffer areas. For sequential access, the buffer size is fixed at 128 bytes. This 128-byte block of data is called a record. The buffer permits BASIC to be more efficient in accessing information on a disk. Information to be written to the disk is temporarily stored in the buffer until the buffer is full. Then this information is transferred to DOS. Each disk access is relatively slow because mechanical devices are involved. Therefore, using a buffer in this manner requires fewer disk accesses, resulting in improved program execution time. A buffer is used in a similar manner whenever a read operation is performed.

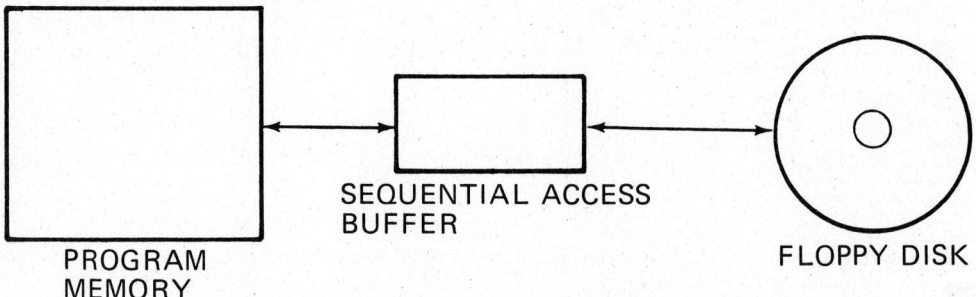

Figure 13.2—illustration of sequential access buffer area

Writing and Reading Sequential Files

The following steps are required to write data to a sequential file, and then read those data from the file:

Write to a Sequential File	Read from a Sequential File
1. Specify the file name.	1. Specify file name.
2. Specify output or append mode.	2. Specify input mode.
3. Specify the file number.	3. Specify file number.
4. Write data to the file.	4. Read data from file.
5. Close file.	5. Close file.

In BASIC, the first three of the above steps are performed with one statement, OPEN. For example, to OPEN the file DESIGN.JAN residing on disk B for output, we use a statement of the form:

```
10 OPEN "B:DESIGN.JAN" FOR OUTPUT AS #1
```

The #1 is a reference number which serves two purposes. First, it designates a buffer area in memory where a record resides before a write to the disk occurs. In the case of a read from the disk, this buffer area contains the record read from the disk.

The second function of this number is to reduce cumbersome file specifications. As long as the file DESIGN.JAN is open, you may refer to it using only the reference number (#1). The following alternate form of the OPEN statement may be used to open a file for output:

```
10 OPEN "O", #1, "DESIGN.JAN"
```

BASIC provides three modes for accessing a sequential disk file: OUTPUT, INPUT, and APPEND. If you are going to write to a sequential disk file, you must open the file using the output or the append mode. If you later want to read (input) from this file within the same program, you must close the file and then reopen it using the input mode. The latter two statements might look like these:

```
20 CLOSE #1
30 OPEN "DESIGN.JAN" FOR INPUT AS #1
```

There is an alternate form for input:

```
30 OPEN "I", #1, "DESIGN.JAN"
```

Closing a file lets you change the access mode. If the file had been open for output or append, closing the file causes the final buffer contents to be written to disk. Executing an END statement will automatically close all open files. If a file will not be used continuously, it is a good practice to close it using a CLOSE statement. You thus insure that should a power failure occur, the contents of the output buffer will not be lost.

The append mode applies only to sequential output. Using this mode causes the file to be positioned at the end of the data existing in the file (sometimes called end of file). In this way, additional data can be added to the end of a file. If the file does not exist (the file name is not in the diskette directory), it is created and positioned at the beginning, as it is when using the output mode.

It is important to remember that opening a file using the output mode destroys any existing data in the file. Therefore, if you want to add a file, you must use the append mode. A good rule to follow for writing to a sequential file is always to use the append mode unless you specifically want to destroy the file contents.

There is no alternate form of the open statement for append operations. You must use a statement of the form:

```
10 OPEN "DESIGN.JAN" FOR APPEND AS #1
```

After you have opened a file for output, you may write data to the file using the PRINT#, WRITE#, or PRINT#,USING statements. For example, assume that you want to output to a file the string variable "EXPERIMENT$" and the numeric variable "VALUE." This could be done using the instruction:

```
10 WRITE#1, EXPERIMENT$, VALUE
```

Here the #1 is the reference number assigned to the file when it was opened.

The WRITE# instruction inserts quotation marks to delimit strings. In addition, a carriage return/line feed sequence is inserted after the last item in the list (WRITE# file number, list) is written (this will be illustrated with an example). In contrast, the PRINT# statement writes data to a file as it would be displayed on the screen with a PRINT statement.

For instance, let X$="EXAMPLE" and Y$="ONE". Using the statement

```
10 PRINT#1,X$;Y$
```

the following characters would be written in file #1:

EXAMPLEONE

This is the same image that would have been displayed on your monitor if a PRINT statement had been used. The problem occurs when you try to input this image from the file. Because there are no delimiters between the two words, the computer assumes there is only one word. This problem can be avoided by including delimiters in the PRINT# statement such as:

```
10 PRINT#1,X$;",";Y$
```

Therefore, if you use a PRINT# statement to write to a file, care must be taken to properly delimit the data on the file.

Executing this statement:

```
10 WRITE#1,X$;Y$
```

will result in the following image being written to the file:

```
              CL
"EXAMPLE","ONE"RF
           C L
```

The symbols R and F represent carriage return and line feed. Consequently, when writing data to a sequential file, most of the time you will want to use a WRITE# statement.

EXAMPLE 1. Create a sequential data file consisting of date and high temperature for each day during the month of January. Let DAY$ denote the date and

13.2 Sequential Access Files

TEMP denote the high temperature for that date. The program should continue asking for data until you type "END" for the date.

SOLUTION.

```
10 OPEN "B:HITEMP.JAN" FOR APPEND AS #1
20 INPUT "DATA";DAY$
30 IF DAY$="END" GOTO 70
40 INPUT "TEMPERATURE";TEMP
50 WRITE#1,DAY$,TEMP
60 GOTO 20
70 CLOSE#1
80 END
```

Assume that when this program is run, the following data are input:

```
1/3/83
33.56
1/4/83
18.32
END
```

After writing this information to the diskette, the file ("HITEMP.JAN") would contain the following information:

```
              CL            CL
"1/3/83",33.56RF"1/4/83",18.32RF
```

Remember that a WRITE# statement delimits strings with quotation marks, places commas between items, and inserts a carriage return/line feed after the last item in the list. In the above example, the WRITE# statement was executed two times; consequently, there are two sets of carriage return/line feed entries in the file. All information in a sequential file is stored as ASCII characters. Even a number such as 33.56 is stored as five ASCII characters rather than the binary number representation of the numeric value.

EXAMPLE 2. Write a program that will search the data file created in Example 1 for the date 1/4/83 and print on the screen the temperature for this date. If the specified date is not found, the program should print a message indicating this condition.

SOLUTION.

```
110 OPEN "B:HITEMP.JAN" FOR INPUT AS #1
120 INPUT "DATE TO SEARCH FOR:"; MATCH$
130 INPUT#1,DAY$,TEMP
140 IF MATCH$=DAY$ THEN 170 ELSE 150
150 IF EOF(1) THEN 200
160 GOTO 130
170 CLS
180 PRINT "DATE: ";DAY$," HIGH TEMP ";TEMP
190 GOTO 210
200 PRINT "NO MATCH FOUND"
```

```
210 CLOSE#1
220 END
```

Statement 150 is used to check for an end of file (EOF mark) on file 1. This mark denotes that the file contains no more information. The disk operating system automatically places this indicator after the last data element in a sequential file.

Some applications may involve large amounts of data. Consequently, you need to know how much data you can store on a diskette. Each sector on a diskette can store 512 bytes of information. However, when you are using BASIC and the sequential access mode, each sector will contain 4 records, because the sequential access mode uses a record size of 128 bytes (which is the size of the buffer).

Looking at Example 1 above, we can determine how many days of temperature data will fit into one sector. For each day, the following information is written to the diskette:

```
              CL
"1/3/83",33.56RF
```

Counting characters, we see that 16 characters are required. In a sequential file, each character requires 1 byte. Assuming that each day requires 16 characters, data for 8 days can be stored in 1 record (128/16=8). Therefore, data for 32 days can be stored in one sector (512/128=4 records/sector).

Space on a diskette is allocated in blocks of 512 bytes (one sector). If a file uses less than a full sector, the remaining space is not available to be assigned to other files. Consequently, this unused space is wasted.

Sequential File Management Program

It is difficult, if not impossible, to write a sequential file management program that can be used for all sequential access applications. Also, programming is largely an art; therefore, we learn from seeing how others handle a given problem. As a result, the approach used here is to present an example that illustrates many of the concepts we have been discussing as well as some new ideas.

EXAMPLE 3. All work received by Specialty Engineering Co. is obtained by submitting a bid to prospective clients. At any time, this company may have 80 to 100 jobs in one of the following stages:

1. Working on bid.
2. Waiting to start work.
3. Work in process.
4. Complete.
5. Past due.

Write a program that will maintain a job inventory in a sequential disk file. The following information should be maintained for each bid:

1. Job number.
2. Job description.
3. Job due date.

13.2 Sequential Access Files

4. Bid dollar amount.
5. Status (bidding, complete, late, etc.)

Include in the program the ability to create a sequential file, append to the file, modify the file contents, and list the file contents.

SOLUTION. Interactive computer programs often are developed using a "menu" consisting of different options. When executing the program, the user must specify which option is to be performed. This concept was used to develop this program.

A menu is provided consisting of five options:

1—Create (Create a new job inventory data file)
2—Append (Add new jobs to this file)
3—Modify (Modify job information in this file)
4—List (List job inventory data)
5—Quit (Exit from program)

A subroutine is used for each option except the last, QUIT. A listing of this program appears in Figure 13.3.

```
1000 REM PROGRAM:SEQMAN
1010 REM Example sequential file data management program
1020 REM Maintains a job inventory using the sequential access method.
1030 REM The following information is stored for each job:
1040 REM     1-Job number
1050 REM     2-Job description
1060 REM     3-Due date
1070 REM     4-Bid dollar amount
1080 REM     5-Job status
1090 REM
1100 INPUT "ENTER DISK:FILE NAME TO BE USED FOR JOB INVENTORY";NAM$
1110 PRINT:PRINT TAB(5) "MENU:"
1120 PRINT TAB(10) "1- CREATE"
1130 PRINT TAB(10) "2- APPEND"
1140 PRINT TAB(10) "3- MODIFY"
1150 PRINT TAB(10) "4- LIST"
1160 PRINT TAB(10) "5- QUIT"
1170 INPUT "OPTION"; OPT%
1180 IF(OPT%<1) OR (OPT%>5) GOTO 1210
1190 ON OPT% GOSUB 1250,1430,1770,1570,1220
1200 GOTO 1110
1210 PRINT:PRINT "INVALID OPTION, TRY AGAIN!":GOTO 1170
1220 PRINT:PRINT TAB(20) "**END OF PROGRAM**"
1230 CLOSE:END
1240 REM****************************************************************
1250 REM SUBROUTINE: CREATE A SEQUENTIAL FILE AND ACCEPT DATA
1260 REM Determine if file name is unique
1270 ON ERROR GOTO 1350
1280 OPEN NAM$ FOR INPUT AS #1
1290 CLOSE #1
1300 ON ERROR GOTO 0
1310 PRINT "FILE ALREADY EXISTS WITH THIS NAME: ", NAM$
```

```
1320 PRINT "CREATING NEW FILE WILL DESTROY CONTENTS OF OLD FILE"
1330 INPUT "ABORT (Y/N)"; AN$
1340 IF AN$="Y" THEN RETURN ELSE 1400
1350 IF ERR=53 THEN RESUME 1390
1360 RESUME 1370
1370 ON ERROR GOTO 0
1380 RETURN
1390 ON ERROR GOTO 0
1400 OPEN NAM$ FOR OUTPUT AS #1
1410 GOTO 1460
1420 REM*******************************************************************
1430 REM SUBROUTINE: APPEND TO FILE
1440 OPEN NAM$ FOR APPEND AS #1
1450 REM Request data
1460 PRINT:INPUT "JOB NO."; JNO$
1470 REM End of input?
1480 IF JNO$="END" THEN 1550
1490 INPUT "JOB DESCRIPTION"; DESC$
1500 INPUT "DUE DATE";DAT$
1510 INPUT "BID $ AMOUNT";AMOUNT
1520 INPUT "JOB STATUS";STATUS$
1530 WRITE #1, JNO$, DESC$, DAT$, AMOUNT,STATUS$
1540 GOTO 1460
1550 CLOSE #1:RETURN
1560 REM*******************************************************************
1570 REM SUBROUTINE: LIST DATA
1580 REM Open file for input
1590 GOSUB 2210
1600 IC%=1
1610 LINENO%=1
1620 PRINT:PRINT "JOB NO.", "DESCRIPTION", "DUE DATE", "$
     AMOUNT","STATUS"
1630 LINENO%=LINENO%+1
1640 IF LINENO%<>(IC%*20) THEN 1680
1650 REM If lines on CRT=20, stop execution to read data
1660 IC%=IC%+1
1670 PRINT:INPUT "PRESS ENTER TO CONTINUE";Y$:PRINT
1680 INPUT #1,JNO$,DESC$,DAT$,AMOUNT,STATUS$
1690 PRINT JNO$,DESC$,DAT$,AMOUNT,STATUS$
1700 REM Stop reading data from file when reach EOF
1710 IF EOF(1) THEN PRINT:PRINT "END OF FILE":GOTO 1730
1720 GOTO 1630
1730 CLOSE#1
1740 PRINT
1750 RETURN
1760 REM*******************************************************************
1770 REM SUBROUTINE: MODIFY DATA
1780 REM Open file for input
1790 GOSUB 2210
1800 INPUT "ENTER JOB TO BE MODIFIED"; JNO$
1810 REM Find specified data ID for modification
1820 REM Open scratch file
```

```
1830 OPEN "B:SEQTEMP.DAT" FOR OUTPUT AS #2
1840 INPUT #1, TJNO$,TDESC$,TDAT$,TAMOUNT,TSTATUS$
1850 IF TJNO$=JNO$ THEN 1990
1860 IF EOF(1) THEN 1920
1870 REM If job number is blank, do not keep job information
1880 IF TJNO$="" THEN 1910
1890 REM Write to scratch file
1900 WRITE #2, TJNO$,TDESC$,TDAT$,TAMOUNT,TSTATUS$
1910 GOTO 1840
1920 PRINT "END OF FILE-NO MATCH FOUND FOR:"; JNO$
1930 CLOSE #2
1940 REM Erase scratch file because match not found, original file ok
1950 KILL "B:SEQTEMP.DAT"
1960 CLOSE #1
1970 RETURN
1980 REM Modify data for specified job
1990 PRINT:INPUT "JOB NO."; JNO$
2000 INPUT "JOB DESCRIPTION"; DESC$
2010 INPUT "DUE DATE"; DAT$
2020 INPUT "BID $ AMOUNT"; AMOUNT
2030 INPUT "JOB STATUS";STATUS$
2040 REM If job number is blank, do not keep job information
2050 IF JNO$="" THEN 2090
2060 REM Write modified data to scratch file
2070 WRITE #2, JNO$,DESC$,DAT$,AMOUNT,STATUS$
2080 REM Write remaining data to scratch file
2090 IF EOF(1) THEN 2130
2100 INPUT #1, JNO$,DESC$,DAT$,AMOUNT,STATUS$
2110 WRITE #2, JNO$,DESC$,DAT$,AMOUNT,STATUS$
2120 GOTO 2090
2130 CLOSE #2
2140 CLOSE #1
2150 REM Erase original file
2160 KILL NAM$
2170 REM Rename scratch file to name of original file
2180 NAME "B:SEQTEMP.DAT" AS NAM$
2190 RETURN
2200 REM*****************************************************************
2210 REM SUBROUTINE: OPEN FILE FOR INPUT
2220 OPEN NAM$ FOR INPUT AS #1
2230 REM Is input file empty?
2240 IF NOT EOF(1) THEN 2280
2250 PRINT:PRINT "INPUT FILE IS EMPTY"
2260 CLOSE #1
2270 GOTO 1110
2280 RETURN
```

Figure 13.3—sequential file management program

Option 1, CREATE, will create a file having the name specified by the user at the beginning of the program. Before a file is created, the program determines if the file name specified is unique. This is done because the open statement used to create the new file specifies the output mode. Therefore, creating this file will destroy the contents of a file having the same name. If a file with the same name is found, the user is given the option of aborting or continuing with the creation of this new file. If the file is created, the program asks for data. The user denotes end of data by typing "END" for the JOB NO. The data file is then closed and control is returned to the menu.

You might note that an ON ERROR GOTO statement is used in the the CREATE subroutine. To determine if a file name is unused, we first try to open a file by that name using the input mode. If the file does not exist, BASIC returns the error message

```
File not found
```

This is error message number 53. When this error occurs, program execution is terminated unless the error is "trapped." The ON ERROR statement permits us to trap the error and provide a means for dealing with it. In this case, the occurrence of the error is good because it means that no file exists having the name we want to use. The statement

```
ON ERROR GOTO 0
```

disables error trapping.

The second option, APPEND, permits us to add data to our file whenever we desire. Like option 1, this option terminates whenever "END" is entered for a JOB NO. The data file is then closed and control is returned to the menu.

As requested, we provided a modification option, called MODIFY. The job number is used as the key to finding the appropriate data in the file. We assumed that the job numbers are unique. The only way data in a sequential file can be modified is by rewriting the entire file. Therefore, this option used two files, the existing data file and a temporary "scratch" file named "B:SEQTEMP.DAT." Data are read from the existing data file and written to the temporary file until a match is made on the job number that was specified to be changed. At this point, the user is asked (prompted) for all of the information associated with this job. Next, this information is written to the temporary file followed by the remainder of the data that were in our original file. When the end of file (EOF) is encountered, the original file is KILLed (erased). The temporary file is then renamed with the name we had assigned to our original file. In this way we satisfy the requirement of rewriting the entire file whenever the contents of the sequential file are changed. At this point, control is returned to the menu.

Jobs can be removed from the data file by simply pressing the enter key when the program prompts for the job number. This results in a job number having all blanks. The MODIFY option checks for a blank job number. Whenever one is found, the data associated with this job are not written to the new file.

One other subroutine is used in the program. This subroutine opens the specified file using the input mode and checks to see if the file is empty. Options 3 and 4 (APPEND and MODIFY) open the designated file using this subroutine. If the file is empty, we know we should not proceed with these options. Therefore, when an empty file is detected, execution of these options is terminated and control is returned to the menu.

You might note that at the end of each option all open files are closed. This causes the disk operating system to write all data in the buffer to the disk file. Therefore, if power to your computer is interrupted during program execution, the only data you will lose are those in a buffer associated with the option you are currently executing.

13.3 Random Access Files

The second type of file supported by BASIC on your Personal Computer is random access. Direct access is another term for this type of file. Random access files permit you to access data anywhere in a file without first reading all the preceding information in the file.

Besides randomly accessing data, this access method differs from sequential access in how information is stored (coded) on a diskette. The sequential method stored all information as ASCII characters; however, random files are stored in a packed binary format. Consequently, a random file can require less disk space than a sequential file. Remember that storing the number 33.56 in a sequential file required five bytes. In a random access file, this number can be stored in four bytes.

Writing to a Random Access File

Although random access files offer more flexibility, they require more programming steps to use. The following steps are required to write data to a random file:

1. Specify the file name.
2. Specify the file number.
3. Specify the record length.
4. Allocate space in a random buffer for the variables that will be written to the random files.
5. Convert numerical values to string values.
6. Move data into the random buffer.
7. Write data from the buffer to the desired file record.
8. Close file.

The first three of the above steps are specified using on OPEN statement. For example, consider the following:

```
10 OPEN "B:COMPUTER.DAT" AS #1 LEN=30
```

This statement assigns the file reference number one (#1) to the file named COMPUTER.DAT, residing on disk drive B. A record length of 30 characters (bytes) is also specified.

For random access files, a *record* is the number of bytes that are moved in and out of the random buffer at one time (the amount of data transferred with one GET or one PUT statement). The *random buffer* is a temporary storage area that is like the sequential file buffer. Similarly, it improves the efficiency of the disk input and output by reducing the number of disk accesses.

Comparing this statement to the OPEN statement used for sequential files, you will note two differences. First, the mode specification is omitted (output, input, and append). When the mode is omitted, random access is assumed. The second difference is the record length specification (LEN=30). Here, a record length of 30 characters (bytes) is specified. You cannot specify the record length for sequential files; the length is fixed at 128 bytes. If you omit this specification for random files, the default length is 128 bytes.

An alternative form of the above OPEN statement is:

```
10 OPEN "R", 1, "B:COMPUTER.DAT", 30
```

In this form, R denotes random access. Specifying the random access mode in an OPEN statement does not destroy data in a file as would using the output mode of the sequential access method.

It is best to specify the record length; the default record size will usually result in wasted space. For example, if we do not specify a record size and only move 30 characters at a time into the random buffer, BASIC writes the data onto the file using a 128 character record (the default record size). Consequently, in each record on the disk, we would waste 98 bytes (128−30). The method for calculating the record size will be illustrated with an example.

The next step in creating a random file is to allocate space in the random buffer. As noted above, this buffer is used in all random file reads and writes. As an example, consider the following:

```
20 FIELD #1, 2 AS N$, 12 AS D$, 4 AS C$, 12 AS L$
```

This statement allocates space in the record specified in the OPEN statement for four variables (N$, D$, C$, and L$). The first 2 bytes are assigned to the string variables N$, 12 bytes to D$, 4 bytes to C$, and 12 bytes to L$, for a total of 30 bytes. The space allocated to a particular variable in a record is called *field*. Thus, we have defined four fields in the above FIELD statement, one field for each variable. The total number of bytes allocated to the fields cannot exceed the record length specified in the OPEN statement when the file was opened. As a result, one way of determining the appropriate record length is to total the number of characters specified in a FIELD statement.

A FIELD statement does not move data into the random file buffer. This statement only allocates space to the specified variable. It is also important to note that the variable names in the FIELD statement (N$, D$, etc.) point to locations (specific character positions) within the random file buffer. Therefore, you must be careful when using these names on the left side of an assignment statement or in an input statement. These operations redefine the locations of respective variables from the random buffer space to program space.

Looking at the above FIELD statement, you will note that all of the variables are specified as string variables. Only string variables can be used in I/O statements accessing a random file. This may appear to be a severe limitation because a numeric value is not considered a string.

However, BASIC provides three functions (MKI$, MKS$, MKD$) for converting numeric type values to string type values. MKI$ converts an integer value to a 2-byte string. Similarly, MKS$ converts a single precision value to a 4-byte string, and MKD$ converts a double precision value to an 8-byte string. Although this conversion requires an extra programming step, one major advantage over sequential files is that, in most cases, numeric data will not require as much disk space. For instance, the number 12345 would require 5 bytes to store in a sequential file. However, this number would require only 2 bytes in a random access file.

Once space is allocated in a random buffer, string data must be moved into this buffer before it can be stored on a diskette. This can be done using an LSET or RSET statement. For instance,

```
50 LSET N$=MKI$(NOM%)
60 LSET C$=MKS$(COST)
```

MKI$ in statement 50 converts the integer variable NOM into a 2-byte string, and then LSET assigns this string to the 2-byte location N$ in the random buffer. Similarly, the next statement converts the single precision variable COST to a 4-byte string and then assigns this string to location C$. The difference between LSET and RSET is that LSET left-justifies a string in a field in the random buffer while RSET right justifies the string. To left-justify a string means to place the string as far left in the field as possible. Any space unused on the right side is padded with blanks. Right-justify has a similar meaning.

Character string data must also be placed in the random buffer using LSET or RSET. However, the MKI$, MKS$, and MKD$ functions are not used because a conversion to character string is not required. Thus, if DESC$ represents a character string, the following statement could be used to place this string in the random buffer defined by the above FIELD statement:

```
70 LSET D$=DESC$
```

If DESC$ contains more than 12 characters (D$ is defined for 12 characters in the FIELD statement), characters are dropped from the right. If DESC$ contains fewer than 12 characters, blanks are added to the right.

Only one more step remains in the process of storing information on a random access file, that of actually writing a record from the random buffer to the file. This is accomplished using a PUT statement such as:

```
80 PUT#1,REC%
```

This statement writes the contents of the random file buffer into the file specified by the file reference number one (#1). The integer variable REC% denotes the record number in a file where the contents of the buffer will be written. Specifying the record number is optional; if it is omitted, the record is written in the file at the position of the next available record. Since the record pointer is automatically advanced one record after each read or write statement is executed, the next available record is the record following the last record accessed. When a random file is opened, the record pointer is set equal to the first record in the file.

EXAMPLE 4. Write a program which will create a file for experimental data using a random access file. This file should provide space for the following types of data: parameter description, date, and parameter value.

SOLUTION.

```
100 OPEN "B:EXPDATA" AS #1 LEN=32
110 FIELD #1, 20 AS N$, 8 AS D$, 4 AS VA$
120 RECNO%=0
130 INPUT "ENTER PARAMETER";NAM$
140 IF NAM$="END" THEN CLOSE #1:END
150 INPUT "ENTER DATE"; DAT$
160 INPUT "PARAMETER VALUE";VALUE
170 REM Place data in random buffer
180 LSET N$=NAM$
190 LSET D$=DAT$
200 LSET VA$=MKS$(VALUE)
210 RECNO%=RECNO%+1
220 REM Write data to random disk file
230 PUT#1,RECNO%
240 GOTO 130
```

We specified a record length of 32 characters. Within this record, a field of 20 characters was allocated in the random buffer for the string variable N$, 8 characters for D$, and 4 characters for VA$. The variable RECNO% was used to denote the next available record. If END is typed when the program prompts for name, the program terminates execution. Note that the single precision variable VALUE is converted to a four byte string in line 190. Lines 180 through 200 place the data in the random buffer. Then line 230 writes (PUTs) the data in the random buffer (a record of 32 bytes) to the random file names PERSONEL.

Reading From a Random Access File

The steps required to read data from a random file are very similar to those for writing data. These steps are:

1. Specify the file name.
2. Specify the file number.
3. Specify the record length.
4. Allocate space in random buffer.
5. Move the desired record into the random buffer.
6. Convert numeric values back to numbers.
7. Close the file.

The first four steps are usually performed with the same statements (OPEN and FIELD) that were used to write data to the random file. Consequently, if input and output to the same random file are performed in the same program, only one OPEN and one FIELD statement are required. One good rule to follow is always to read data from a file using the same format used to write it. You will, therefore, want to use the same OPEN and FIELD statements.

To move the desired record from a random access disk file into the random buffer, we use the GET statement.

```
10 GET#1,REC%
```

13.3 Random Access Files

The logical record number (REC%) is optional, similar to the PUT statement. After executing a PUT statement, it is not necessary to close the file before executing a GET statement. These statements may be executed in any order after a file is opened.

Conversion of numeric values from character strings to numbers is accomplished by using the three functions CVI, CVS and CVD. The first function CVI converts a 2-byte string to an integer; CVS converts a 4-byte string to a single precision number; and CVD converts an 8-byte string to a double precision number. After this conversion process, the data are now ready to be used by your program.

EXAMPLE 5. Write a program which will access any desired record in the experiment data file created in Example 4 and print the contents of this record on your video display.

SOLUTION.

```
100 OPEN "B:EXPDATA" AS #1 LEN=32
110 FIELD #1, 20 AS N$, 8 AS D$, 4 AS VA$
120 INPUT "ENTER NUMBER OF RECORD TO BE RETRIEVED";RECNO%
130 IF RECNO%=0 THEN CLOSE#1: END
140 GET#1,RECNO%
150 REM Convert string variable to single precision variable
160 VALUE=CVS(VA$)
170 PRINT:PRINT "RECORD NO. =";RECNO%
180 PRINT "PARAMETER : ";N$
190 PRINT "DATE : ";D$
200 PRINT "VALUE :"; VALUE
210 GOTO 120
```

Note that the records can be retrieved in any order. When we input a record number of 0 the program terminates. Also, note that line 160 converts the string variable VA$ to a single precision variable, VALUE. Remember that we had converted VALUE to a string variable in Example 4.

EXAMPLE 6. Combine the programs from Examples 4 and 5 such that after all the experiment data have been input, the user can print out the contents of any record.

SOLUTION.

```
100 OPEN "B:EXPDATA" AS #1 LEN=32
110 FIELD#1, 20 AS N$, 8 AS D$, 4 AS VA$
120 RECNO%=0
130 INPUT "ENTER PARAMETER";NAM$
140 IF NAM$="END" THEN 250
150 INPUT "ENTER DATE";DAT$
160 INPUT "PARAMETER VALUE";VALUE
170 REM Place data in random buffer
180 LSET N$=NAM$
190 LSET D$=DAT$
200 LSET VA$=MKS$(VALUE)
```

```
210 RECNO%=RECNO%+1
220 REM Write data to random file
230 PUT#1,RECNO%
240 GOTO 130
250 INPUT "ENTER NUMBER OF RECORD TO BE RETRIEVED";RECNO%
260 IF RECNO%=0 THEN CLOSE#1: END
270 GET#1,RECNO%
280 REM Convert string variable to single precision variable
290 VALUE=CVS(VA$)
300 PRINT:PRINT "RECORD NO. =";RECNO%
310 PRINT "PARAMETER : ";N$
320 PRINT "DATE : ";D$
330 PRINT "VALUE : ";VALUE
340 GOTO 250
```

Note that only one OPEN and one FIELD statement are required. Also, only one CLOSE statement was used.

Random File Management Program

It would require several thousand BASIC program statements to write a general random file access management program. Therefore, we will use a specific example that illustrates many of the concepts that are considered. The example will be the same one used to illustrate sequential access file management.

Before the details of the program are presented, we need to discuss how to locate a specific record. In Example 3, the Specialty Engineering Co. problem, we developed a program to store job related information. Assume that we have stored this information in a random access file. To locate a record with a particular job number, we will have to read one record at a time until we find the right job number. This is not an improvement over the sequential access method.

There are techniques which will improve this search process. One is the keyed access method. With this method we build a directory (table) containing job numbers and associated record numbers. The record number specifies where the job information for a given job number is stored in the direct access file (See Figure 13.4). When a job is to be stored in the file, the directory is searched for an unused record. The job information is then written to this location, and the directory is updated to show which job information is stored at this location. Figure 13.4(a) and 13.4(b) depict this process. If we want to locate a specific record (such as the record containing information for job B106) we can search the directory and see which record (record two for job B106) contains the desired information. This record can then be directly accessed.

One other problem remains. What do we do with the directory? It can be stored in a disk file or it can be reconstructed each time we use the program. Often it is stored at the beginning of the random access file for which it serves as a directory.

EXAMPLE 7. Reconsider Example 3 such that the data are stored in a random access file. The same types of data will be stored. The key to each record is the job number.

13.3 Random Access Files

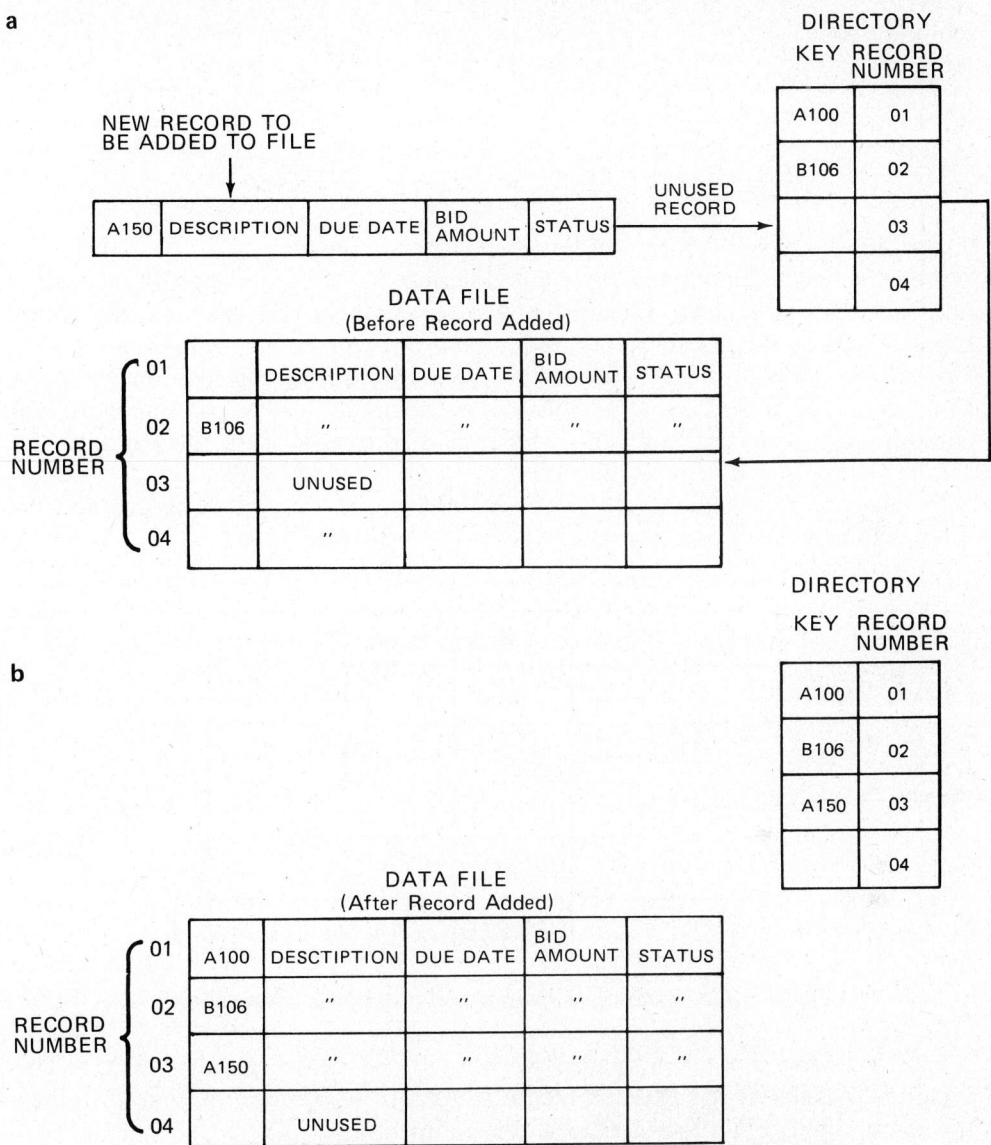

Figure 13.4 (A&B)—keyed access method illustrated

SOLUTION. Given the same information as in Example 3, we need to decide what record length should be used. This can be determined by assigning fields to each data type that is to be stored. Assume that the following field sizes will be used:

Field	Number of Characters	Variable Type
Job Number	4	Character
Job Description	24	Character
Due Date	8	Character (MM-DD-YY)
Bid Amount	4	Single Precision
Status	8	Character
Total	48	

Therefore, the record length will be 48 characters (bytes).

We will slightly modify the procedure described above for storing the data record number in the directory. Looking at Figure 13.4, we can see that the record number corresponds to the position of a job number in the column denoted KEY. For instance, job number B106 was in position two in this column, and the associated data record used for this job is record number two. Therefore, the column used for record numbers is redundant; the position in the column KEY can be used to specify the record number.

In the data file in Figure 13.5, the first record is a directory record and the following records are used for data.

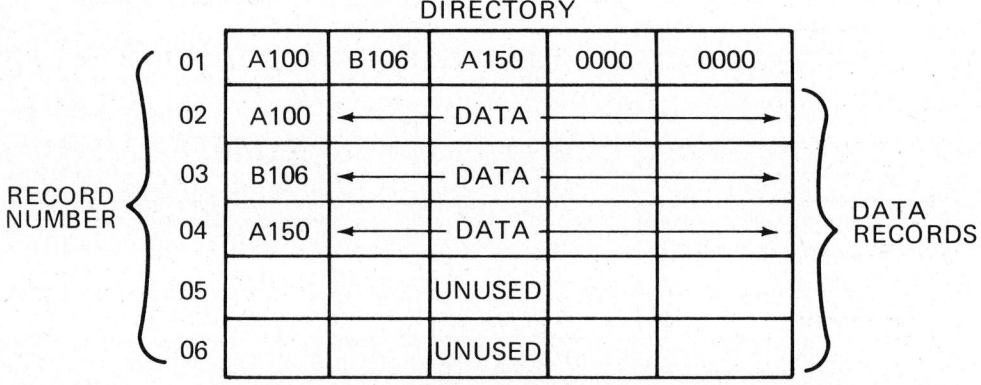

Figure 13.5—example directory and associated data records

First, the directory is initialized so that all the fields contain "0000." This indicates that the associated data record is unused. As job information is entered, the appropriate job number is placed in the directory.

In Figure 13.5, the first field in the directory contains the job number A100. Since this is located in field number one of the directory, the associated data are found in the first data record, record number two in the file. The specific file record is found by adding:

Number of   Field number
directory records + in directory = File record number

1 + 1 = 2.

In a similar manner we can calculate the file record number for job number A150:

1 + 3 = 4

Maintaining the directory in this manner saves space in the file.

For the Specialty Engineering Co. example, we will use a directory to maintain the location of each job related record as well as the location of each unused record. The directory will be stored at the beginning of the random access file, and the field number in the directory will designate the associated data record.

Now that a file index method has been defined, we need to determine how many directory records will be required, assuming that as many as 100 jobs may be active at one time. We specified that the size of the job number is 4 characters. Consequently, each directory record can contain 12 fields:

48 characters/record ÷ 4 characters/field = 12 fields/record

We will use 10 directory records; therefore, we can maintain indices for 120 jobs. This number provides some margin for growth.

We will use a menu, like the one used in Example 3. The menu will consist of the following options:

1—Initialize Directory
2—New Job
3—Modify Job Data
4—Delete a Job
5—List Data
6—List Directory Contents
7—Exit Program

In the random access file management program, each of these options is a subroutine. The program appears in Figure 13.6. Each of these subroutines will be explained.

```
1000 REM PROGRAM RANMAN
1010 REM Example random file data management program
1020 REM Maintains a job inventory using the random access method.
1030 REM The following information is stored for each job:
1040 REM    1-Job number
1050 REM    2-Job description
1060 REM    3-Due date
1070 REM    4-Bid dollar amount
1080 REM    5-Job status
1090 REM     Variable Name      Variable Name      Description
1100 REM    in Random Buffer   in Program Space   of Variable
1110 REM         NO$                JNO$           JOB NUMBER
1120 REM         DS$                DESC$          JOB DESCRIPTION
1130 REM         DA$                DAT$           DUE DATE
1140 REM         AM$                AMOUNT         BID $ AMOUNT
1150 REM         ST$                STATUS$        JOB STATUS
1160 REM         DI$(I)             DIR$(I)        DIRECTORY FIELD I
1170 REM
1180 DEFINT I
1190 NOFIELD%=12              'No. fields in dir. record
1200 NODIR%=10                'No. of directory records
1210 DIM DIR$(NOFIELD%),DI$(NOFIELD%)
1220 REM Headings for CRT output
```

Chapter 13—Disk Data Files

```
1230 HEAD1$="JOB NO.        DESCRIPTION                    DUE DATE"
1240 HEAD2$="     BID AMOUNT        STATUS"
1250 INPUT "INPUT DISK:FILE NAME TO BE USED FOR JOB INVENTORY";NAM$
1260 PRINT:PRINT TAB (5) "MENU:"
1270 PRINT TAB(10) "1 - INITIALIZE DIRECTORY"
1280 PRINT TAB(10) "2 - NEW JOB"
1290 PRINT TAB(10) "3 - MODIFY JOB DATA"
1300 PRINT TAB(10) "4 - DELETE A JOB"
1310 PRINT TAB(10) "5 - LIST DATA"
1320 PRINT TAB(10) "6 - LIST DIRECTORY CONTENTS"
1330 PRINT TAB(10) "7 - EXIT PROGRAM"
1340 INPUT "OPTION"; OPT%
1350 IF(OPT%<1) OR (OPT%>7) GOTO 1440
1360 IF OPT%=1 THEN GOSUB 1460       'Initialize directory
1370 IF OPT%=2 THEN GOSUB 1760       'New job
1380 IF OPT%=3 THEN GOSUB 1930       'Modify job data
1390 IF OPT%=4 THEN GOSUB 2120       'Delete a job
1400 IF OPT%=5 THEN GOSUB 2320       'List data
1410 IF OPT%=6 THEN GOSUB 2690       'List directory contents
1420 IF OPT%=7 THEN PRINT:PRINT TAB(20) "**END OF PROGRAM**":END
1430 GOTO 1260
1440 PRINT'PRINT "INVALID OPTION, TRY AGAIN!":GOTO 1260
1450 REM****************************************************************
1460 REM SUBROUTINE: INITIALIZE DIRECTORY
1470 REM Determine if file name is unique
1480 ON ERROR GOTO 1560
1490 OPEN NAM$ FOR INPUT AS #1
1500 CLOSE#1
1510 ON ERROR GOTO 0
1520 PRINT:PRINT "FILE ALREADY EXISTS WITH THIS NAME: ";NAM$
1530 PRINT "INITIALIZING DIRECTORY WILL DESTROY OLD FILE!"
1540 INPUT "ABORT (Y/N)";AN$
1550 IF AN$<>"N" THEN RETURN ELSE 1640
1560 IF ERR=53 THEN RESUME 1630
1570 REM Invalid error
1580 RESUME 1590
1590 ON ERROR GOTO 0
1600 PRINT "ERROR OCCURRED WHILE ATTEMPTING TO OPEN FILE"
1610 RETURN
1620 REM Valid error
1630 ON ERROR GOTO 0
1640 GOSUB 2860                 'Open file
1650 FOR I=1 TO NOFIELD%
1660 DIR$(I)="0000"
1670 NEXT I
1680 GOSUB 2940                 'Place dir. in random buffer
1690 REM Initialize all directory fields to 0000
1700 FOR DIRECT%=1 TO NODIR%
1710 PUT #1,DIREC%
1720 NEXT DIREC%
1730 CLOSE #1
1740 RETURN
```

```
1750 REM****************************************************************
1760 REM SUBROUTINE: ADD A JOB TO FILE
1770 GOSUB 2860                 'Open file
1780 GOSUB 3180                 'Ask for new data
1790 MATCH$="0000"              'Match this value in dir.
1800 GOSUB 3430                 'Obtain unused record no.
1810 IF IFL<>0 THEN PRINT "FILE FULL":RETURN
1820 GOSUB 3100                 'Place data in random buffer
1830 PUT #1, FILREC%            'Write data to file
1840 DIR$(IREC)=JNO$            'Update directory
1850 GOSUB 3000                 'Write directory to disk
1860 IF IFL<>0 THEN RETURN 'IFL<>0-invalid record no.
1870 INPUT "ADD ANOTHER JOB (Y/N)"; Y$
1880 IF Y$="Y" THEN 1780
1890 CLOSE#1
1900 RETURN
1910 REM****************************************************************
1920 REM SUBROUTINE: MODIFY JOB INFORMATION
1930 GOSUB 2860                 'Open file
1940 INPUT "ENTER JOB TO BE MODIFIED";JNO$
1950 MATCH$=SPACE$(4):LSET MATCH$=JNO$ 'Match this value in dir.
1960 GOSUB 3430                 'Obtain job record no.
1970 IF IFL<>0 THEN RETURN 'IFL<>0-no match
1980 GOSUB 3280                 'Read designated record
1990 GOSUB 3370                 'Print record contents
2000 PRINT "ENTER MODIFIED JOB INFORMATION"
2010 GOSUB 3180                 'Ask for information
2020 GOSUB 3100                 'Place data in random buffer
2030 PUT#1, FILREC%             'Write modified record to file
2040 DIR$(IREC)=JNO$            'Update directory
2050 GOSUB 3000                 'Write directory to disk
2060 IF IFL<>0 THEN RETURN 'IFL<>0-invalid record no.
2070 INPUT "ANY MORE JOBS TO BE MODIFIED (Y/N)";Y$
2080 IF Y$="Y" GOTO 1940
2090 CLOSE #1
2100 RETURN
2110 REM****************************************************************
2120 REM SUBROUTINE: DELETE A JOB
2130 GOSUB 2860                 'Open file
2140 INPUT "JOB TO BE DELETED"; JNO$
2150 MATCH$=SPACE$(4):LSET MATCH$=JNO$ 'Match this value in dir.
2160 GOSUB 3430                 'Obtain job record no.
2170 IF IFL<>0 THEN RETURN 'IFL<>-no match
2180 GOSUB 3280                 'Read designated record
2190 PRINT:PRINT "INFORMATION OF JOB TO BE DELETED"
2200 PRINT:PRINT HEAD1$;HEAD2$
2210 PRINT JNO$,DESC$,DAT$,AMOUNT,STATUS$
2220 INPUT "DELETE THIS RECORD (Y/N)";Y$
2230 IF Y$<>"Y" THEN 2270
2240 DIR$(IREC)="0000"          'Update directory
2250 GOSUB 3000                 'Write directory to disk
2260 IF IFL<>0 THEN RETURN
```

```
2270 INPUT "DELETE ANOTHER JOB (Y/N)";Y$
2280 IF Y$="Y"THEN 2140
2290 CLOSE#1
2300 RETURN
2310 REM**********************************************************************
2320 REM SUBROUTINE: LIST DATA
2330 GOSUB 2860                    'Open file
2340 INPUT "ENTER JOB TO BE LISTED";JNO$
2350 IF JNO$="ALL" THEN 2450 'List all jobs?
2360 MATCH$=SPACE$(4):LSET MATCH$=JNO$ 'Match this value in dir.
2370 GOSUB 3430                    'Obtain job record no.
2380 IF IFL<>0 THEN GOSUB 2860:GOTO 2410 'IFL<>0-no match
2390 GOSUB 3280                    'Obtain data
2400 GOSUB 3370                    'Print record contents
2410 INPUT "LIST ANOTHER JOB (Y/N)";Y$
2420 IF Y$="Y" THEN 2340
2430 GOTO 2660
2440 REM List all jobs-stop every 20 jobs to read data
2450 ILINE=1
2460 IC=1
2470 PRINT HEAD1$;HEAD2$
2480 FOR DIREC%=1 TO NODIR%
2490 GET #1, DIREC%
2500 REM Save directory record in program space
2510 FOR I=1 TO NOFIELD%
2520 DIR$(I)=DI$(I)
2530 NEXT I
2540 FOR IREC=1 TO NOFIELD%
2550 IF DIR%(IREC)="0000" THEN 2640 'Record unused?
2560 REM Calculate file record no.
2570 FILREC%=NOFIELD%*(DIREC%-1)+IREC+NODIR%
2580 ILINE=ILINE+1
2590 IF ILINE<>(IC*20) THEN 2620
2600 IC=IC+1
2610 PRINT:INPUT "PRESS ENTER TO CONTINUE"; Y$:PRINT
2620 GOSUB 3280                           'Obtain data record
2630 PRINT JNO$,DESC$,DAT$,AMOUNT,STATUS$
2640 NEXT IREC
2650 NEXT DIREC%
2660 CLOSE #1
2670 RETURN
2680 REM**********************************************************************
2690 REM SUBROUTINE: LIST DIRECTORY CONTENTS
2700 GOSUB 2860                            'Open file
2710 PRINT:PRINT TAB (20) "DIRECTORY CONTENTS"
2720 FOR DIREC%=1 TO NODIR%
2730 PRINT:PRINT TAB(7) "DIRECTORY RECORD NO.";DIREC%
2740 PRINT "FIELD NO.","FIELD VALUE"
2750 GET #1, DIREC%
2760 FOR IREC=1 TO NOFIELD%
2770 FIELDNO%=NOFIELD%*(DIREC%-1)+IREC
```

```
2780 PRINT FIELDNO%, DI$(IREC)
2790 NEXT IREC
2800 REM Stop after every directory record to read data
2810 PRINT:INPUT "PRESS ENTER TO CONTINUE"; Y$:PRINT
2820 NEXT DIREC%
2830 CLOSE #1
2840 RETURN
2850 REM****************************************************************
2860 REM SUBROUTINE: OPEN FILES
2870 OPEN NAM$ AS #1 LEN=48
2880 REM Field specifications for job data record
2890 FIELD #1,4 AS NO$, 24 AS DS$, 8 AS DA$, 4 AS AM$, 8 AS ST$
2900 REM Field specifications for directory record
2910 FIELD #1,4 AS DI$(1),4 AS DI$(2),4 AS DI$(3),4 AS DI$(4), 4 AS
     DI$(5),4 AS DI$(6),4 AS DI$(7),4 AS DI$(8),4 AS DI$(9),4 AS
     DI$(10),4 AS DI$(11),4 AS DI$(12)
2920 RETURN
2930 REM****************************************************************
2940 REM SUBROUTINE: WRITE DIRECTORY TO RANDOM BUFFER
2950 FOR I=1 TO NOFIELD%
2960 LSET DI$(I)=DIR$(I)
2970 NEXT I
2980 RETURN
2990 REM****************************************************************
3000 REM SUBROUTINE: WRITE DIRECTORY RECORD TO DISK
3010 IFL=0                             'Indicates no error
3020 GOSUB 2940                        'Place dir. in random buffer
3030 IF(DIREC%>0)AND(DIREC%<(NODIR%+1)) THEN 3070
3040 PRINT "INVALID DIRECTORY RECORD"
3050 IFL=1:CLOSE#1
3060 RETURN
3070 PUT #1, DIREC%
3080 RETURN
3090 REM****************************************************************
3100 REM SUBROUTINE: WRITE DATA TO RANDOM BUFFER
3110 LSET NO$=JNO$
3120 LSET DS$=DESC$
3130 LSET DA$=DAT$
3140 LSET AM$=MKS$(AMOUNT)           'Convert integer to string
3150 LSET ST$=STATUS
3160 RETURN
3170 REM****************************************************************
3180 REM SUBROUTINE: ASK FOR JOB INFORMATION
3190 PRINT:INPUT "ENTER JOB NUMBER:";JNO$
3200 IF LEN(JNO$)>4 THEN PRINT "ONLY 4 CHARACTERS ALLOWED":GOTO 3190
3220 INPUT"ENTER JOB DESCRIPTION";DESC$
3230 INPUT "ENTER DUE DATE"; DAT$
3240 INPUT "ENTER AMOUNT BID"; AMOUNT
3250 INPUT "JOB STATUS";STATUS$
3260 RETURN
3270 REM****************************************************************
3280 REM SUBROUTINE: READ A DATA RECORD FROM RANDOM FILE
```

```
3290 GET #1, FILREC%
3300 JNO$=NO$
3310 DESC$=DS$
3320 DAT$=DA$
3330 AMOUNT=CVS(AM$)            'Convert string to integer
3340 STATUS$=ST$
3350 RETURN
3360 REM*******************************************************************
3370 REM SUBROUTINE: PRINT RECORD CONTENTS ON MONITOR
3380 PRINT:PRINT HEAD1$;HEAD2$     'Print headings
3390 PRINT JNO$, DESC$, DAT$, AMOUNT,STATUS$
3400 PRINT
3410 RETURN
3420 REM*******************************************************************
3430 REM SUBROUTINE: MATCH JOB NO. IN DIRECTORY
3440 IFL=0                          'Indicates no error
3450 FOR DIREC%=1 TO NODIR%
3460 GET #1, DIREC%
3470 FOR IREC=1 TO NOFIELD%
3480 DIR%(IREC)=DI$(IREC)
3490 IF DIR$(IREC)<>MATCH$ THEN 3570
3500 REM Calculate file record no.
3510 FILREC%=NOFIELD%*(DIREC%-1)+IREC+NODIR$
3520 IF IREC>=NOFIELD% THEN 3620
3530 FOR IREM=IREC+1 TO NOFIELD%
3540 DIR$(IREM)=DI$(IREM)
3550 NEXT IREM
3560 GOTO 3620
3570 NEXT IREC
3580 NEXT DIREC%
3590 PRINT "NO MATCH FOUND FOR JOB ID: "; MATCH$
3600 CLOSE #1
3610 IFL=1                          'Indicates error
3620 RETURN
```

Figure 13.6—random file management program

Option 1, INITIALIZE DIRECTORY, should only be used when you want to create a new job inventory file. This subroutine checks for a unique file name because the initialization procedure will destroy the contents of an existing file. If a file already exists, the user is given an opportunity to abort this option so that the contents of the existing file will not be destroyed. The directory is initialized by writing '0000' in each field of each directory record. This value denotes that the associated data record is unused. After the directory is initialized, control is returned to the menu.

The second option, NEW JOB, is used to add new jobs to the file. This option prompts you for the data for a job and then locates an unused record in the data file. The job information is written into this record and the appropriate directory field is updated to denote the job data stored in the data record.

Job information could change (job status could change from working to completed, for example) or the original information could have been entered incorrectly. Option 3, MODIFY JOB DATA, may be used to change the data in the disk file. This option prompts you for the job number associated with the data to be changed. Using this number, the directory is searched to find where the job data record is stored. This search is performed by reading one directory record at a time and comparing each field in the directory with the specified job number. If no match is found in any of the directory records, the following message is displayed:

```
NO MATCH FOUND FOR JOB ID: XXXX
```

where XXXX is the job number. When the specified job number is found in the directory, the program prompts you for the data that will replace the existing job data.

Option 4, DELETE JOB, permits you to remove a job from the file. You are prompted for the job number. When the record containing this job information is located, the information is displayed. The program then asks if you want to delete this information. In this way you get a chance to review the data before they are deleted. If you reply with a "yes," the value '0000' is written into the directory field associated with this data record. Consequently, the status of this data record is now unused.

The fifth option, LIST DATA, will list the contents of the data records. The job information for a particular job number can be displayed. If you respond with 'ALL' when prompted for the job number, the job information for all jobs in the file will be listed. The contents of the directory can be listed using option 6, LIST DIRECTORY CONTENTS. This option displays all fields in the directory one record at a time.

The last option, EXIT PROGRAM, lets you terminate the program. You should note that as each option is completed, the open file is closed. This avoids the problem of determining if the data file is open each time a new option is specified. Using this procedure, we know that the file is not open. This also causes the contents of the random buffer to be written to the data file.

The subroutine starting at line 2860 is used to open the data file and to define the field sizes. Note that two field statements are used for the same file. This was necessary because we have two types of records in the file: directory records and data records.

A modular approach was used to develop this program: each option consists of one or more subroutines. This made the program easier to write and easier to debug. Also note that the number of jobs that can be accommodated is now 120 (10 directory records × 12 job fields per record). This value can be changed by changing only one variable, NODIR, which specifies the number of directory records. It now has a value of 10.

In the last example, a keyed index method was used to maintain the location of the data records. Other methods, such as hashing, can be used. Space limitations do not permit an explanation of this technique. The interested reader might refer to reference [1]. Data base management systems (DBMS) are also available. These systems are designed to handle a wide range of sequential and random file applications. Initially DBMS were available only for minicomputers and main frame computers. They are now becoming available for microcomputers. For a general description of DBMS, refer to reference [3].

13.4 Summary

Two methods for storing data in disk files were presented in this chapter. We showed how to create and use sequential and random access disk files. A keyed index method was discussed that could be used to quickly locate records in a random access file. Although the disk input/output programs presented are not very general, these programs will be useful guides as you develop programs of your own.

References

Baase, Sara, *Computer Algorithms: Introduction to Design and Analysis*. Addison-Wesley, Massachusetts, 1978.

Bradley, James, *File and Data Base Techniques*. Holt, Reinhart and Winston, New York, 1981.

Ross, Ronald G., *Data Base Systems Design, Implementation and Management*. AMACOM, New York, 1978.

Exercises

1. Give some examples of primary storage devices. Give a few examples of secondary storage devices.
2. What is a *sector* on a floppy disk? How many sectors are on one side of an IBM Personal Computer floppy diskette?
3. What is the primary function of the DOS diskette directory?
4. What is the sequential access method? What is the random access method?
5. Describe the input/output modes that can be used with the sequential access method.
6. For random access files, define the following terms:

   a. record
   b. random buffer
   c. file

7. Data from several experimental tests along with the date of the tests are to be stored in a disk file. At this time, the following data are available:

Date	Test 1	Test 2	Test 3
2/15/83	100.4	96.7	105.2
2/18/83	151.0	159.3	165.8
3/1/83	125.9	117.4	110.6

   Write a program which will

   (a) create a sequential data file named TESTDATA on disk drive B, and
   (b) store the above test data in this file.

8. Write a program which will add the following data to the file created in Exercise 7.

Date	Test 1	Test 2	Test 3
3/15/83	172.4	185.9	190.6
3/22/83	200.6	205.0	203.7

9. Write a program which will

   a. create a random access file named OUTCOMES on disk drive B, and
   b. store the data from Exercise 7 in this file.

10. Write a program which will display on your printer the contents of the file created in Exercise 7.
11. Write a program which will display on your printer the contents of the file created in Exercise 9.

14

Data Structures

Small amounts of data can be stored on top of a desk; however, as the amount of information increases, a file cabinet will be required. Decisions must then be made regarding how information will be stored. If the data are well organized, a given item can be located in less time and less storage space is required.

Working with data within a computer is similar. When working with a small amount of data, we usually are not concerned with how the data are stored (structured). However, when substantial amounts of data are involved, we learn that the data structure can dramatically affect the program execution time, amount of storage required, ease of use, and ease of programming.

14.1 Arrays

Several techniques have been developed to aid in structuring data. Some of the simplest to implement will be discussed in this chapter. One such structure is the *array*. An array is a group of data elements referenced by the same name, A. Each element is referenced by specifying its associated subscript; thus A(3) specifies the third element in the array A. The array A(I) is called a one-dimensional array. Likewise, A(I,J) is a two-dimensional array. In your Personal Computer, an array can have 255 dimensions; however, arrays having more than three dimensions are seldom used.

Use of an array can greatly simplify a programming task. When initializing the value of a series of variables, using an array reduces the number of statements required. For example, consider the problem of initializing 20 variables to the value of 100, using an array:

```
10 DIM A(20)
20 FOR I = 1 TO 20
30 A(I) = 100
40 NEXT I
```

Without an array, 20 assignment statements would have been required to perform this task:

A1 = 100
A2 = 100

-
-
-

A20 = 100

An array is the simplest of data structures, and it is the structure most often used. In fact, arrays are used in most of the programs in this book.

14.2 Stacks

The second data structure we will consider is called a *stack*. The operation of a stack can be visualized by considering a springloaded cafeteria tray holder. When trays are placed in a tray holder, the trays descend until only the top tray is visible. As a tray is removed from the top, the remaining trays move up so that the top tray is visible. This is also an example of last-in, first-out (LIFO) storage. A stack data structure works in a similar manner. All data additions and deletions to the stack occur at one end. Adding data to the stack is often called "pushing" the stack; removing data is called "popping" the stack.

In a BASIC program, an array is often used to implement a stack. Data elements are stored in the array. Moving the data within an array each time an element is added or deleted (similar to the cafeteria tray holder) can be avoided by using a variable (sometimes called a stack pointer) to denote the "top of stack." Consider the example in Figure 14.1 where three data elements are added to an empty stack and then two elements are removed.

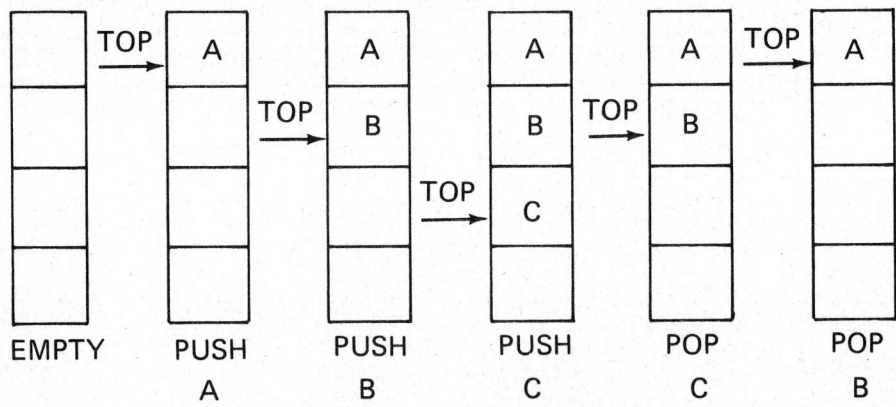

Figure 14.1—stack structure

A stack data structure is used by the central processing unit of most computers to store information temporarily as instructions are executed. This structure is widely used because it requires very few program statements to implement. Within a program, two of the more common applications of a stack data structure are the maintenance of unused space in an array and unused records in a disk file. In a following section of this chapter, there is a program which uses a stack to maintain a list of unused space in an array. Figure 14.2 contains three subroutines which will assist you in using a stack data structure. Note the small number of BASIC statements required.

14.2 Stacks

```
24500 REM STACK SUBROUTINES
24510 REM These subroutines will initialize a stack and
24520 REM perform data additions and deletions to
24530 REM array SK. TOP is the stack pointer. SIZ is
24540 REM the maximum size of the stack. Data to be
24550 REM placed in the stack resides in IDAT.
24560 REM ****************************************************************
24570 REM SUBROUTINE: STACK INITIALIZE
24580 DIM SK(SIZ%)                    'Define array size
24590 TOP%=-1                         'Denote empty stack
24600 RETURN
24610 REM ****************************************************************
24620 REM SUBROUTINE: STACK PUSH
24630 IFL=0                           'Set no error condition
24640 IF(TOP%>=SIZ%) THEN 24680       'Stack full?
24650 TOP%=TOP%+1                     'Push data
24660 SK(TOP%)=IDAT
24670 RETURN
24680 PRINT:PRINT "STACK OVERFLOW"
24690 IFL=1:RETURN                    'Set error condition
24700 REM ****************************************************************
24710 REM SUBROUTINE: STACK POP
24720 IFL=0                           'Set no error condition
24730 IF(TOP%<=-1) THEN 24770         'Stack empty?
24740 IDAT=SK(TOP%)                   'Pop data
24750 TOP%=TOP%-1
24760 RETURN
24770 PRINT: "STACK UNDERFLOW"
24780 IFL=1:RETURN                    'Set error condition
```

Figure 14.2—stack subroutines

In these subroutines the stack is maintained in the array SK and the maximum size (SIZ) of this array is supplied by the main program. In the subroutine STACK INITIALIZATION, all the elements in the array SK are equated to 0 by the statement:

DIM SK(SIZ)

In BASIC the minimum value for an array subscript is (0); therefore, to show that the stack is empty, the stack pointer TOP is initially set equal to (−1). Subroutine STACK POP will remove a data element from the stack and place the value in IDAT. However, before an element is removed, the stack pointer is evaluated to see if the stack is empty. If it is empty, an error indicator, IFL, is set equal to (1) and a return from subroutine occurs. Otherwise, IFL remains equal to (0), indicating no error, and an element is removed.

Subroutine STACK PUSH will add data to the stack. Before this occurs, the stack pointer is checked to determine if adding one more data element will exceed the maximum size of the array. If so, IFL is set equal to (1) and a return from subroutine occurs.

EXAMPLE 1. Write a program that will use the subroutines in Figure 14.2 to store the three data elements A=10, B=5, and C=25 in a stack. At this point, your program should print out the contents of the stack; then elements C and B are to be removed, printing each element as it is removed. Next, print the final contents of the stack. Assume that all data stored in the stack will be integers.

SOLUTION. Your response to Example 1 might look like the program in Figure 14.3. There are many ways to write this program. The maximum size (SIZ) of the stack array (SK) was set as 6. Also, all variables beginning with I were defined to be integers.

```
100 REM STACK EXAMPLE PROGRAM
110 DEFINT I                        'Define integer type variable
120 SIZ% = 6                        'Set array size
130 GOSUB 24560                     'Initialize stack
140 A=10
150 B=5
160 C=25
170 IDAT=A
180 GOSUB 24620                     'Push A on stack
190 IF IFL<>0 THEN 410
200 IDAT=B
210 GOSUB 24620                     'Push B on stack
220 IF IFL<>0 THEN 410
230 IDAT=C
240 GOSUB 24620                     'Push C on stack
250 IF IFL<>0 THEN 410
260 FOR I=0 TO TOP%
270 PRINT "STACK", I, SK(I)         'Print stack contents
280 NEXT I
290 GOSUB 24710                     'Pop C from stack
300 IF IFL<>0 THEN 410
310 C=IDAT
320 PRINT "C", C
330 GOSUB 24710                     'Pop B from stack
340 IF IFL<>0 THEN 410
350 B=IDAT
360 PRINT "B", B
370 FOR I=0 TO TOP%
380 PRINT "STACK", I, SK(I)         'Print stack contents
390 NEXT I
400 STOP
410 PRINT:PRINT "ERROR"
420 END
```

Figure 14.3—using stack subroutines

14.3 Queues

A data structure that processes data in a first-in, first-out (FIFO) order is called a queue. Additions to data are made at one end of the structure and all deletions are made at the other end. You might relate this structure to a waiting line of people (a queue in front of a bank teller). New arrivals join the *rear* of the line, and as people are serviced, they leave the *front* of the line. We can use an array to implement a queue structure. In addition, two pointers are needed; one to denote the front of the queue (F) and one to denote the rear (R). Figure 14.4 illustrates data elements being added and removed from a queue.

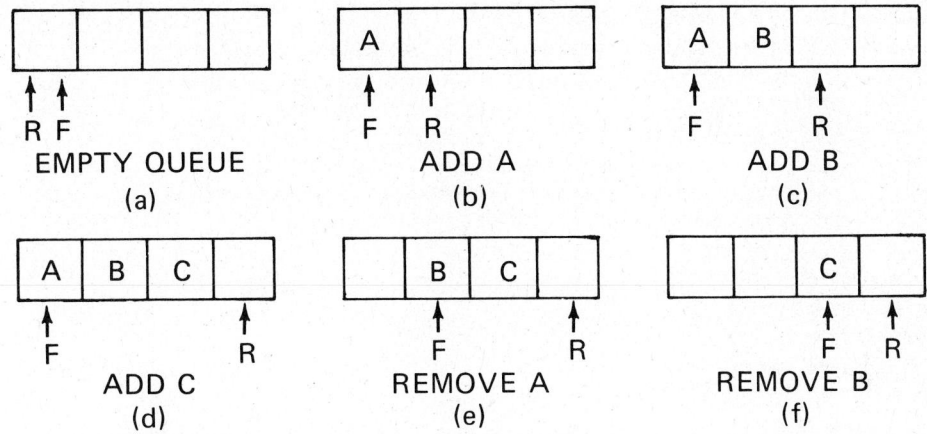

Figure 14.4—queue structure

Note that the ends of the data structure move from left to right across the array. When the end of the array is reached, the next data element to be added to the structure will be placed in the first cell of the array, if that location is empty. Figure 14.5 illustrates this process.

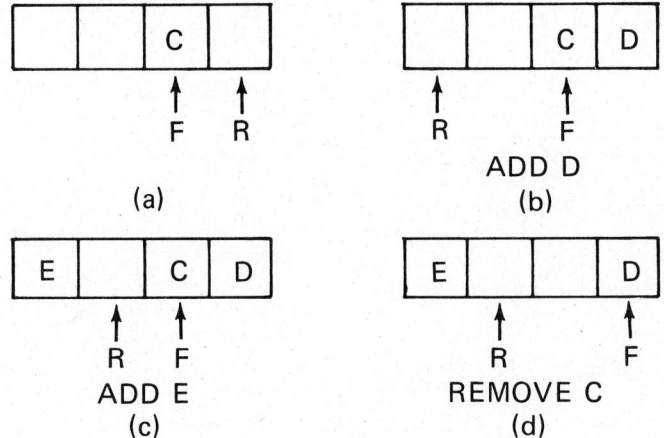

Figure 14.5—additional queue structure illustrations

Before writing the programs to implement a queue structure, we need to decide how to determine when the array holding the data is empty or full. Looking at Figure 14.4, you can see that an empty queue occurred when the rear pointer equaled the front pointer (R=F). If we adopt this convention, we could also say that the queue is full when R=F−1, such as in Figure 14.5c. These conventions mean that only (n−1) data elements can be stored in an array with n cells; however, maintaining the simplicity is worth losing one memory location.

Figure 14.6 contains three subroutines that will assist you in using a queue data structure. Like the stack related routines, these programs will initialize a queue structure and perform additions and deletions to a queue.

```
24800 REM QUEUE SUBROUTINES
24810 REM These subroutines will initialize a queue
24820 REM and perform data additions and deletions
24830 REM to a queue. The queue is maintained in
24840 REM the array Q. FQ% is the front of queue
24850 REM pointer, and RQ% is the rear of queue pointer.
24860 REM ****************************************************************
24870 REM SUBROUTINE: QUEUE INITIALIZE
24880 DIM Q(QSIZ%)
24890 FQ%=0                              'Denote empty queue
24900 FQ%=RQ%
24910 RETURN
24920 REM ****************************************************************
24930 REM SUBROUTINE: QUEUE ADD
24940 IFL=0                              'Set no error condition
24950 IF RQ%=(FQ%−1) THEN 25010          'Is queue full?
24960 IF (FQ%=0) AND (RQ%=QSIZ%) THEN 25010
24970 Q(RQ%)=IDAT                        'Add to queue
24980 RQ%=RQ%+1
24990 IF RQ%>QSIZ% THEN RQ%=0
25000 RETURN
25010 PRINT:PRINT "QUEUE FULL"
25020 IFL=1:RETURN                       'Set error condition
25030 REM ****************************************************************
25040 REM SUBROUTINE: QUEUE DELETE
25050 IFL=0                              'Set no error condition
25060 IF RQ%=FQ% THEN 25110               'Is queue empty
25070 IDAT=Q(FQ%)
25080 FQ%=FQ% +1
25090 IF FQ%>QSIZ% THEN FQ%=0
25100 RETURN
25110 PRINT:PRINT "QUEUE EMPTY"
25120 IFL=1:RETURN
```

Figure 14.6—queue subroutines

EXAMPLE 2. Write a program that will use the subroutines in Figure 14.6 to store the three data elements A=10, B=5, and C=25 in a queue data structure. At this point, print out the contents of the queue. Then remove elements A and B. Next, print the contents of the queue. Assume that all data stored in the queue will be integers.

14.3 Queues

SOLUTION. Figure 14.7 contains a program that is one possible solution. This program first defines all variables starting with I to be integer type. The queue data array, Q, is dimensioned to six by setting QSIZ=6. In this program, data to be input to the queue are read into the variable IDAT using a subroutine residing at line 230 (called INPUT TO QUEUE). Each time a data element is input, this subroutine is used. Following the read statement in this subroutine, subroutine QUEUE ADD from Figure 14.6 is referenced. This latter subroutine places the data into the queue data array.

A subroutine has also been written to print the queue (called QUEUE PRINT) contents; this subroutine starts at line 350. A third subroutine (called TAKE FROM QUEUE) has been written for this example. It starts at line 300 and is referenced every time a data element is removed from the queue. This subroutine references subroutine QUEUE DELETE from Figure 14.6. A data element removed from a queue is placed in the variable IDAT.

```
100 REM QUEUE EXAMPLE PROGRAM
110 DEFINT I                    'Define integer type variables
120 QSIZ%=6                     'Set array size
130 GOSUB 24870                 'Initialize queue
140 GOSUB 230                   'Place data in queue
150 GOSUB 350                   'Print queue contents
170 GOSUB 300                   'Take data from queue
180 GOSUB 300
190 GOSUB 350                   'Print queue contents
200 DATA 10,5,25,-9999
210 END
220 REM *************************************************************
230 REM SUBROUTINE: INPUT TO QUEUE
240 READ IDAT
250 IF IDAT=-9999 THEN RETURN
260 GOSUB 24930
270 IF IFL<>0 THEN PRINT "ERROR":END
280 GOTO 240
290 REM *************************************************************
300 REM SUBROUTINE: TAKE FROM QUEUE
310 GOSUB 25040
320 IF IFL<>0 THEN PRINT "ERROR":END
330 RETURN
340 REM *************************************************************
350 REM SUBROUTINE: QUEUE PRINT
370 PRINT "QUEUE CONTENTS-FIFO ORDER"
380 PRINT "ARRAY LOCATION","VALUE"
390 IF FQ%=RQ% THEN 460
400 I=FQ%
410 PRINT I,Q(I)
420 I=I+1
430 IF I>QSIZ% THEN I=0
440 IF I=RQ% THEN PRINT :RETURN
450 GOTO 410
460 PRINT "QUEUE EMPTY":PRINT:RETURN
```

Figure 14.7—using queue subroutines

14.4 Linked Lists

A *list* is an ordered collection of data elements. In a *linked list*, each element has an accompanying *pointer*, which specifies the next element in the list. As a result, although the data is not physically stored in a specified order, it can be logically processed in the proper order using the pointers.

Figure 14.8 contains an example of a linked list ordered alphabetically. Using the pointers contained in the box labeled FRONT and the array LINK, the data elements in the array AY can be processed in alphabetical order.

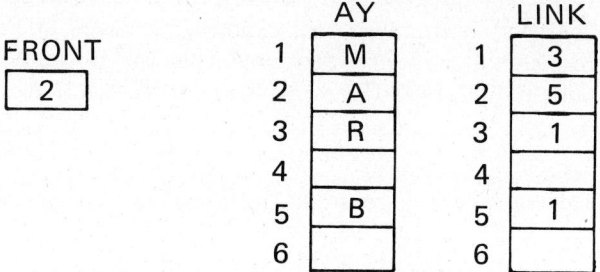

Figure 14.8—linked list structure

FRONT denotes the location of the first data element in the list (located in cell 2 of the array AY). The number (5) in cell 2 of the array LINK (adjacent to position 2 of the array AY) points to the location of the next data element in alphabetical order. This process continues until the pointer value is a (−1), which indicates "end of list." In Figure 14.9, a new data element has been added to the list.

```
           AY         LINK
FRONT   1 [ M ]    1 [ 4 ]
[ 2 ]   2 [ A ]    2 [ 5 ]
        3 [ N ]    3 [ -1 ]
        4 [ R ]    4 [ 3 ]
        5 [ B ]    5 [ 1 ]
        6 [   ]    6 [   ]
```

Figure 14.9—additional linked list illustrations

Most sets of data are dynamic in that over time data are added and deleted. Maintaining data in a specific order, such as alphabetical, can involve moving the data elements as deletions and additions occur. However, if a linked list data structure is used, we need not move any of the data elements when the contents of the set of data changed. Instead, the linked list pointers are modified to reflect any changes.

It is also advantageous to use a linked list when a limited amount of storage space is available for data lists containing varying numbers of elements. An example would be a block of storage space allocated to storing the names of employees in each department of some firm. Each employee in a particular department must

14.4 Linked Lists

be associated with the right department. However, the number of employees in a department may vary from small to large depending on the size of the department. Another consideration is the dynamic nature of the data; new employees must be added and terminated employees must be removed. If we reserve a block of space in an array for each department, each space must be large enough to handle the largest department. This will result in unused space for the small departments. If we use a linked list data structure for this type of problem, we can efficiently utilize our storage space while keeping our program execution times small.

A linked list data structure permits more than one list to be stored in a common block of storage space. Figure 14.10 illustrates an array that holds data for two linked lists. Note the FRONT pointer associated with each list in the array. A stack data structure has been utilized to provide a list of unused cells in the data array in this figure.

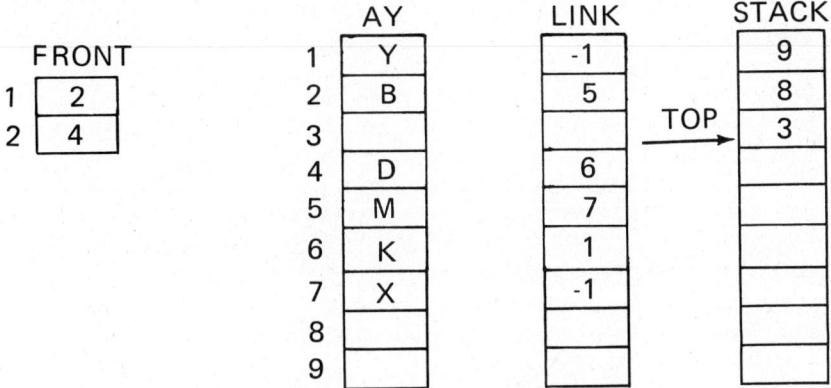

Figure 14.10—multiple linked lists

A stack is a convenient method of keeping track of unused space in a block of storage space (memory or disk). As an element is deleted from the block of storage, the unused space is added to the top of the stack. Another data element is added to the data array in the location specified by the number at the top of the stack. For example, if another data element were added to the data array in Figure 14.10, it would be placed in row 3, and the top of stack would then point to the number 8 in the stack.

Linked Queues

We have been discussing linked lists, saying that each list is an ordered collection of data elements. How the lists are ordered (LIFO, FIFO, alphabetical, etc.) depends upon the particular application. Also associated with each element in a linked list might be several other data fields. An example might be an alphabetical list of employees in a department. With each name, we may want to maintain the social security number, address, and age.

Figure 14.11 contains programs that can be used to implement linked lists when each list is a queue. The data elements are maintained in the array LQDAT; each element in an individual queue is linked using the array, LINK. The front and rear of each queue are maintained in the arrays FRNT and REAR.

Because more than one queue may be stored in the array LQDAT, the convention described in Section 14.4 cannot be used to denote an empty or full queue (R=F and R=F−1). Instead, the programs in Figure 14.11 use the stack routines in Figure 14.2 to maintain a list of available space in the array LQDAT. Initially the stack will contain all cells in the data array LQDAT because the data array is empty. The queue initialization subroutine places all of the cell numbers available in LQDAT in the stack. No more data can be added to a queue if the stack indicates that no space is available in the array LQDAT. We will know that queue I is empty when the variable FRNT(I) = −1.

```
25200 REM SUBROUTINES FOR LINKED LISTS (QUEUES)
25210 REM These routines will initialize, perform additions and
25220 REM deletions to linked lists of queues. The data reside in
25230 REM an array called LQDAT%; these data are linked into FIFO
25240 REM queues using the array LINK%. FRNT%(I) points to front
25250 REM of queue I; REAR% points to end of queue I. Location in
25260 REM array LQDAT% of data to be added or removed is indicated
25270 REM by PT%. A list of available space in the data array is
25280 REM maintained in stack array, SK.
25290 REM ****************************************************************
25300 REM SUBROUTINE: INITIALIZE LINKED QUEUES AND STACK
25310 DIM LQDAT%(SIZ%), LINK%(SIZ%)
25320 DIM FRNT%(NQ%), REAR%(NQ%)
25330 REM Must set front and rear pointers to −1 to
25340 REM indicate empty queues.
25350 FOR I=1 TO NQ%
25360 FRNT%(I)=−1
25370 REAR%(I)=−1
25380 NEXT I
25390 REM Place all available data locations in stack.
25400 REM Want to use lower numbered cells first (LIFO)
25410 FOR I=SIZ% TO 0 STEP −1
25420 IDAT=I
25430 GOSUB 24620            'Place data element in stack
25440 NEXT I
25450 RETURN
25460 REM SUBROUTINE: ADD DATA TO A LINKED QUEUE
25470 IFL=0                  'Set no error condition
25480 GOSUB 24710            'Obtain available data cell location
25490 IF IFL<>0 THEN PRINT "ERROR STACK ROUTINE":END
25500 PT%=IDAT               'Location obtained from stack
25510 IF(PT%<0)OR(PT%>SIZ%)THEN 25610    'Does location exist?
25520 REM If queue not empty, link to rear of list
25530 IF REAR%(IQ)<>−1 THEN LINK%(REAR%(IQ))=PT%
25540 LINK%(PT%)=−1          'Denote end of list
25550 REAR%(IQ)=PT%
25560 REM If queue was empty, must set front pointers
25570 IF FRNT%(IQ)=−1 THEN FRNT%(IQ)=PT%
25580 LQDAT%(PT%)=LDAT%
```

14.4 Linked Lists

```
25590 RETURN
25600 IFL=1                      'Set error condition
25610 PRINT:PRINT "INVALID LOCATION IN DATA ARRAY":RETURN
25620 REM ****************************************************
25630 REM SUBROUTINE: DELETE DATA FROM A LINKED QUEUE
25640 IFL=0                      'Set no error condition
25650 IF FRNT%(IQ)=-1 THEN 25760           'Is queue empty?
25660 REM Reset front pointer and retrieve data
25670 PT%=FRNT%(IQ)
25680 FRNT%(IQ)=LINK%(PT%)
25700 LDAT%=LQDAT$(PT%)
25710 REM If queue is empty, set rear pointer
25720 IF FRNT%(IQ)=-1 THEN REAR%(IQ)=-1
25730 IDAT=PT%                   'Available location
25740 GOSUB 24620                'Put new available location on stack
25750 RETURN
25760 IFL=1                      'Set error condition
25770 PRINT:PRINT "QUEUE IS EMPTY;CAN'T DELETE ELEMENT":RETURN
```

Figure 14.11—linked list subroutines

EXAMPLE 3. Write a program which will put the following data in a linked list of queues:

Queue Number	Data
1	50
1	10
2	75
3	113

After these data have been placed in the respective queues, your program should print out the contents of each queue; delete one data element from queue 1 and queue 3; and then print out the queue contents. Assume that no more than three queues are required and that the maximum size of the data array is 6. Also assume that all data are of integer type.

SOLUTION. Figure 14.12 contains one possible solution to Example 3. You should note that the data needed by the example program (queue number, data values) are supplied in DATA statements at the end of the main program. All variables starting with the letter I are defined to be integers. The maximum number of queues is set to 3, and the size of the data array is set to 6.

Data are read into the program using subroutine INPUT. The data to be input to a particular queue must reside in a DATA statement in the following format: queue number, data value. This subroutine terminates reading data when it encounters the value −9999 for the queue number. With subroutine REMOVE we can specify from what queue a data element is to be removed. The queue number is read from a DATA statement for each element to be removed. When the number −9999 is encountered in the DATA statement, this subroutine terminates

Chapter 14—Data Structures

reading data and no more data are removed from any of the queues. A subroutine has also been written to print the contents of each queue.

```
100 REM MAIN PROGRAM ILLUSTRATING USING LINKED QUEUES
110 DEFINT I
120 SIZ%=6                    'Maximum size of arrays
130 NQ%=3                     'Maximum number of queues
140 REM Initialize stack to 0
150 GOSUB 24570
160 REM Initialize linked queues to 0 and place available
165 REM space in stack
170 GOSUB 25300
180 REM Add data elements to linked queues
190 GOSUB 290
200 REM Print queue contents
210 GOSUB 440
220 REM Delete data elements from linked queues
230 GOSUB 370
240 REM Print queue contents
250 GOSUB 440
260 DATA 1,50,2,75,1,10,3,113,-9999
270 DATA 1,3,-9999
280 END
285 REM ****************************************************************
290 REM SUBROUTINE: INPUT
300 READ IQ
310 IF IQ=-9999 THEN RETURN
320 IF (IQ<1) OR (IQ>NQ%) THEN PRINT "INVALID QUEUE":END
330 READ LDAT%
340 GOSUB 25460                'Add to linked queue
350 IF IFL<>0 THEN PRINT "ERROR IN ADDING TO QUEUE":END
360 GOTO 300
365 REM ****************************************************************
370 REM SUBROUTINE: REMOVE
380 READ IQ
390 IF IQ=-9999 THEN RETURN
400 IF (IQ<1) OR (IQ>NQ%) THEN PRINT "INVALID QUEUE":END
410 GOSUB 25630                'Delete from linked queue
420 IF IFL<>0 THEN PRINT "ERROR IN DELETING FROM QUEUE:END
430 GOTO 380
435 REM ****************************************************************
440 REM SUBROUTINE: PRINT LINKED QUEUE CONTENTS
450 PRINT
460 FOR I=1 TO NQ%
470 PRINT "CONTENTS QUEUE NO.",I
480 IF FRNT%(I)=-1 THEN PRINT "QUEUE EMPTY":GOTO 540
490 LLOC%=FRNT%(I)
500 PRINT LQDAT%(LLOC%)
510 LLOC%=LINK%(LLOC%)         'Location of next element in queue
520 IF LLOC%=-1 THEN 540
530 GOTO 500
```

```
540 NEXT I
550 PRINT:PRINT
560 RETURN
```

Figure 14.12—linked list subroutines

Bidirectional Linked Lists

The linked list structure discussed in the previous section works well with LIFO and FIFO ordered lists. Additions and deletions to these types of lists occur only at the ends. However, many other types of ordered lists may involve additions and deletions somewhere between the ends of the lists. For example, deleting a name from inside an alphabetical list requires changing the pointer that linked this name in the list to the next (successor) name in the list. The successor name is easily found because the pointer to this name is associated with the name being deleted. However, the preceding (predecessor) name must be located. Locating this name requires searching through the list.

Bidirectional lists make random deletions and insertions to linked lists easier. Associated with each data element in a list are two sets of pointers. One set points forward to each *successor* in the list; another set points backward to each *predecessor* in the list. Figure 14.13 illustrates a bidirectional list in which employee names are maintained in alphabetical order by department.

	FRONT		NAME	PREDECESSOR	SUCCESSOR
1	7	1	HALL	5	4
2	3	2	ZINK	6	-1
		3	ABLE	-1	9
		4	WINK	1	-1
		5	BRADY	7	1
	REAR	6	WOLFE	8	2
1	4	7	BAKER	-1	5
2	2	8	SMITH	9	6
		9	JONES	3	8

Figure 14.13—bidirectional lists

Using Figure 14.13, we can obtain the following lists of employees ordered alphabetically:

Department 1
Baker
Brady
Hall
Wink

Department 2
Able
Jones
Smith
Wolfe
Zink

With the successor and predecessor pointers, this type of list can be processed in the forward or backward direction. Deleting the name Smith, in Department 2, is easy once the name is located:

	Predecessor	Successor
Smith	9	6

From this information, we know that the predecessor to Smith is located in row 9 and the successor is located in row 6. Therefore, the successor pointer for the name in row 9 (Jones) must be changed from 8 to 6 (pointing to Wolfe). Now the predecessor for the name in row 6 (Wolfe) must be changed from 8 to 9 (pointing to Jones). The result of these changes is shown in Figure 14.14.

FRONT
1	7
2	3

REAR
1	4
2	2

	NAME	PREDECESSOR	SUCCESSOR
1	HALL	5	4
2	ZINK	6	1
3	ABLE	-1	9
4	WINK	1	-1
5	BRADY	7	1
6	WOLFE	9	2
7	BAKER	-1	5
8			
9	JONES	3	6

Figure 14.14—additional bidirectional list illustrations

Assuming that the predecessor and successor pointers are maintained in the respective arrays PRD and SUC, the required changes in a program would be:

SUC%(PRD%(8)) = SUC%(8),   giving SUC%(9) = 6; and

PRD%(SUC%(8)) = PRD%(8),   giving PRD%(6) = 9.

Consequently, you can see that deleting a name required two statements to make the needed pointer changes. This is much less time consuming than searching a list to find the immediate predecessor to the name deleted, then changing the predecessor pointer.

Figure 14.15 contains programs that utilize a bidirectional list structure. It is assumed that each data element is a string. Previous programs had assumed that all data were integer type. However, the programs in Figure 14.11 can be easily modified to work with other data types (integer and real) by changing the variable declaration from QDAT$ and DAT$ to the desired data type, such as DAT.

These routines initialize the bidirectional list and place all available storage locations on a stack. Additions and deletions can be made to the bidirectional lists. Strings are maintained in alphabetical order. However, deletions from a specified list can be made in three different ways depending upon the value of the control parameter ITY:

ITY=1, delete the first string.

14.4 Linked Lists

ITY=2, delete the matching string.

ITY=3, delete the last string.

The value for ITY and the matching string are supplied by the main program. In addition, the number of lists (NQ) and the maximum size (SIZ) of the arrays are supplied by the main program.

```
25800 REM SUBROUTINES FOR BIDIRECTIONAL LINKED LISTS
25810 REM These subroutines will initialize, perform additions
25820 REM and deletions to bidirectional linked lists. The lists
25830 REM reside in an array called QDAT$; forward pointers
25840 REM reside in the array PRD%; backward pointers reside
25850 REM in the array SUC%. FRNT%(I) points to front of list
25860 REM I; REAR%(I) points to rear of list I. Location in array
25870 REM QDAT$ where data is to be added or deleted is indicated
25880 REM by PT%. Variable DAT$ contains string to be added or
25890 REM string that was removed. Stack array SK contains a
25900 REM list of available space in the data array.
25910 REM ***************************************************************
25920 REM SUBROUTINE: INITIALIZE BIDIRECTIONAL LISTS
25930 DIM QDAT$(SIZ%),PRD%(SIZ%),SUC%(SIZ%)
25940 DIM FRNT%(NQ%),REAR%(NQ%)
25950 FOR I=1 TO NQ%
25960 FRNT%(I)=-1              'Denotes that list is empty
25970 REAR%(I)=-1
25980 NEXT I
25990 FOR I=SIZ% TO 0 STEP -1   'Load stack with available locations
26000 IDAT=I
26010 GOSUB 24620               'Push available location on stack
26020 NEXT I
26030 RETURN
26040 REM ***************************************************************
26050 REM SUBROUTINE: ADD TO A BIDIRECTIONAL LIST
26060 IFL=0                     'Set no error condition
26070 GOSUB 24710               'Obtain available data cell
26080 IF IFL<>0 THEN PRINT "ERROR STACK ROUTINE":END
26090 PT%=IDAT                  'PT% is available data cell
26100 IF(PT%<0)OR(PT%>SIZ%)THEN 26180     'Does location exist?
26110 IF  (FRNT%(IQ)=-1)THEN 26190        'List is empty
26120 SEQ%=FRNT%(IQ)
26130 IF DAT$<QDAT$(SEQ%)THEN 26240       'Insert at front
26140 IF REAR%(IQ)=SEQ% THEN 26300        'Insert at rear of list
26150 SEQ%=SUC%(SEQ%)
26160 IF DAT$<QDAT$(SEQ%)THEN 26360       'Insert in list
26170 GOTO 26140
26180 PRINT:PRINT"INVALID LOCATION FOR DATA ELEMENT":END
26190 FRNT%(IQ)=PT%             'Empty list, insertion at front
26200 REAR%(IQ)=PT%             'Empty list, insertion at front
26210 PRD%(PT%)=-1:SUC%(PT%)=-1
26220 QDAT$(PT%)=DAT$
26230 RETURN
```

```
26240 SUC%(PT%)=FRNT%(IQ)      'Insertion at front of list
26250 PRD%(PT%)=-1
26260 PRD%(FRNT%(IQ))=PT%
26270 FRNT%(IQ)=PT%
26280 QDAT$(PT%)=DAT$
26290 RETURN
26300 SUC%(SEQ%)=PT%            'Insertion at front of list
26310 SUC%(PT%)=-1
26320 PRD%(PT%)=SEQ%
26330 REAR%(IQ)=PT%
26340 QDAT$(PT%)=DAT$
26350 RETURN
26360 SUC%(PRD%(SEQ%))=PT%   'Insertion in list
26370 PRD%(PT%)=PRD%(SEQ%)
26380 PRD%(SEQ%)=PT%
26390 SUC%(PT%)=SEQ%
26400 QDAT$(PT%)=DAT$
26410 RETURN
26420 REM ****************************************************************
26430 REM SUBROUTINE: DELETE STRING FROM BIDIRECTIONAL LIST
26440 REM Will delete first, matching, or last string from list.
26450 REM      ITY=1 Delete first string
26460 REM      ITY=2 Delete matching string
26470 REM      ITY=3 Delete last string
26480 IFL=0                      'Set no error condition
26490 IF FRNT%(IQ)=-1 THEN 26860    'Is list empty?
26500 ON ITY GOTO 26510, 26710, 26630
26510 REM Delete first string from list
26520 SEQ%=FRNT%(IQ)
26530 DAT$=QDAT$(SEQ%)
26540 IF REAR%(IQ)<>SEQ% THEN 26600       'Is list empty?
26550 FRNT%(IQ)=-1:REAR%(IQ)=-1           'List is empty
26560 IDAT=SEQ%
26570 GOSUB 24620               'Put available space on stack
26580 IF IFL<>0 THEN PRINT"ERROR IN STACK ROUTINE":END
26590 RETURN
26600 FRNT%(IQ)=SUC%(SEQ%)    'List is not empty
26610 PRD%(SUC%(SEQ%))=-1
26620 GOTO 26560
26630 REM Delete last string from bidirectional list
26640 SEQ%=REAR%(IQ)
26650 DAT$=QDAT$(SEQ%)
26660 IF FRNT%(IQ)<>SEQ% THEN 26680       'Is list empty?
26670 GOTO 26550               'Empty list, push location on stack
26680 REAR%(IQ)=PRD%(SEQ%)    'List is not empty
26690 SUC%(PRD%(SEQ%))=-1
26700 GOTO 26560               'Push location on stack
26710 REM Delete matching string from bidirectional list
26720 SEQ%=FRNT%(IQ)
26730 IF DAT$=QDAT$(SEQ%) THEN 26770      'Match found
26740 IF REAR%(IQ)=SEQ% THEN 26830        'No match found
```

```
26750 SEQ%=SUC%(SEQ%)
26760 GOTO 26730
26770 REM Delete string
26780 IF SEQ%=FRNT%(IQ)THEN 26510        'Go to delete first string
26790 IF SEQ%=REAR%(IQ)THEN 26630        'Go to delete last string
26800 SUC%(PRD%(SEQ%))=SUC%(SEQ%)
26810 PRD%(SUC%(SEQ%))=PRD%(SEQ%)
26820 GOTO 26560                         'Push location on stack
26830 REM No match found
26840 PRINT:PRINT"NO MATCH FOUND FOR STRING:",DAT$
26850 RETURN
26860 IFL=1                              'Set error condition
26870 PRINT:PRINT"CANNOT REMOVE AN ELEMENT-LIST IS EMPTY":END
```

Figure 14.15—bidirectional list subroutines

EXAMPLE 4. Using a bidirectional list, store the following names in the respective lists:

List 1	List 2
Baker	Able
Brady	Jones
Hall	Smith
Wink	Wolfe
	Zink

After loading the lists, print the contents of each list. Then delete the first string (name) from List 1, the last string from List 2, and a string matching Wolfe from List 2. Now, print the contents of the remaining lists.

SOLUTION. Figure 14.16 contains a program that provides one possible solution to Example 4. This program has been written using subroutines (INPUT, REMOVE, etc.) like the solution to Example 3. The INPUT subroutine expects data to be in a specific format:

string, list number

The DATA statements in lines 560 and 570 conform to this format. Thus:

BAKER, 1

means that the string BAKER is to be placed in list 1. Also, the value -9999 will terminate input. This value must be in the position a string would occupy in the DATA statement.

The DATA statement in line 580 contains information that specifies what type of deletions are to occur. The following format should be used for deletion types 1 (delete first string) and 3 (delete last string):

deletion type, list number

If deletion type 2 (matching string) is specified, the following format should be used:

deletion type, list number, matching string

The value −9999 appearing in the deletion type field will terminate the deletions.

```
100 REM MAIN PROGRAM ILLUSTRATING BIDIRECTIONAL LINKED LISTS
110 REM STRINGS WILL BE STORED IN LISTS
120 DEFINT I
130 SIZ%=12                        'Maximum size arrays
140 NQ%=3                          'Maximum number of lists
150 REM Initialize stack to 0
160 GOSUB 24570
170 REM Initialize lists to blank and place available space in stack
180 GOSUB 25880
190 REM ****************************************************************
200 REM SUBROUTINE: INPUT
210 READ DAT$
220 IF DAT$="-9999" THEN 260       'End of data
230 READ IQ                        'List number
240 GOSUB 26050                    'Go to add subroutine
250 GOTO 210                       'Read next data
260 REM Print contents of lists
270 GOSUB 430                      'Go to print subroutine
280 REM ****************************************************************
290 REM SUBROUTINE: DELETE
300 READ ITY                       'Type of deletion
310 IF ITY=-9999 THEN 390          'End of data
320 IF ITY=2 THEN 360
330 READ IQ                        'List number
340 GOSUB 26430                    'Go to delete subroutine
350 GOTO 300                       'Read next data
360 READ IQ,DAT$                   'List number, string match
370 GOSUB 26430
380 GOTO 300                       'Read next data
390 REM Print contents of lists
400 GOSUB 430                      'Go to print subroutine
410 GOTO 590
420 REM ****************************************************************
430 REM SUBROUTINE: PRINT BIDIRECTIONAL LIST CONTENTS
440 PRINT
450 FOR I=1 TO NQ%
460 PRINT "CONTENTS OF LIST NO."I
470 IF FRNT%(I)=-1 THEN PRINT"LIST EMPTY":GOTO 530
480 LLOC%=FRNT%(I)
490 PRINT QDAT$(LLOC%)
500 LLOC%=SUC%(LLOC%)
510 IF LLOC%=-1 THEN 530           'End of list?
520 GOTO 490
530 NEXT I
540 PRINT:PRINT
550 RETURN
560 DATA BAKER,1,ABLE,2,JONES,2,BRADY,1,HALL,1
```

```
570 DATA SMITH,2,WINK,1,WOLFE,2,ZINK,2,-9999
580 DATA 1,1,3,2,2,2,WOLFE,-9999
590 END
```

Figure 14.16—using bidirectional list subroutines

14.5 Summary

In this chapter, we discussed several types of data structures. Familiarity with these structures can be very beneficial when you are developing your own programs. Subroutines were provided that will aid you in implementing structures such as stacks, queues, linked queues, and bidirectional linked queues. Reference [2] describes other structures not discussed here, such as trees.

Although the techniques presented in this chapter were illustrated using arrays in memory, many of the techniques can be used to structure disk files. For example, consider the bidirectional linked lists appearing in Figure 14.13. This structure can be implemented in a disk file, substantially reducing the time needed to access these records by department. Each record would contain a pointer field that holds the number of the next record in the list. Thus, linked lists and other data structure techniques can be applied to disk files. A set of programs called data base management systems are used to manage data structures in disk files. Systems of this type are widely used on minicomputers and mainframe computers. Some data base management systems are becoming available for microcomputers. Reference [1] provides a discussion of the concepts used to develop a data base management system.

References

Bradley, James, *File and Data Base Techniques*. Holt Reinhart and Winston, New York, 1981.
Standish, Thomas A.; *Data Structure Techniques*. Addison-Wesley, Reading, Mass., 1980.

Exercises

1. What is a stack structure?
2. What is a queue data structure?
3. Assume that A is a single-dimensional array [DIM A(50)]. You are to use a stack to manage the available storage space in this array. The largest array elements are to be used first [A(50), A(49), etc.]. As an element becomes available to be used to store data, that element is to be "pushed" onto the stack. Write a program that will satisfy the specifications of this exercise. You are also to use the stack subroutine provided in this chapter.
4. A series of tests are to be performed over an extended time period. Each time a test is performed, a numeric value in the range of 0 to 499.99 is recorded. Altogether, 200 tests are to be performed. However, only the most recent 25 readings will be maintained. Therefore, this data will be maintained on a FIFO (first-in, first-out) basis. After more than 25 test results are compiled, the oldest result will be discarded as the new result is saved. Write a program which will store this data in an array having no more than 25 elements. If possible, use subroutines from this chapter in developing this program.
5. The following data must be saved each time an inspection is performed on a product prior to shipping.

inspector's name
date
value 1
value 2
value 3

You are to write a program which will maintain 100 sets of inspection results in memory. All results for an inspector are to be linked together with a bidirectional linked list. There are 4 different inspectors.

6. Write a program which stores the data from exercise 5 in a random access disk file. Each record in the file should be bidirectionally linked.
7. Sometimes an NxM array must be sorted using one of the columns as a "key" for the sort procedure. For instance, a 50×3 array (50 rows × 3 columns) may contain an individual's social security number in column 1, his age in column 2, and his weight in column 3. Such an array might contain data for 50 people (50 rows). We might want to sort this data on age. Using the routines provided in this chapter can mean moving all data in a row when an age must be moved to another location in the array. A better way is to use a linked list where only the pointers in the list are changed. You are to modify the bubble sort algorithm presented in Chapter 12 so that an array can be sorted using any column as the key. You are also to use a linked list to implement your algorithm.

15

Random Numbers & Simulation

BASIC has a numeric function named RND which can generate "random" numbers between 0 and 1. Each number in this interval is equally likely to occur and successively generated values are independent. The RND function can be used for such things as games, simulation models, and statistical studies. Every time RND is executed, a number between 0 and 1 is generated. The number generated is called a *pseudo-random number* because an algorithm is used to generate it. This is not equivalent to performing an experiment such as throwing a die to generate a number between 1 and 6. However, functions like RND permit us to generate numbers that emulate random numbers.

15.1 Using the Function RND

The function RND will always generate the same sequence of random numbers unless we vary the *random number seed*. Using the BASIC statement RANDOMIZE, we can control the sequence of numbers generated. A program with the statement

```
10 RANDOMIZE
```

will result in the following question on your monitor:

```
Random Number Seed (-32768 to 32767)?
```

Your response should be a number within the indicated interval. Assume that your response was 100; the computer would use this number to generate the first random number. The algorithm used in RND utilizes the last random number generated to generate the next random number. Consequently, by specifying the seed for the first random number, we are in a way controlling all random numbers that will be generated until the seed is reset. This will also permit us to generate a given random number sequence as many times as desired. Control such as this can be very useful in validating a simulation program or other computer programs using random numbers.

Consider the following program:

```
10 FOR I = 1 TO 5
20 PRINT RND
30 NEXT I
40 END
```

After running this program, the following numbers will be displayed:

.6291626

.1948297

.6305799

.8625749

.736353

Every time you run the above program you will get this sequence of numbers. Now add a RANDOMIZE statement to the above program:

```
5 RANDOMIZE
10 FOR I=1 TO 5
20 PRINT RND
30 NEXT I
40 END
```

If you run this program and input a 100 as a response to the request for the random number seed, the following numbers will be displayed:

.1851404

.9877729

.806621

.8573399

.6208935

The same random numbers will be displayed each time this program is run with a seed of 100.

A random number between 0 and 1 will not always satisfy our needs. For example, we may be writing a program that simulates rolling a die. If a die is unbiased, throwing it several times will result in a series of uniformly distributed integers in the range of 1 to 6. Consequently, we need to be able to generate a random integer having the value of 1, 2, 3, 4, 5, or 6.

The function RND generates a random number between 0 and 1. Specifically, the random X is in the range,

$0 \leq X < 1$

Therefore, the expression

15.1 Using the Function RND

```
X = RND*6
```

will generate a number in range of

$0 \le X < 6$.

Now we must convert these numbers to integers:

```
I = INT(RND*6),
```

This expression produces an integer in the range of

$0 \le I \le 5$.

Each integer in this range has an equal likelihood of occurring. Consequently, if we add 1 to the above expression we can simulate rolling a die:

```
I = INT(RND*6) + 1
```

EXAMPLE 1. Write a program that will simulate tossing two dice. After each toss, display the value for each die and the total of the dice. Write your program so that you can control how many times the dice are tossed.

SOLUTION. In the following program, the variables D1 and D2 represent the individual die.

```
10   CLS
20   RANDOMIZE
30   D1 = INT(RND*6) + 1
40   D2 = INT(RND*6) + 1
50   TOTAL = D1 + D2
60   PRINT "DIE 1"; D1, "DIE 2"; D2
70   PRINT "TOTAL"; TOTAL
80   INPUT "ROLL AGAIN (Y/N)?"; Y$
90   IF Y$="Y" THEN PRINT:GO TO 30
100  END
```

You might run this program several times using different random number seeds to determine if the integers generated for the individual die are uniformly distributed between 1 and 6.

15.2 Developing a Random Number Generator

One of the most popular methods of generating pseudo-random numbers, the congruential method, will be described in this section. The interested reader might refer to [3] for additional material on this subject. The congruential method involves multiplying a constant, C, by another number, L, that varies in value, to obtain a product. Another number, R (called the residual), is created from the value of the product by keeping the last x digits of the product. An expression for this process is:

$$R = L*C - M*INT(L*C/M),$$

where   R = residual
              C = constant multiplier

L = varying multiplier
M = modulus (used to obtain residue)

Consider an example where we want to keep the last two digits of the product, L*C; therefore, we will set M=100. Also, let C=33 and L=13. Using the above expression,

R=13*33−100*INT(13*33/100)
 =429−100*INT(4.29)
 =429−100*4
 =429−400
 =29

The product of C times L is 429. Keeping the residual, composed of the last two digits, we get 29.

The next number in the sequence would be generated by replacing the old value of L, 13, with the new residual, 29. Using this value for L and substituting into the above expression, we get:

R=29*33−100*INT(29*33/100)
 =57

Continuing this process until we have generated ten numbers results in the following pseudo-random number series:

29,57,82,73,9,97,1,33,89,37

If these numbers are all divided by M, the resulting series is between 0 and 1. You should be careful in choosing the values for C and M. Experience has shown that M should be large, C should not be too small relative to M, and M and C should have no common factors.

Figure 15.1 contains a subroutine called ORGRND, which will generate random numbers using the congruential method. We selected M=8,388,608 (2^{23}) and C=2,893 (an odd number approximately the square root of 2^{23}). To use this subroutine, you must specify a seed in the main program. The seed must be negative and within the range −8388608 ≤ SEED ≤ 0. You may reseed the generator at any time by specifying a negative seed.

```
27000 REM ***************************************************************
27010 REM SUBROUTINE: ORGRND RANDOM NUMBER GENERATOR
27020 REM CONGRUENTIAL METHOD
27030 REM User must initially specify SEED (−8388608<seed<0)
27040 REM in MAIN program, setting SEED equal to a negative
27050 REM value will reseed generator.
27060 M=8388608!            'Modulus value
27070 C=2893                'Constant multiplier
27080 IF SEED >0 THEN 27140
27090 IF SEED=0 THEN L=C:SEED=L:GOTO 27140
27100 REM CHECK FOR VALID SEED
27110 IF (SEED< −8388608!) THEN PRINT "INVALID SEED":END
27120 SEED=(−1)*SEED
27130 L=SEED
```

```
27140 L=L*C-M*INT(L*C/M)
27150 RNO=L/M
27160 RETURN
```

Figure 15.1—ORGRND random number generator

15.3 Testing a Random Number Series

It is possible to statistically check a series of numbers to determine if they are uniformly distributed. One such test is called a *chi-square goodness of fit test*. To apply this test, we must divide the range of the numbers generated into I equal intervals. A count is then made of the random numbers that fall within each interval. If the numbers are uniformly distributed, the count for each interval should be approximately the same.

Using the function RND, we can generate random numbers in the range of 0 to 1. Figure 15.2 contains a program which will generate such a series.

```
100 REM PROGRAM RANDOM NUMBERS USING RND
110 REM Array AY contains random number series
120 REM N is the length of the series
130 RANDOMIZE
140 INPUT "ENTER NUMBER FOR LENGTH OF SERIES"; N
150 DIM AY(N)
160 PRINT:PRINT "GENERATING;N; " random numbers"
170 FOR I=1 TO N
180 RNO=RND
190 AY(I)=RNO
200 NEXT I
220 END
```

Figure 15.2—generating random numbers

A natural way to segment this range is to specify 10 intervals. Thus the first interval would be from 0 to .1. A count would be made of all the numbers that had a value within this range. The same procedure would be followed for the interval of .1 to .2, etc. We would expect the count in each of the intervals to be equal to N/10, where there are N numbers.

Having determined the count for each interval, the chi-square statistic S can be determined:

$$S = \frac{(C(1)-A)^2}{A} + \frac{(C(2)-A)^2}{A} + \ldots + \frac{(C(10)-A)^2}{A}$$

where C(1) is the count for the first interval and A=N/10. When 10 intervals are used and S≥16.92, we can be 95% confident that the series is not uniformly distributed. In that case, our generator would need some improvements. The value of 16.92 is only valid for a test involving 10 intervals. This value was obtained from a table of percentage points for the chi-square distribution; 16.92 is a chi-square value for the 95 percentile with 9 degrees of freedom. Several other statistical tests can be used to evaluate the random behavior of a series of pseudo-random numbers. Reference [3] contains a brief summary of the most common tests applied.

In Figure 15.3 you will find a subroutine that will perform a chi-square goodness of fit test on a series of numbers. The main program must supply the series in an array, AY, and the length of the series, N. The program uses 10 intervals and compares the chi-square statistic to 16.9.

```
27200 REM ****************************************************************
27210 REM SUBROUTINE: CHI-SQUARE GOODNESS OF FIT TEST
27220 REM For a uniformly distributed series of
27230 REM numbers in the range of 0<X<1.
27240 REM Array AY contains a series of numbers.
27250 REM N is the length of the series
27260 DIM C(10)
27270 CHI=16.9
27280 K=10
27290 FOR I=1 TO N
27300 INTV=0!
27310 FOR J=1 TO K
27320 IF(AY(I)>INTV)AND(AY(I)<=(INTV+.1))THEN 27360
27330 INTV=INTV+.1
27340 NEXT J
27350 PRINT:PRINT "NUMBER OUT OF RANGE":END
27360 C(J)=C(J)+1
27370 NEXT I
27380 S=0:A=N/K
27390 PRINT
27400 PRINT:PRINT TAB(7) "RESULTS OF CHI-SQUARE TEST":PRINT:
27410 PRINT "INTERVAL","ACTUAL","EXPECTED","CHI-SQUARE"
27420 PRINT "NUMBER","OBSERVATIONS","OBSERVATIONS","COMPONENT"
27430 FOR J=1 TO K
27440 S1=((C(J)-A)^2)/A
27450 PRINT J,C(J),A,S1
27460 S=S+S1
27470 NEXT J
27480 PRINT
27490 PRINT "CHI-SQUARE SUM=";S
27500 IF S>=CHI THEN 27540
27510 PRINT "We cannot reject the hypothesis that the number"
27520 PRINT "series is uniformly distributed"
27530 RETURN
27540 PRINT "We are 95% confident that the number series"
27550 PRINT "is not uniformly distributed."
27560 RETURN
```

Figure 15.3—chi-square goodness of fit subroutine

15.3 Testing a Random Number Series

EXAMPLE 2, PART A. Using the chi-square goodness of fit test subroutine in Figure 15.3, test a series of 100 numbers generated using the RND function.

PART B. Likewise, test a series of 100 numbers generated using the ORGRND subroutine in Figure 15.1.

SOLUTION, PART A. The program in Figure 15.4 generates 100 random numbers using RND and then calls the chi-square goodness of fit subroutine. A random number seed of 100 was used to get the results shown in Figure 15.5.

```
100 REM PROGRAM TO TEST RANDOM NUMBER GENERATOR
110 REM Array AY contains number series
120 REM N is the length of the series
130 DIM AY(500)
140 RANDOMIZE
150 INPUT "ENTER LENGTH OF RANDOM NUMBER SERIES";N
160 PRINT:PRINT "Generating ";N;" random numbers"
170 FOR I = 1 TO N
180 RNO=RND
190 AY(I)=RNO
200 NEXT I
210 GOSUB 27210
220 END
```

Figure 15.4—generating 100 numbers using RND

RESULTS OF CHI-SQUARE TEST

Interval Number	Actual Observations	Expected Observations	Chi-Square Components
1	11	10	.1
2	13	10	.9
3	10	10	0
4	10	10	0
5	8	10	.4
6	13	10	.9
7	7	10	.9
8	11	10	.1
9	12	10	.4
10	5	10	2.5

CHI-SQUARE SUM = 6.2

Figure 15.5—chi-square results, RND function

PART B. Figure 15.6 contains a program which performs a similar analysis using the ORGRND subroutine. Note that only three statements were changed (statements 140, 145, and 180). A seed of −100 was used to get the results shown in

Figure 15.7. From these results, we cannot reject the hypothesis that RND and ORGRND generate random numbers in the interval of 0 to 1.

```
100 REM PROGRAM TO TEST RANDOM NUMBER GENERATOR
110 REM Array AY contains number series
120 REM N is the length of the series
130 DIM AY(500)
140 INPUT "ENTER RANDOM NUMBER SEED (-8388608 TO 0)";SEED
145 IF SEED>0 THEN PRINT "INVALID":GOTO 140
150 INPUT "ENTER LENGTH OF RANDOM NUMBER SERIES";N
160 PRINT:PRINT "Generating ";N;" random numbers"
170 FOR I = 1 TO N
180 GOSUB 27000
190 AY(I)=RNO
200 NEXT I
210 GOSUB 27210
220 END
```

Figure 15.6—generating 100 numbers using ORGRND

RESULTS OF CHI-SQUARE TEST

Interval Number	Actual Observations	Expected Observations	Chi-Square Component
1	10	10	0
2	12	10	.4
3	14	10	1.6
4	8	10	.4
5	5	10	2.5
6	10	10	0
7	10	10	0
8	11	10	.1
9	9	10	.1
10	11	10	.1

CHI-SQUARE SUM = 5.2

Figure 15.7—chi-square results, ORGRND subroutine

15.4 Random Number Generators for Specific Distributions

Computers are often used to simulate real world activities such as the services of a bank or a particular inventory replenishment policy. Simulation is often used because a mathematical model of the physical process may be too complicated to solve; therefore, the physical activities might be modeled using a computer program. An example of a simple simulation model will be presented in the next section. Another reason for using simulation is to introduce uncertainty (random-

15.4 Random Number Generators for Specific Distributions

ness) into the analysis. Analytical solutions to a model may only give expected values (averages); simulation studies can provide additional insight into the variation about the expected value.

One way of introducing uncertainty into simulation studies is by means of random number generators. However, to do this we need to be able to generate random numbers from several types of distributions. Subroutines will be presented in this section that generate random numbers from some of the most commonly used distributions. A main program should initially RANDOMIZE the seed before using these subroutines because the RND function is used in each subroutine.

Each of these subroutines was tested by generating a sample of 500 numbers using a random number seed of 1000. Then for each sample, the mean and standard deviation were calculated, and a frequency histogram was plotted. These calculations and plots were made using the subroutine provided in Chapter 2 (Data Reduction). The results are included in the description of each random number generator.

Uniform Distribution

The uniform distribution has a continuous probability density function over some interval, a to b. This distribution is described by:

$$f(x) = \frac{1}{b-a}$$

where the mean is $(b+a)/2$ and the variance is $(b-a)^2/12$.

Since the pseudo-random number generator produces numbers uniformly distributed in the interval of 0 to 1, it is a simple matter to transform these numbers to the interval a to b:

$$x = a + (b-a)*RND, \quad 0 \le RND < 1$$

Figure 15.8 contains a subroutine that will generate a random number in the interval A to B. Values for A and B must be supplied by the main program. The random number generated resides in the variable RUFM. Figure 15.9 contains a frequency histogram of a sample of 500 numbers generated using this subroutine.

```
28000 REM ***************************************************************
28010 REM SUBROUTINE: UNIFORM DISTRIBUTION GENERATOR
28020 REM Main program should initially RANDOMIZE seed
28030 REM Main program must supply values for limits of
28040 REM interval, A and B
28050 REM Random number generated is RUFM
28060 RUFM=A+(B-A)*RND
28070 RETURN
```

Figure 15.8—uniform distribution generator

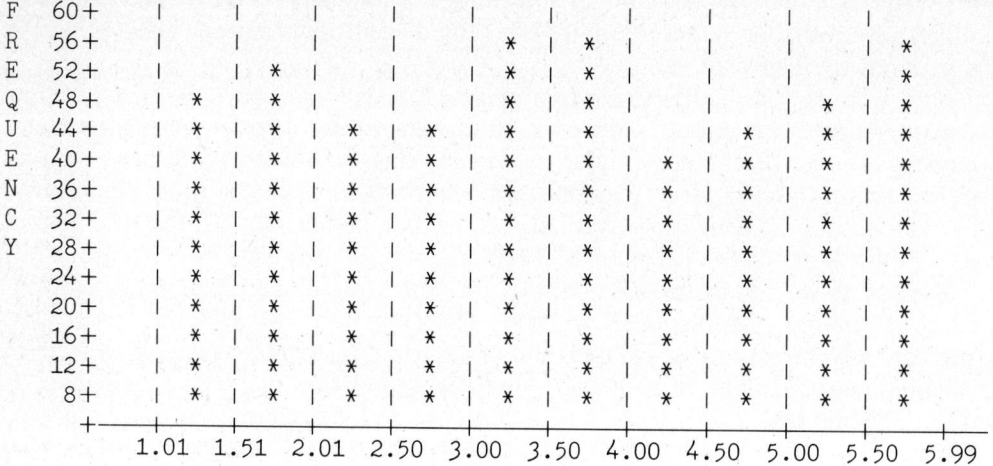

Figure 15.9—frequency histogram uniform distribution
A=1, B=6, N=500

Normal Distribution

The normal (Gaussian) distribution may be the most widely used distribution. It is a continuous distribution characterized by its mean and its variance. This distribution is symmetrical about its mean. To simplify generation of normally distributed numbers, the distribution is usually transformed to have a mean of 0 and a variance of 1. The transform used is

$$z = (x-u)/\sigma$$

where u is the mean and σ is the standard deviation. After a standard normal deviate z is generated, it is transformed to the desired normal distribution with the desired mean and variance using

$$x = u+z\sigma.$$

Using the standard normal transform, the normal distribution is described by

$$f(z) = \frac{e^{-z^2/2}}{2\pi}$$

Several methods have been devised to generate normal variates. The one used in the subroutine in Figure 15.10 was proposed by Marsaglia and Bray (1964). Values for the mean, XM, and the standard deviation, SD, must be supplied by the main program. The variable RNOR contains the random number generated. A frequency histogram of a sample of 500 numbers appears in Figure 15.11.

```
28080 REM ************************************************************
28090 REM SUBROUTINE: NORMAL DISTRIBUTION GENERATOR
28100 REM Main program should initially RANDOMIZE seed
28110 REM Main program must supply mean, XM,
```

15.4 Random Number Generators for Specific Distributions

```
28120 REM and standard deviation, SD
28130 REM Random number generated is RNOR
28140 IF NRN=1 THEN 28240
28150 R1=2*RND-1
28160 R2=2*RND-1
28170 S=R1*R1 + R2*R2
28180 IF S>=1 THEN 28150
28190 RNN1=R1*SQR((-2*LOG(S))/S)
28200 RNN2=R2*SQR((-2*LOG(S))/S)
28210 RNOR=XM+RNN1*SD
28220 NRN=NRN+1
28230 RETURN
28240 RNOR=XM+RNN2*SD
28250 NRN=0
28260 RETURN
```

Figure 5.10—normal distribution generator

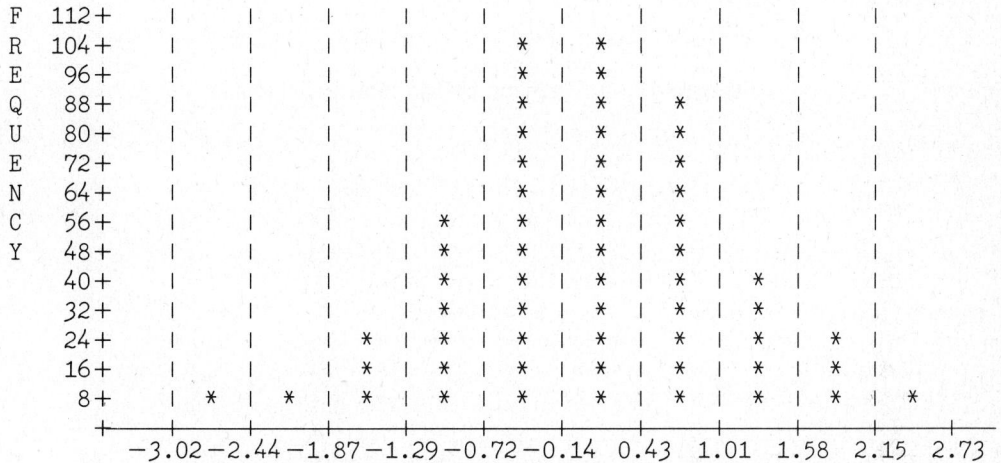

Figure 15.11—frequency histogram normal distribution
$\mu=0$, $\sigma=1$, $N=500$

Poisson Distribution

The Poisson distribution is often used in statistical and computer simulation studies. This distribution is described by:

$$f(x) = \frac{\lambda^x e^{-\lambda}}{x!}$$

where the mean is λ, the variance is λ, and x is an integer value ≥ 0.

The probability function f(x) describes the probability that an event occurs x times in some interval of time. Consequently, x must be a positive integer value.

This distribution might be used to describe the number of arrivals to a bank or the number of car accidents per hour.

The subroutine in Figure 15.12 will generate Poisson variates (for an additional explanation see reference [1]). The main program should initialize the mean of the distribution, XM. The random number generated resides in the variable RPOI. Figure 15.13 contains a frequency histogram of a sample of 500 numbers.

```
28270 REM ***************************************************************
28280 REM SUBROUTINE: POISSON DISTRIBUTION GENERATOR
28290 REM Main program should initially RANDOMIZE seed
28300 REM Main program must supply mean, XM
28310 REM Random number generated is RPOI
28320 RPOI=0
28330 E=EXP(-XM)
28340 R=1!
28350 URN=RND
28360 R=R*URN
28370 IF R<E THEN RETURN
28380 RPOI=RPOI+1
28390 GOTO 28350
```

Figure 15.12—Poisson distribution generator

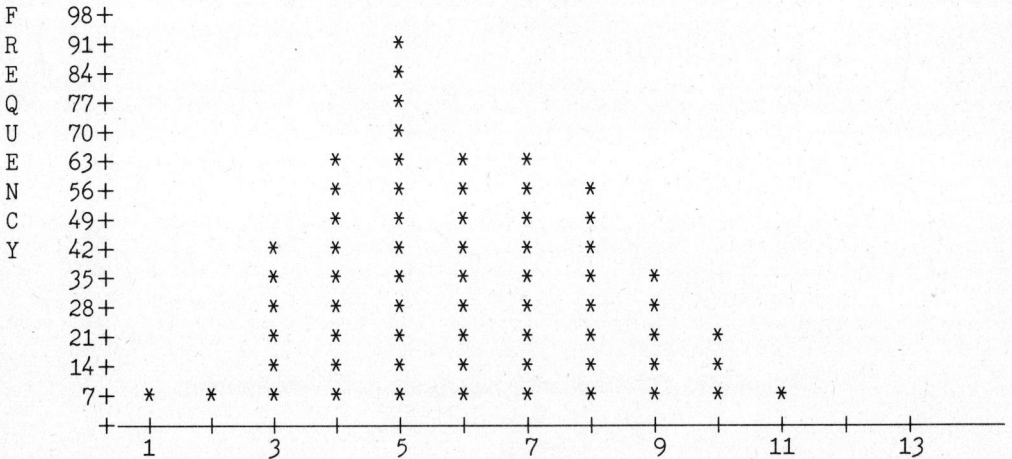

Figure 15.13—frequency histogram Poisson distribution
$\mu=6$, N=500

Exponential Distribution

The probability is large that no phone calls will occur in the next minute, if the average number of phone calls a day is five. However, the probability is small that no phone calls will occur during the day. If the phone calls are independent, the

15.4 Random Number Generators for Specific Distributions

time between phone calls can be represented by an exponential distribution. This distribution, which is continuous, is described by

$$f(x) = \lambda e^{-\lambda x}$$

where the mean is $1/\lambda$ and the variance is $1/\lambda^2$.

Computer simulation studies often use this distribution because it can represent so many types of phenomena. Some of these are: time between arrival of customers, time between occurrence of automobile wrecks, and life of electric parts. Figure 15.14 contains a subroutine that can be used to simulate an exponential distribution. The main program must specify the mean, XM. The variable REXP contains the sample generated. Figure 15.15 contains a frequency histogram of a sample of 500 numbers.

```
28400 REM ****************************************************************
28410 REM SUBROUTINE: EXPONENTIAL DISTRIBUTION GENERATOR
28420 REM Main program should initially RANDOMIZE seed
28430 REM Main program must supply mean, XM
28440 REM Random number generated is REXP
28450 URN=RND
28460 REXP=-XM*LOG(URN)
28470 RETURN
```

Figure 15.14—exponential distribution generator

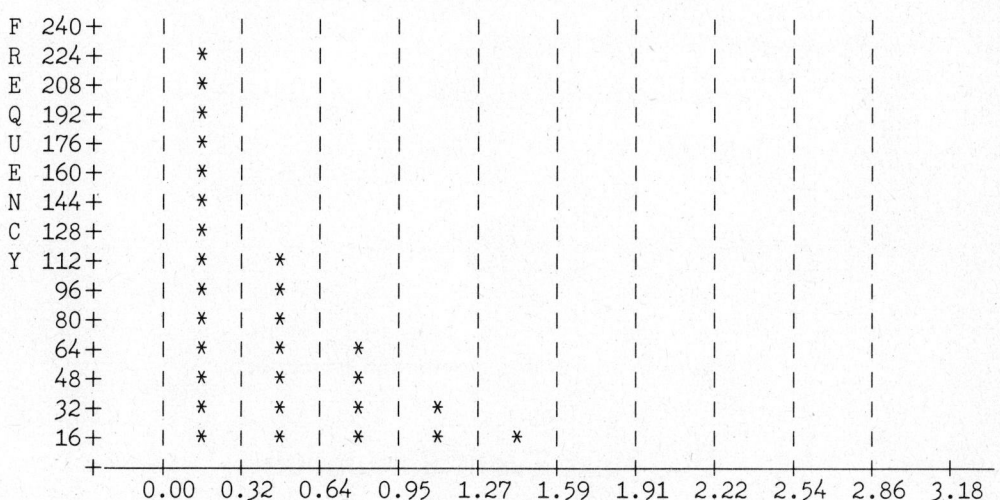

Figure 15.15—frequency histogram exponential distribution
$\mu=.5$, $\sigma^2=.25$, N=500

Erlang Distribution

If the random numbers must be positive and generally follow a unimodal distribution, then the phenomenon might be represented by an Erlang distribution. The density function is:

$$f(x) = \begin{cases} \dfrac{1}{(k-1)!} \lambda^K (\lambda x)^{K-1} e^{-\lambda x} & x > 0 \\ 0 & \text{Otherwise} \end{cases}$$

where k must be an integer, the mean is k/λ, and the variance is k/λ^2. The parameter k is shown as a shape parameter. The exponential distribution is a special case of the Erlang distribution when k=1. The Erlang distribution can be generated by summing k-independent and identically distributed exponential random variables.

Figure 15.16 contains a subroutine that will generate Erlang variates. The main program must supply a value for the shape parameter, K1, and a value for the mean of the exponential distribution, XM. The variable RERL contains the sample generated. A frequency histogram of a sample of 500 numbers appears in Figure 15.17.

```
28480 REM ****************************************************************
28490 REM SUBROUTINE: ERLANG DISTRIBUTION GENERATOR
28500 REM Main program should initially RANDOMIZE seed
28510 REM Main program must supply values for
28520 REM shape parameter, K1, and mean of
28530 REM exponential distribution, XM
28540 REM Random number generated is RERL
28550 IF K1<0 THEN PRINT "INVALID SHAPE PARAMETER":END
28560 RV=1!
28570 FOR I1=1 TO K1
28580 RV=RV*RND
28590 NEXT I1
28600 RERL=-XM*LOG(RV)
28610 RETURN
```

Figure 15.16—Erlang distribution generator

15.5 Next Event Simulation

Computer simulation is often used to analyze a real world physical system. This is done by developing a mathematical-logical model within a computer program that represents the important features of the real system. Once the model is developed, experiments can be performed varying parameters of the model to provide a better understanding of the system. Since models are usually a simplification of the real world, computer simulation is often less expensive and more convenient than manipulation of a real-world system.

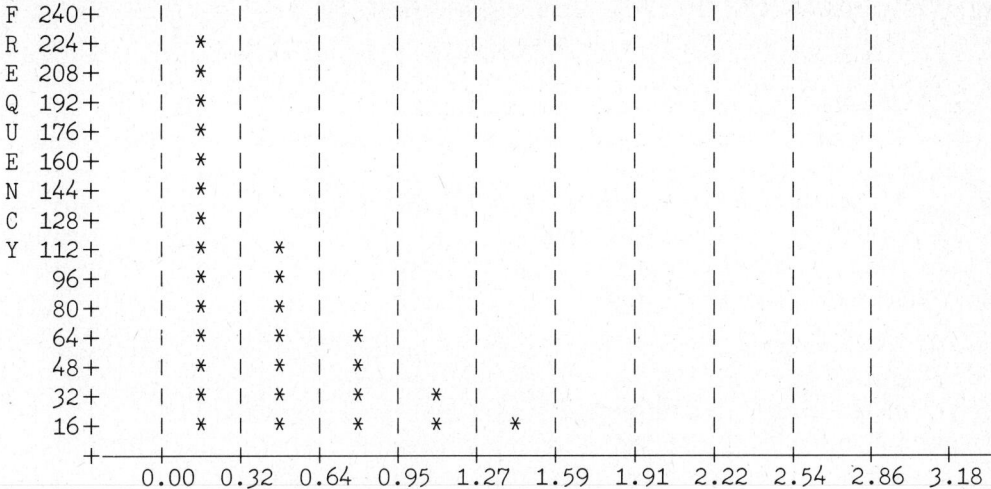

Figure 15.17—frequency histogram Erlang distribution
u=.5 σ²=.25, N=500, K=1

Simulation models can be very simple or very complex. An example of a simple simulation model is one that emulates tossing a coin or throwing a die. An example of a complex model is one that simulates the weather. This latter example also illustrates the fact that computer simulation models are simplifications of real-world phenomena. As such, they may not always provide accurate results if the underlying physical relationships are not represented correctly. However, computer simulation is recognized as one of the most powerful techniques we have for analyzing a wide variety of real-world systems. Developing simulation models is an art; consequently, we improve with experience. Some of the programs contained in this text can be useful in developing simulation models. Three of these subroutines will be used in the next example presented.

EXAMPLE 3. Develop a program which will simulate a waiting line (queue) with a single server. If the server is busy a new arrival to the system will enter the back of the queue. The queue discipline will be first-in, first-out (FIFO). Assume that the time between arrivals is exponentially distributed with a mean of 5.5 minutes. Also, assume that the service time is normally distributed with a mean of 8 minutes and a standard deviation of 1.5 minutes. The program developed should generate random samples for the arrival and service times (round event times to nearest minute). Because customers do not like to wait in long lines, assume that new arrivals balk (turn away) whenever the queue contains four customers. Therefore, the maximum queue size is four. Use next-event simulation methodology to simulate this system for 200 minutes. During this time, collect the following statistics:

a) Number of customers served.
b) Average service time.

320   Chapter 15—Random Numbers & Simulation

c) Average waiting time for customers served.
d) Number of customers who balked.

SOLUTION. Figure 15.18 contains a program that will simulate the system described. Before explaining the program, we will go through the logic used in developing the model.

In this model there are two types of events that occur: arrival of a customer and the completion of service. Consequently, the model must determine each arrival time and each service completion time. These times are determined in the following manner:

a) When an arrival occurs, determine the time interval to the next arrival. In this example the interval is a random time generated from an exponential distribution with a mean of 5.5 minutes between arrivals.
b) When service is initiated, determine the required service time. This will be a random service time generated from a normal distribution having a mean of 8 minutes and a standard deviation of 1.5 minutes.

The program in Figure 15.18 was developed using the following logic:

Initialize variables.
Schedule first arrival.
If TIME is less than limit STOPSIM,
    Then determine the next event;
        If it is an arrival,
            Then update TIME.
            If number in queue is ≤ 4,
                Then insert arrival in queue.
                Else turn arrival away.
            Determine next arrival time.
            If server is idle
                Then take arrival from queue
                and schedule service completion time
        Else it is a service completion,
            Update TIME and remove
            customer from service
            If number in queue > 0
                Then take a customer from queue
                and schedule service completion time.
    Else stop simulation.
        Print simulation results.

```
1000 REM NEXT EVENT SIMULATION OF A QUEUE WITH A SINGLE SERVER
1010 REM Time between arrivals is exponentially distributed with mean
1020 REM of EMU minutes. Service time is normally distributed with mean
1030 REM of NMU minutes and standard deviation of SD minutes.
1040 REM     ARRIVE-next arrival time
1050 REM     SERCOM-next service completion time
1060 REM
```

15.5 Next Event Simulation

```
1070 DEFINT I
1080 RANDOMIZE
1090 REM INITIALIZE VARIABLES
1100 SER%=0                          'Number served
1110 TOT%=0                          'Total service time
1120 MXQ%=4                          'Maximum queue length
1130 TWAT%=0                         'Total waiting time
1140 AWAY%=0                         'Number customers turned away
1150 TIME%=0                         'System time
1160 IDLE=0                          'Server idle
1170 NOQ%=0                          'Number in queue
1180 EMU=5.5                         'Mean arrival rate
1190 NMU=8                           'Mean service rate
1200 SD=1.5                          'Service standard deviation
1210 QSIZ%=10                        'Set queue array size for Q sub.
1220 INPUT "ENTER TIME TO TERMINATE SIMULATION";STOPSIM%
1230 GOSUB 24880                     'Initialize queue array
1240 GOSUB 1530                      'Schedule first arrival
1250 REM Initialize service completion to large number so next
1260 REM event will be an arrival
1270 SERCOMP%=STOPSIM%+1
1280 INPUT "TRACE MODEL PROGRESS ON PRINTER (Y/N)";Y$
1290 PRINT:PRINT TAB(25) "SIMULATION IN PROGRESS":PRINT
1300 REM Determine next event
1310 IF ARRIVE%>SERCOMP% THEN 1390
1320 REM Next event is an arrival
1330 TIME%=ARRIVE%
1340 IF TIME%>=STOPSIM% THEN 1450    'Simulation over
1350 GOSUB 1680
1360 IF IDLE=<>0 THEN 1300           'Determine next event
1370 GOSUB 1840
1380 GOTO 1300                       'Determine next event
1390 REM Next event is a service completion
1400 TIME%=SERCOMP%
1410 IF TIME%>=STOPSIM% THEN 1450    'Simulation over
1420 GOSUB 1980                      'Process service completion
1430 GOTO 1300                       'Determine next event
1440 REM ****************************************************************
1450 REM SIMULATION RESULTS
1460 PRINT
1470 PRINT,"NUMBER","NUMBER","AVERAGE","AVERAGE"
1480 PRINT "TIME", "SERVED", "TURNED AWAY", "SERVICE TIME", "WAIT TIME"
1490 PRINT TIME%, SER%, AWAY%, TOT/SER%, TWAT/SER%
1500 PRINT:PRINT TAB(25) "END OF SIMULATION"
1510 END
1520 REM ****************************************************************
1530 REM SUBROUTINE: GENERATE ARRIVALS
1540 XM=EMU
1550 GOSUB 28410                     'Generate a random exponential
                                      deviate
1560 ARRIVE%=TIME%+INT(REXP+.5)      'Time next arrival rounded to
                                      integer
```

```
1570 IF Y$<>"Y" GOTO 1590
1580 LPRINT "NEXT ARRIVAL TIME";ARRIVE%
1590 RETURN
1600 REM ****************************************************************
1610 REM SUBROUTINE: GENERATE SERVICE COMPLETIONS
1620 XM=NMU
1630 GOSUB 28090                     'Generate a random normal deviate
1640 WORK%=INT(RNOR+.5)              'Service time rounded to integer
1650 SERCOMP%=TIME%+WORK%            'Completion time
1660 RETURN
1670 REM ****************************************************************
1680 REM SUBROUTINE: PROCESS NEW ARRIVAL
1690 IF NOQ%=MXQ% THEN 1770          'Turn customer away
1700 IDAT=TIME%
1710 GOSUB 24940
1720 IF IFL<>0 THEN PRINT "ERROR":END
                                     'Error in queue subroutine
1730 NOQ%=NOQ%+1
1740 IF Y$<>"Y" GOTO 1760
1750 LPRINT "NO IN QUEUE AT ARRIVAL";NOQ%,"TIME";TIME%
1760 GOTO 1800
1770 AWAY%=AWAY%+1
1780 IF Y$<>"Y" GOTO 1800
1790 LPRINT "NUMBER TURNED AWAY";AWAY%,"TIME";TIME%
1800 REM Schedule new arrival
1810 GOSUB 1530                      'Arrival subroutine
1820 RETURN
1830 REM ****************************************************************
1840 REM SUBROUTINE: INITIATE SERVICE COMPLETION
1850 IDLE=1                          'Not idle
1860 SER%=SER%+1                     'Number served
1870 GOSUB 25050                     'Remove customer from queue
1880 NOQ%=NOQ%-1
1890 TWAT%=TWAT%+TIME%-IDAT          'Total waiting time
1900 GOSUB 1610                      'Obtain service interval
1910 TOT%=TOT%+WORK%                 'Total service time
1920 IF Y$<>"Y" GOTO 1960
1930 LPRINT "SERVICE COMPLETION TIME";SERCOMP%,"TIME";TIME%
1940 LPRINT "NO. IN Q "; NOQ%;", WAIT TIME ";TIME%-IDAT;
1950 LPRINT ", NUMBER SERVED";SER%
1960 RETURN
1970 REM ****************************************************************
1980 REM SUBROUTINE PROCESS SERVICE COMPLETIONS
1990 IF NOQ%<>0 THEN 2060
2000 IDLE=0                          'Server is idle
2010 REM Make sure a completion cannot occur before an arrival
2020 SERCOMP%=STOPSIM%+1
2030 IF Y$<>"Y" GOTO 2050
2040 LPRINT "QUEUE IS EMPTY:TIME", TIME%
2050 RETURN
2060 GOSUB 1840                      'Initiate service
2070 RETURN
```

15.5 Next Event Simulation

```
2080 REM ****************************************************************
24800 REM QUEUE SUBROUTINES
24810 REM These subroutines will initialize a queue
24820 REM and perform data additions and deletions
24830 REM to a queue. The queue is maintained in
24840 REM the array Q. FQ% is the front of queue
24850 REM pointer, and RQ% is the rear of queue pointer
24860 REM ****************************************************************
24870 REM SUBROUTINE: QUEUE INITIALIZE
24880 DIM Q(QSIZ%)
24890 FQ%=0                      'Denote empty queue
24900 FQ%=RQ%
24910 RETURN
24920 REM ****************************************************************
24930 REM SUBROUTINE: QUEUE ADD
24940 IFL=0                      'Set no error condition
24950 IF RQ%=(FQ%-1) THEN 25010
                                 'Is queue full?
24960 IF (FQ%=0) AND RQ%=QSIZ%) THEN 25010
24970 Q(RQ%)=IDAT
24980 RQ%=RQ%+1
24990 IF RQ%>QSIZ% THEN RQ%=0
25000 RETURN
25010 PRINT:PRINT "QUEUE FULL"
25020 IFL=1:RETURN               'Set error condition
25030 REM ****************************************************************
25040 REM SUBROUTINE: QUEUE DELETE
25050 IFL=0                      'Set no error condition
25060 IF RQ%=FQ% THEN 25110
                                 'Is queue empty
25070 IDAT=Q(FQ%)
25080 FQ%=FQ% +1
25090 IF FQ%>QSIZ% THEN FQ%=0
25100 RETURN
25110 PRINT:PRINT "QUEUE EMPTY"
25120 IFL=1:RETURN
28080 REM ****************************************************************
28090 REM SUBROUTINE: NORMAL DISTRIBUTION GENERATOR
28100 REM Main program should initially RANDOMIZE seed
28110 REM Main program must supply mean, XM,
28120 REM and standard deviation, SD
28130 REM Random number generated is RNOR
28140 IF NRN=1 THEN 28240
28150 R1=2*RND-1
28160 R2=2*RND-1
28170 S=R1*R1 + R2*R2
28180 IF S>=1 THEN 28150
28190 RNN1=R1*SQR((-2*LOG(S))/S)
28200 RNN2=R2*SQR((-2*LOG(S))/S)
28210 RNOR=XM+RNN1*SD
28220 NRN=NRN+1
```

```
28230 RETURN
28240 RNOR=XM+RNN2*SD
28250 NRN=0
28260 RETURN
28400 REM ****************************************************************
28410 REM SUBROUTINE: EXPONENTIAL DISTRIBUTION GENERATOR
28420 REM Main program should initially RANDOMIZE seed
28430 REM Main program must supply mean, XM
28440 REM Random number generated is REXP
28450 URN=RND
28460 REXP=-XM*LOG(URN)
28470 RETURN
```

Figure 15.18—next event simulation program

The QUEUE subroutine (Figure 14.6) from Chapter 14 is used in the above program to store the arrival times of customers waiting for service. In addition, the EXPONENTIAL GENERATOR and NORMAL GENERATOR subroutines from this chapter were used to generate random arrival and service times.

The technique of next-event simulation is illustrated in the example. Although the program will simulate 200 minutes of activity (STOPSIM=200), the system time (TIME) does not advance by seconds or minutes. Instead, TIME is reset each time an event occurs (a customer arrival or a service completion). Since only two events can occur, it is not necessary to look at the time between the occurrence of one of these events because nothing can happen. With practice, next-event simulation models are relatively easy to develop and they execute faster than a "clock oriented" model in which time is incremented by fixed intervals.

The program in Figure 15.18 contains LPRINT statements that will permit you to trace the progress of a simulation run. You are given the option of skipping these print statements. Also, you may easily change the parameters of the model, such as maximum queue number (MXQ), simulation run length (STOPSIM), mean arrival rate (EMU), and mean service rate and standard deviation (NMU and SD). Varying these parameters lets you study the system being modeled. The final results from running this model for 200 minutes with a random number seed of 1000 may be seen in Figure 15.19.

Time	Number Served	Number Turned Away	Average Service Time	Average Wait Time
200	25	5	8	27.76

Figure 15.19—simulation results

15.6 Summary

At the beginning of this chapter, we explained how to use the BASIC numeric function RND. We then presented a subroutine (named ORGRND) which would generate random numbers using the congruential method. A chi-square goodness of fit subroutine was developed to determine if the numbers generated by the RND and ORGRND subroutines were not random. Both subroutines passed the chi-square test. Next, random number generators (subroutines) were provided for the uniform, normal, Poisson, exponential, and Erlang distributions.

In the last section of this chapter, we presented the next-event simulation technique. This is one of the most useful computer techniques available. In fact, simulation is so useful that several general purpose simulation languages have been developed for minicomputers and mainframe computers. These languages reduce the programming effort required to develop a model and to generate statistical results. One popular language is SLAM [2]. To date, most of these languages have been developed using the FORTRAN programming language. However, the memory addressing capability of your Personal Computer is making it feasible to develop some simulation models on a microcomputer.

References

Marsaglia, G., and T. A. Bray, "A Convenient Method for Generating Normal Variables," *SIAM Review*, Vol. 6, No. 3, pp. 260–64, 1964.

Pritsker, A. Alan B. and Claude Dennis Pegden, *Introduction to Simulation and SLAM*. Halsted Press, New York, 1979.

Shannon, Robert E., *Systems Simulation the Art and Science*. Prentice-Hall, Englewood Cliffs, N.J., 1975.

Exercises

For exercises 1–5, you can use the subroutines in Chapter 2 to calculate the mean and variance of the numbers generated and to plot a histogram.

1. Write a program that will simulate the tossing of a coin 20 times. Keep track of the number of heads and tails that occur. Do you think that your program realistically simulates a coin tossing event?
2. Generate 50 uniformly distributed random numbers in the interval of $50 \leq x \leq 99$. Use the chi-square goodness of fit test to determine of the resulting numbers are not uniformly distributed.
3. Generate 300 normally distributed numbers having a mean of 15 and variance of 3.
4. Using the Poisson distribution, generate 200 numbers that have a mean of 20.
5. Generate 400 numbers using the Erlang Distribution Generator subroutine. Set the shape parameter equal to 3 and use a exponential distribution mean of 2.
6. Simulate the process of a blind man crossing the street. While he walks across the street several events can occur. These events and the associated probabilities of occurring are:

    a. Walks across the street without falling down; probability of .67.

    b. Falls down while walking across the street; probability of .33.

    c. Hit by a truck while walking in street; probability of .15.

    d. Hit by a truck while lying in the street; probability of .33.

    Simulate the man crossing the street 30 times keeping track of how many times he successfully crosses the street. What is the probability of the blind man successfully crossing the street?

Chapter 15—Random Numbers & Simulation

7. Using the next-event simulation model of a single queue, single server system provided in this chapter, simulate 500 minutes of activity. The average time between arrivals is to be 9 and the average service time is to be 9 with a variance of 4.
8. Develop a next-event simulation model of a single queue with two servers. Run this model for 300 time periods. The time between arrivals is to be exponentially distributed with a mean of 8. The service time for each server is to be normally distributed with a mean of 4 and a variance of 2.

Answers to Exercises

Answers to selected exercises are provided below. A few exercises have been left to the reader to answer.

Chapter 1

1. A source program is a program written in a higher language such as BASIC that must be translated to machine language before it can be executed by a computer.

   An object program is a program in machine language that can be executed without additional translation.

2. A BASIC interpreter is a program that translates a source (BASIC) statement into machine language statements. These machine language statements are then executed before another source statement is translated.

   A BASIC compiler translates the entire source program into a machine language program (object program), after which, the machine language program is executed.

3. A compiler should be used when program execution time is to be minimized.

   An interpreter should be used when memory is limited and when ease of program debugging is important.

4. Significant features of the IBM Personal Computer:
   1. Amount of memory that can be addressed.
   2. MS-DOS is easy to use.
   3. The special function keys.
   4. Full screen editing of programs.
   5. Special keys that aid in program editing.
   6. High resolution monitor.
   7. Provision for the 8087 numeric data processor.
   8. Large character set (256 characters).
   9. Supports a color monitor.
   10. High resolution graphics.
   11. Enhanced BASIC language.
   12. Parity bit included at each memory location.
   13. Large instruction set (includes multiply and divide).
   14. Ability to list on the printer what is currently displayed on the monitor.

5. CHAIN MERGE "B:EXSMOOTH", 1000

Answers to Exercises

6. MERGE "QUEUE"

7. ```
    5 PRINT
   10 N%=40
   20 X=3/10
   30 FOR I%=1 TO N%
   40 X=X/10
   50 PRINT I%,SIN(X),X
   60 NEXT I%
   70 END
   ```

I	SIN(X)	X
1	.0299955	.03
2	2.999996E-03	.003
3	.003	.0003
4	.00003	.00003
5	.000003	.000003
6	.0000003	.0000003
7	3E-08	3E-08
8	3E-09	3E-09
9	3E-10	3E-10
10	3E-11	3E-11
11	3E-12	3E-12
12	3E-13	3E-13
13	3E-14	3E-14
14	3.000001E-15	3.000001E-15
15	3E-16	3E-16
16	3.000001E-17	3.000001E-17
17	3.000001E-18	3.000001E-18
18	3.000001E-19	3.000001E-19
19	3E-20	3E-20
20	3.000001E-21	3.000001E-21
21	3E-22	3E-22
22	3.000001E-23	3.000001E-23
23	3.000001E-24	3.000001E-24
24	3.000001E-25	3.000001E-25
25	3E-26	3E-26
26	3.000001E-27	3.000001E-27
27	3E-28	3E-28
28	3E-29	3E-29
29	3E-30	3E-30
30	3E-31	3E-31
31	3E-32	3E-32
32	3E-33	3E-33
33	3E-34	3E-34
34	3E-35	3E-35
35	3E-36	3E-36
36	3E-37	3E-37
37	3E-38	3E-38
38	3E-39	3E-39
39	0	0
40	0	0

8. A%=2   A%=2
   B=3   B=3
   X=.6666667   X%=1
   ```

Chapter 1

```
A=2              A=2
B%=3             B%=3
X=.6666667       X%=1

A%=2             A%=2
B%=3             B%=3
X=.6666667       X%=1
```

9. Using the program below, an incorrect answer was obtained when the difference of X and X1 was equal to 0 (at I=6).

   ```
   30  N%=10
   40  FOR I%=1 TO N%
   50  X1=X*10
   60  X=X1+1
   70  DIF=X-X1
   80  PRINT I%,X,X1,DIF
   90  NEXT I%
   100 END
   ```

I	X	X1	DIFFERENCE
1	501	500	1
2	5011	5010	1
3	50111	50110	1
4	501111	501110	1
5	5011111	5011110	1
6	5.011111E+07	5.011111E+07	0
7	5.011111E+08	5.011111E+08	0
8	5.011111E+09	5.011111E+09	0
9	5.011111E+10	5.011111E+10	0
10	5.011111E+11	5.011111E+11	0

Chapter 2

1. Mode: 3.2 and 4.6 each occurred twice
   Median: 3.2
   Mean: 3.252381
   Range: 7.9
   Variance: 3.867259
   Standard Deviation: 1.966535

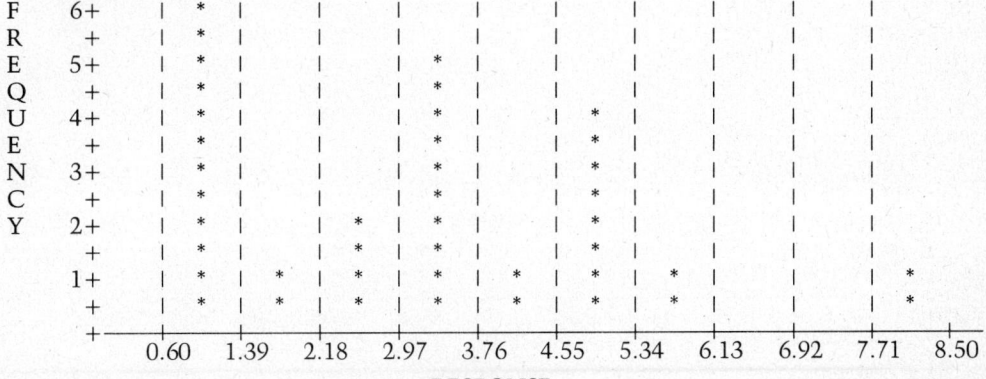

2.   *Metal A*
Mode: No value occurred more than once
Median: 9511.5
Mean: 9542.917
Range: 503
Variance: 16160
Standard Deviation: 127.122

       Metal B
Mode: No value occurred more than once
Median: 9358
Mean: 9439.667
Range: 1560
Variance: 240794.7
Standard Deviation: 490.7083

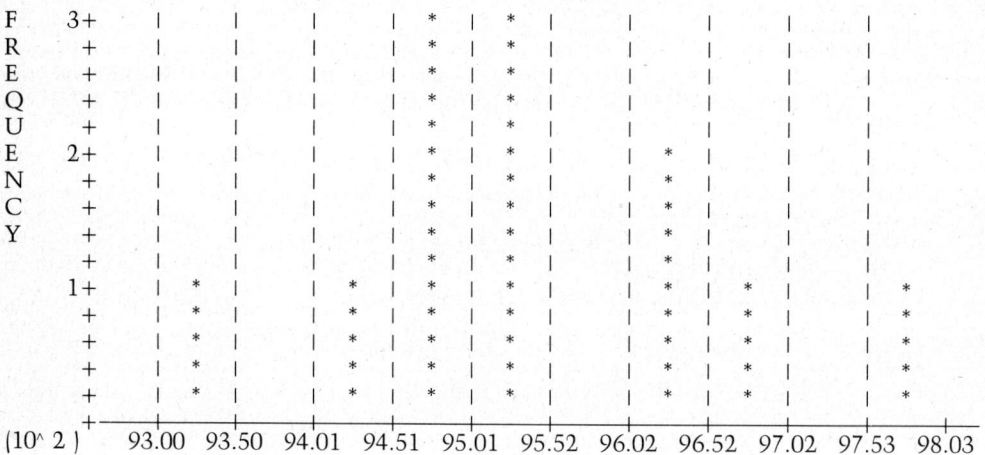

METAL A

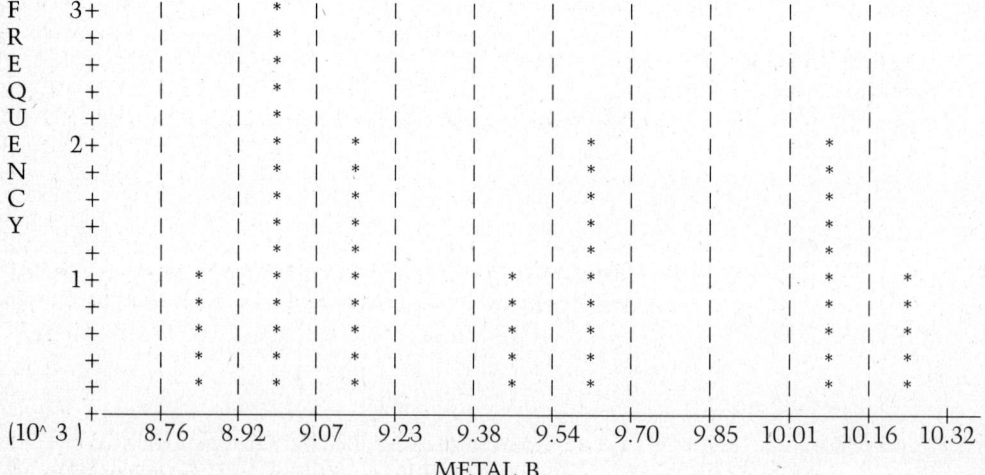

METAL B

3. See plot: Sunlight does not appear to be the only factor relevant to plant growth.

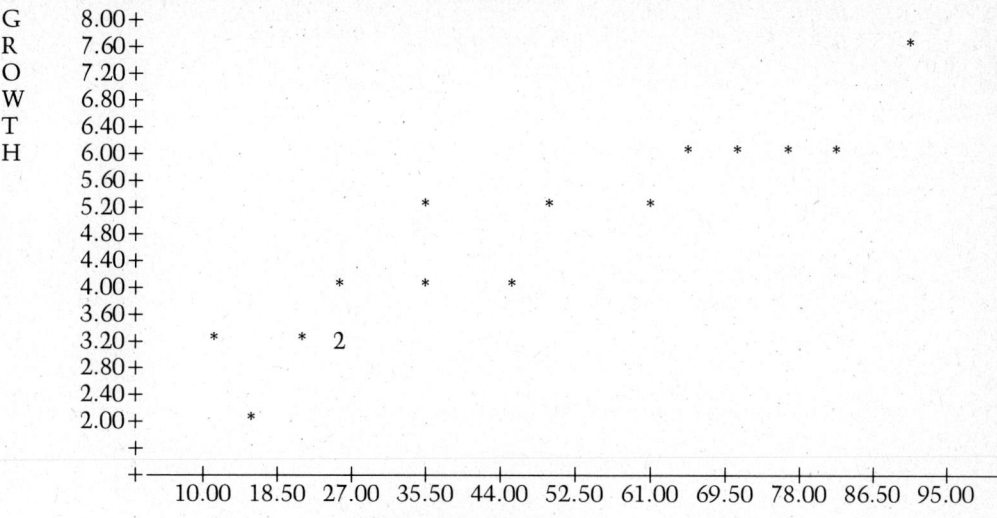

PERCENT SUNLIGHT

Chapter 3

1.
$$A \times B = \begin{bmatrix} 32 & 15 & 37 \\ 64 & 30 & 14 \\ 24 & 13 & 67 \end{bmatrix} \quad B \times A = \begin{bmatrix} 52 & 54 & 20 \\ 6 & 54 & 18 \\ 37 & 61 & 23 \end{bmatrix}$$

For $A \times B = B \times A$, let

$$A = \begin{bmatrix} 3 & 5 & 1 \\ 6 & 0 & 2 \\ 1 & 9 & 3 \end{bmatrix} \quad B = A^{-1} = \begin{bmatrix} 0.225 & 0.075 & -0.125 \\ 0.2 & -0.1 & 0 \\ -0.675 & 0.275 & 0.375 \end{bmatrix}$$

3. a. let $b_1 = 3$, $b_2 = 4$, $b_3 = 5$
   then $x_1 = 0.698$, $x_2 = 2.9070$, $x_3 = -2.0930$
   b. let $b_1 = 9$, $b_2 = 3$, $b_3 = 2$
   then $x_1 = 0.0930$, $x_2 = 1.7442$, $x_3 = -0.2558$

4. Determinant of the resulting matrix is zero.

7. $|kA| = k^2 |A|$

8.
$$A^{-1} = \begin{bmatrix} 0 & 1 & -2 & 0 \\ 0.0909 & -0.7273 & 3.5455 & -0.2727 \\ 0.1212 & -0.6364 & 1.3939 & -0.0303 \\ -0.0909 & -0.2727 & -0.5455 & 0.2727 \end{bmatrix}$$

Chapter 4

1. Participate = 9.704306 + 1.551366 Reagent
   Coefficient of determination = .7340198

2. e^{Distance} = 3.08747 + .03896823 MPH, or
   Distance = 21.92163$e^{.03896823 \text{ MPH}}$

   Coefficient of determination = .9951174
   Predicted distance for 60 MPH = 227.1404 feet

3. Salary = 14786.3 + 1912.329 Years
   Coefficient of determination = .9700621

4. Yield = −11.54382 + 15.53989 *LOG (Lots)
   Coefficient of determination = .9720327
   Predicted yield for lot number 400 is .8156286

5. Viscosity = 5.561412 − .008941539 Temperature
   $e^{\text{Viscosity}}$ = 260.1899$e^{-.008941539 \text{ Temperature}}$
   Coefficient of determination = .9938739

6. Y = −140.1438 + 5.054289 X1 + .666832 X2

   Coefficient of determination = .8383221
   Note that you will not get this answer if you use the stepwise procedure. This is an example of when multiple regression will give better results.

7. Y = 14.98208 + 3.795699 X3
   Coefficient of determination = .9895729
   Only X3 is needed in the regression equation.

8. Price = −1.074853 + 50.73441 Earnings
   Coefficient of determination = .9590872
   Predicted end-of-year stock price is $41.03

Chapter 5

2. a. No solution
   b. Infinite number of solutions
   c. No solution
   d. Infinite number of solutions

4. Input A = 2.1 ounces
   Input B = 1.25 ounces
   Input C = 4.45 ounces

Chapter 6

1. Stationary point: 0.5287065

2. (Some roots are listed below)

   a. x=0.6967, x=1.137866
   b. x=0, x=0.9514, x=−1.2847
   c. x=−4.3598, x=3.3463

5.

	Maximum number of real roots	Roots you may find
a.	1	x=1.0885
b.	0	x=0.8003
		x=1.3684
		x=−0.7301
		x=−3.3904
c.	1	y=0

Chapter 7

1. Trapezoidal Rule for h=0.01   Legendre-Gauss: 2 subintervals

   a. 130.0016   a. 130
   b. 2697.331   b. 2696.394

2. Methods presented automatically handle this; no revision required.

3.

Legendre-Gauss	Trapezoidal with 100 subintervals
a. 0.9247474	0.977186
b. 0.2297294	0.2297441
c. 0.4592219	0.4592056

5. Results found using Legendre-Gauss quadrature

   a. 0.531249
   b. 0.5068681
   c. 14.81073

Chapter 8

1. Amount of substance at t=6 is 9.26744 lbs.

2.

x	0.0	1.0	2.0
y=g(x)	0	0.15524	−0.48046

3. Standard form: $y' = \dfrac{1-xy-y}{x}$

x	1.0	1.5	2.0
y=g(x)	0	0.25969	0.31438

4.

	Order	Could We Solve
a.	2	No
b.	1	Yes
c.	2	No
d.	1	Yes
e.	1	Yes

5. Expression: $\dfrac{dQ}{dt} = kQ$ where k is a constant (in this case, k = −0.02321)

   Amount of Dittmannium left after 24 days = Q (24) = 171.8707

Chapter 9

1. $Z = 34.72$
   $x_1 = 3.2, x_2 = 2.36, x_3 = 0.88$

2. Let x_i = number of barrels of fuel type i produced, where
   $i = 1$:leaded
   $i = 2$:unleaded
   $i = 3$:diesel
   Maximize $Z = 0.12x_1 + 0.09x_2 + 0.10x_3$
   s.t.
   $$0.4x_1 + 0.5x_2 + 0.7x_3 \leq 400$$
   $$0.6x_1 + 0.5x_2 + 0.3x_3 \leq 300$$
   $$x_1, x_2, x_3 \geq 0$$
   Solution:
   $Z = 7600, x_1 = 300, x_2 = 0, x_3 = 400$

3. Let the variables be named so that the letters represent the first letter for each fertilizer/crop combination. That is,
   AC = fertilizer A on corn
   CS = fertilizer C on soybeans, etc.
   unit = bushels
   Maximize $Z = 2(AC+BC+CC) + 1.5(AW+BW+CW) + 1.8(AS+BS+CS)$
   $ -(1.333AC+1.389BC+1.667CC) - (0.645AW+BW+0.571CW)$
   $ -(AS+1.563BS+0.667CS)$
   $ = 0.667AC + 0.611BC + 0.333CC + 0.855AW + 0.5BW$
   $ + 0.929CW + 0.8AS + 0.237BS + 1.133CS$
   s.t.
   $$15AC + 31AW + 20AS \leq 100$$
   $$18BC + 25BW + 16BS \leq 125$$
   $$12CC + 35CW + 30CS \leq 80$$
   all variables ≥ 0
   Solution:
   AC = 6.7   all other variables = 0
   BC = 6.94
   CS = 2.67
   Z = 11.71

4. Let x_1 = number of units of power tool A produced per week
   x_2 = number of units of power tool B produced per week
   Maximize $Z = 2x_1 + x_2$
   s.t.
   $$30x_1 + 40x_2 \leq 2400$$
   $$45x_1 + 30x_2 \leq 2400$$

$$20x_1 + 25x_2 \leq 2400$$
$$x_1, x_2 \geq 0$$

Solution:

Z = 106.67 (although its actual value is irrelevant)

$x_1 = 53.333$

$x_2 = 0$

5. Solution: Produced nothing!
(Can you examine the formulation and tell why?)

Chapter 10

1. a.

Month	Moving Average
1	—
2	—
3	3.43
4	3.50
5	3.70
6	4.23
7	4.70
8	5.10
9	5.43

Estimate for the 10th month is 5.43.

b. Using the average of the first three months as the initial forecast value, then

$F_2 = F_1 + .3(X_1 - F_1)$

$= 3.43 + .3(3.5 - 3.43)$

$= 3.451$

The following values were obtained in a similar manner:

Month	Single Exponential Smoothed Forecast
2	3.451
3	3.496
4	3.407
5	3.495
6	3.71
7	4.03
8	4.35
9	4.67
10	5.01

2. For: season length = 6
initialization periods = 18
lead time = 1
alpha = .3
beta = .1
gamma = .1

Period	Forecast
19	31.52
20	31.50

21	33.63
22	35.88
23	38.25
24	40.60

For the initialization phase, the mean squared error was 2.198 and the mean absolute deviation was 1.167.

3. a.

Period	3-Month Moving Average
1	—
2	—
3	3
4	5
6	9
7	11
8	12
9	15
10	17

b. Using an initial forecast value of $\frac{1+3+5}{3} = 3$, and an alpha = .5, the following results were obtained:

Period	Single Smoothed Value
1	—
2	2
3	2.5
4	3.75
5	5.31
6	7.19
7	9.09
8	11.05
9	13.02
10	15.01

c. Using an initial forecast value of 1 and an initial trend value of 2, the following results were obtained:

Period	Holt's Smoothed Values Alpha = .1 Beta = .1
2	3
3	5
4	7
5	9
6	11
7	13
8	15
9	17
10	19

d. For: season length = 6
initialization periods = 9
lead time = 1
alpha = .1
beta = .1
gamma = .1

Period	Winter's Smoothed Values
2	3
3	5
4	7
5	9
6	11
7	13
8	15
9	17
10	19

Holt's and Winter's methods are the best because these methods were developed to model a process containing a trend.

4.
a. Using an initial smoothed value of $\dfrac{1+4+7+10}{4} = 5.5$

Period	Single Smoothed Value
1	—
2	3.12
3	3.56
4	5.28
5	7.64
6	4.32
7	4.16
8	5.58
9	7.79
10	4.39
11	4.20
12	5.60
13	7.80
14	4.40
15	4.20
16	5.60
17	7.80

b. For: season length = 4
initialization periods = 12
lead time = 1
alpha = .1
beta = .1
gamma = .1

Period	Winter's Smoothed Value
1	—
2	4
3	7
4	10
5	1
6	4
7	7
8	10
9	1
10	4
11	7
12	10
13	4

338   Answers to Exercises

14	4
15	7
16	10
17	1

5. For: season length = 6
     initialization periods = 18
     lead time = 1
     alpha = .3
     beta = .1
     gamma = .1

Period	Winter's Smoothed Value
19	26.60
20	29.43
21	34.86
22	35.70
23	42.57
24	49.23
25	40.97
26	42.23
27	47.98
28	47.36
29	52.11
30	58.30

Chapter 11

1. a.

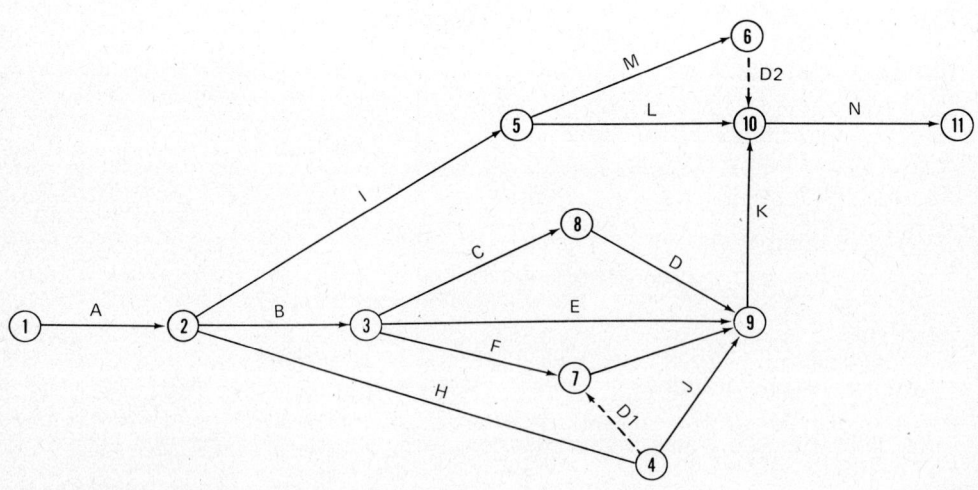

   b. Critical path: A–I–M–D2–N

     Expected project completion date is 19 days.

2. Critical path: B–D–I–J–M–P–O–Q

3. a.

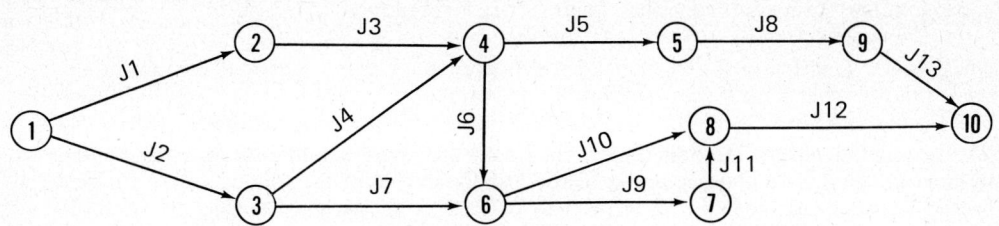

b.

Job	Early Start	Late Start
J1	0	0
J2	0	5
J3	3	3
J4	1	6
J5	8	9
J6	8	8
J7	1	9
J8	12	13
J9	11	12
J10	11	11
J11	12	13
J12	14	14
J13	14	15

c. Critical path: J1–J3–J6–J10–J12

Project duration of 12 days.

4. Critical path: A–C–I–N–K–L–Q–S–R–Y–Z–DC5–END

Project duration of 36 weeks.

Chapter 12

1. Change line 23090 to

    23090 IF X(J+1)<=X(J) GOTO 23120

2. The following lines should be:

    23710 IF (PV<X(R%)) THEN 23780
    23750 IF (PV>X(L%)) THEN 23810

3. Value for time was 55 seconds. This value will vary with the list being sorted.

4. Value for time was 13 seconds.

5. The bubble sort algorithm required 1 second and the quicksort algorithm required 12 seconds. The bubble sort was much faster because this algorithm can determine when a list is in sorted order, and at that time stop the sorting process.

Chapter 13

1. RAM and ROM memory are primary storage devices. Some secondary storage devices are disk and tape devices.

2. Each track on a floppy disk is divided into sectors. The IBM Personal Computer has 8 sectors on one side of a floppy disk. Each sector will hold 512 bytes.

3. The primary function of the DOS diskette directory is to maintain a table containing the names of files that have been saved on the diskette.

4. The sequential access method means that data are written and read in a sequential manner starting from the beginning of the storage device. The random access method means that data records can be written to or read from a storage device in a random order.

5. Sequential access input/output modes:

   a. Input

   b. Output

   c. Append

6. a. A record is the number of bytes that are moved in and out of the random buffer at one time.

   b. A random buffer is where random access records are temporarily stored before being written to a disk file or where these records are stored after being read from a disk file.

   c. A file is a collection of related information that is usually kept on a disk or tape device.

7, 8, and 10. We have left the writing of these programs up to the reader. The programs should be very similar to the SEQMAN program in this chapter.

9 and 11. We have left the writing of these programs up to the reader. The programs should be very similar to the RANMAN program in this chapter.

Chapter 14

1. A stack is a data structure in which data are inserted and deleted in a last-in, first-out manner.

2. A queue is a data structure in which data are inserted and deleted in a first-in, first-out manner.

3.
```
 90 DIM A(50)
100 REM PROBLEM 3, CHAPTER 14
110 DEFINT I
120 SIZ%=50                    'Set array size
130 GOSUB 24560                'Initialize stack
140 REM Place available space in array on stack-LIFO
150 FOR I=1 TO 50
160 IDAT = I
170 GOSUB 24620                'Push I on stack
180 IF IFL< >0 THEN 410
190 NEXT I
200 REM Place first data element in array A
210 INPUT "ENTER DATA ELEMENT"; VALUE
220 GOSUB 24710                'Pop stack
230 IF IFL< >THEN 410
240 A(IDAT)=VALUE
```

```
250 INPUT "CONTINUE(Y/N)"; Y$
260 IF Y$="Y" THEN 210
270 STOP
410 PRINT "ERROR"
420 END
```

4. A queue data structure will be utilized using the subroutines from this chapter. If 25 readings are to be maintained, then the queue size must be set to 26 (one element cannot be used—see explanation of queue subroutines).

```
100 DEFINT I
110 QSIZ%=26                    'Set array size
120 GOSUB 24870                 'Initialize queue
130 REM Data can now be placed in the queue
140 REM When the queue becomes full, the data element at
150 REM front of queue should be removed.
160 REM Then a new data element can be
170 REM added to the rear of the queue.
180 REM This can be done as follows
190 GOSUB 25040                 'Take data from queue
200 IF IFL<>0 THEN 410
210 REM Place new element in queue
220 IDAT=VALUE
230 GOSUB 24930                 'Place data in queue
          •
          •
          •
          •
410 PRINT "ERROR"
420 END
```

5. The writing of this program is left to the reader. The bidirectional linked list subroutines from this chapter can be used. Since the data consists of string and numeric type data, they will be easier to maintain if the numerical values are converted to strings before they are placed into the linked lists. Since there are four inspectors, four linked lists are required.

Chapter 15

1. There are several ways to write this program. We will generate a random number X between 0 and 1. If $0 \leq X < .5$ we will say the toss of the coin resulted in a head; otherwise, the result was a tail.

   ```
   10 RANDOMIZE
   20 INPUT "HOW MANY COIN TOSSES ARE TO BE SIMULATED?; N
   30 FOR I=1 TO N
   40 X = RND
   50 IF X<.5 THEN HEAD=HEAD+1
   60 IF X>=.5 THEN TAIL=TAIL+1
   70 NEXT I
   80 PRINT "HEADS="; HEAD
   90 PRINT "TAILS="; TAIL
   100 END
   ```

2. ```
   10 DIM AY(50)
   20 RANDOMIZE
   30 A=50
   ```
   ```

```
   40 B=99
   50 INPUT "HOW MANY NUMBERS ARE TO BE GENERATED";N
   60 FOR I=1 TO N
   70 GOSUB 2800            'Uniform distribution
   80 AY(I)=RUFM
   90 NEXT I
  100 GOSUB 27200           'Chi-Square Fit Test
  110 END
```

3.
```
   10 RANDOMIZE
   20 DIM X(300)
   30 XM=15
   40 SD=SQR(3)
   50 INPUT "HOW MANY NUMBERS ARE TO BE GENERATED";N
   60 FOR I=1 TO N
   70 GOSUB 28080           'Normal distribution
   80 X(I)=RNOR
   90 NEXT I
  100 END
```

4.
```
   10 DIM X(200)
   20 RANDOMIZE
   30 XM=20
   40 INPUT "HOW MANY NUMBERS ARE TO BE GENERATED";N
   50 FOR I=1 TO N
   60 GOSUB 28270           'Poisson distribution
   70 X(I)=RPOI
   80 NEXT I
   90 END
```

5.
```
   10 DIM X(400)
   20 RANDOMIZE
   30 K1=3
   40 XM=2
   50 INPUT "HOW MANY NUMBERS ARE TO BE GENERATED";N
   60 FOR I=1 TO N
   70 GOSUB 28480           'Erlang distribution
   80 X(I)=RERL
   90 NEXT I
  100 END
```

6. There are several ways to answer this problem.

```
   10 RANDOMIZE
   20 FOR I=1 TO 30
   25 REM Did he fall?
   30 X=RND
   40 IF X<=.66 THEN 100    'Did not fall
   50 REM Fell down
   55 REM Was he hit lying in the street?
   60 X=RND
   70 IF X>.33 THEN 100     'Was not hit after falling
   80 REM Hit while lying in street
   90 HIT=HIT+1:GOTO 130
   95 REM Was he hit walking?
  100 X=RND
  110 IF X<=.15 THEN HIT=HIT+1
```

```
120 IF X>.15 THEN SUCCESS=SUCCESS+1
130 NEXT I
140 PRINT "PROBABILITY OF BEING HIT=";HIT/30
150 PRINT "PROBABILITY OF SUCCESSFULLY CROSSING";SUCCESS/30
160 END
```

7. The following changes must be made:

```
1180 EMU=9
1190 NMU=8
1200 SD=2
```

Index

t = table, f = figure

α, 195
A>B: command, 10
A>BASIC command, 11
A>DIR command, 11
A option, 13
A>prompt, 11
Absolute value, of deviations, 24
Accuracy, desired level of, 150
Activity(ies)
  for CPM program, 233, 234, 235
    listing of, 225, 225t-226t, 226
    slack for, 227, 228
  dummy, 233
Adaptor card, 1
ADD TO DATA option, 82, 96
ADD INDEPENDENT VARIABLE option, 87
ADD NEW ACTIVITIES option, 234
ADD TO TIME SERIES option, 217
Addition(s), 292
  of matrix, 50-51, 51f, 52f
Advanced BASIC, 8, 9, 10
Algebra, linear, 169, 175–177
Algorithm partitions, quicksort, 245
"All," 281
ALL option, 13
Alphabetical information, placing of, 243
  see also Sorting
Alphabetical listing, 293
Analytic solution, 159
Answers, to exercises, 327-343
APPEND mode, 259
Append mode, 259, 260
  omitting of, 268
APPEND option, 263, 266
Applications, types of, 8-9
Arithmetic mean, 20
Array(s), 285
  dimensioning of, 40, 47-48
  dimensions, 285
  in linked lists, 292
  in stack subroutines, 287
  specification of, 48
  storage of matrix in, 47
  use, 285-286, 303
    in queue structure, 289
Array element, 48
Artificial variables, in linear programming solution, 191
ASCII characters, 261, 267
ASCII format, 13
"Average deviation from the mean," 24
AY array, 292

B>A: command, 11

BASIC, 1, 7
  versions of, 8, 9
BASIC COMMANDS, review of, 10-14
BASIC compiler, 10
BASIC editor, 7
BASIC interpreter, 9
  compared to compiler, 9-10
  loss of precision, 14-17
BASIC statement, translation into machine language, 9-10
Basic variables, conversion to non-basic variables, 178
"Biased" estimate, 24
Bidirectional linked lists, 297
  subroutines, 298-303f
  use, 297-298
Bidirectional linked queues, 303
Binary format, packed, 267
Binary number, 261
Bit(s), 1, 2
  incorrect value, 4
Bubble sort, 243-244, 252
  program, 244-245f
Buffer, 258
  random, 267, 268, 269
  type ahead, 7
Buffer area, sequential access file, 258, 258f, 259, 267
Byte(s), 1
  quantity stored, 255, 262, 267
  in random access file, 268
Byte string, 269

C$ variable, 268
Calculus, differential, 66
Capacity limits, in linear programming, 174-175
Caps Lock key, 7
Carriage return, 260
Cassette BASIC, 8, 9, 10
"Center" of data, 20
  see also Central tendency
Central processing unit, 286
Central tendency, measures, 20, 23, 26, 44
  mean, 20, 21
  median, 221-22
  mode, 22-23
CHAIN statement, use, 13
CHANGE EXISTING ACTIVITY DATA option, 234
CHANGE TIME SERIES DATA option, 217
Character(s)
  quantity displayed, 1
  in sequential file, 262

storing of, 7
Character record, 268
Character set, 34-4f
Character strings
  conversion of numeric values to, 271
  length of, 14
Chi-square goodness of fit test, 324
  application, 309-310
  results, 311, 311f
  subroutine, 310f-311
Chi-square statistics, 309
CIRCLE statement, 8
Circuit board(s), 2
CLOSE statement, 259
Closing a file, 259
Coefficient(s), objective, in linear programming solution, 190
Coefficient of correlation, 67
Coefficient of determination, 67, 99, 105, 109, 110
Coefficient matrix
  in Gauss-Seidel method, 132
  in solving simultaneous linear equations, 123
Color display, 1
Color monitor(s), 1
Column(s), 1
Comma(s), 261
Command(s), 7
  BASIC, 10-14
COMMON statement, use, 13
Compiler, 9
  BASIC, 9-10
  compared to interpreter, 9-10
Condenser, charge on, 157
Computation, 8
  accuracy, 6, 7
  speed, 9
    increase in, 5
  time, reduction of, 6
  work, 8
Computer simulation, *see* Simulation
Confidence limits, in linear regression program, 87
CONFIDENCE LIMITS ON PREDICTED VALUES, 109, 110, 112
"Conformable" matrix, 50, 53
Congruential method, for generating pseudo-random numbers, 307-308
Constraint(s), in linear programming, 170, 171, 175-176, 187
Convergence criteria, in Gauss-Seidel method, 132
Corner points, in linear programming solution, 177
CORRECT DATA option, 82, 91
Correlation, coefficient of, 67
Cost effectiveness, 9
CPM, *see* Critical path method
CPU, 9
Create option, 263
CREATE option, 266
Critical path method (CPM), project planning and scheduling with, 225, 241

  exercises, 241-242
  program, 227-234
    examples, 235-240
  technique, 225-227
CRT, 41, 93
  for exponential smoothing forecasting program, 212
  output from regression analysis on, 87
Cursor control keys, 7
Curve, plotting of, 26
Curve fittings, 257
  with linear regression, *see* Linear regression
CVD function, 271
CVI function, 271
CVS function, 271
Cyclical pattern, 195

D$ variable, 268
DAT$, 298
Data
  accessing of, 267
  for dependent variable, 87
  display, 44
  entering of
    in CPM program, 233, 234, 235
    in exponential smoothing forecasting program, 209-210
    for linear regression, 85
    in stepwise regression program, 89
  maintaining of order, 292
  range of, *see* Range
  reading and writing of
    in linked list program, 295
    in random access files, 267-272
    in sequential access files, 259-262
  sorting of, *see* Sorting
  spread of, *see* Dispersion
  storage, 39, 255, 285
    see also Data structures, Storage devices
  variation of, 24
  weighting of, 100
    in linear regression program, 97, 98, 100
Data analysis, plotting subroutines and, *see* Plotting subroutines
Data base management systems (DBMS), 281, 303
Data collection, 19
  device, 8
Data element(s)
  adding of, 287
  change of, 217
  in exponential smoothing, 194
  in linear regression program, 87
  transformation, 87-88
  location, 292
  for stepwise regression program, 90
    changing of, 91
Data file(s), 257
  disk, *see* Disk files
  removal of jobs from, 266
Data management options
  in exponential smoothing forecasting program, 209-210

Index

in linear regression, 82-83
in stepwise regression program, 90, 91, 95-96, 97-98, 103-104, 107-108, 111
Data plot(s), 26
  of subroutines, 26-44
Data point(s), 4, 24, 65
  location and concentration of, 19, 24
  median, 21
  occurrence, 39
Data record(s)
  directory, 274-275
  listing of contents, 281
Data reduction, 19, 26, 44
  basics of, 19-20
  data analysis and plotting subroutines, 26-44
  data plots, 26
  measures of central tendency, 20-23
  measure of dispersion, 23-26
DATA statements, 295, 301
Data storage, in exponential smoothing, 194
Data structures, 285, 303-304
  arrays, 285-286
  linked lists, 292-293
    bidirectional, 297-303f
    linked queues, 293-297f
  queues, 288-291
  stacks, 286-288
Data types, 298
Data values, 295
Date, 10
DAY$, 260
DBMS, see Data base management systems
Debugging, 10
Decision variables, 170, 171, 175
Default disk drive, 10
  changing of, 10-11
Default record, size, 268
Definitional relations, in linear programming, 174
DELETE ACTIVITIES FROM NETWORK, option, 234
DELETE DATA FROM TIME SERIES option, 217
DELETE INDEPENDENT VARIABLE option, 87
DELETE JOB option, 281
DELETE option, use, 13
Deletions, 292
  in linked lists, 297
    bidirectional, 298-299
Density, 101
Dependent variable
  entering name of, 89
  equations with, 71-72
  in linear regression program, 87
  in stepwise regression program, 106
  values required for, 90, 102, 107, 113
Determinant of a matrix, 56-59, 60
Determination, coefficient of, 67, 99, 105, 109, 110
Deviation, about regression line, 65, 66f
  see also Standard deviation
Diagnostic checks, 4, 5
Differential calculus, 66

Differential equations
  numerical solution to, 157, 158-161, 166
    application of Runge-Kutta method, 161-166
  overview, 157-158
Differentiation, inverse of, 147
  see also Integration
DIM statement
  for array size, 47-48
  for defining of vectors, 63
  in linear regression program, 68
DIMENSION statement, see DIM statement
DIR (directory) command, 11
Directory, 272
  DOS diskette, 256-257
  storing data record in, 274-275, 281
Directory command, see DIR command
Disk
  inputting data from, 211, 233
  storage space, 9
Disk BASIC, 8, 9, 258
  use, 9, 10
Disk drive, 2
  default, 10-11
  double sided, 255, 257
Disk files, 92, 255, 282-283
  data storage in, 282
  entering data from, 209, 234
  random access files, 267-282
  recording information on diskette, 255-258
  sequential access files, 258-267
  storing data on, 235
  structure, 303
Disk operating system, see DOS
Disk space, for random file, 267
Diskette, recording information on, 255-258
Diskette file, valid names for, 257
Dispersion, 20, 20f
  measurement, 20, 23, 26, 44
    range, 23-24
    variance and standard deviation, 24-26
Display devices, 1
Display lines, 1
Distribution, use of random number generator for, 312-317f
Divergence, in Gauss-Seidel method, 132
Division, 5
DOS, 7, 256
  diskette, 257
  diskette dictionary, 256-257
  information in, 256
  MS, 1, 7
  review of, 10-14
Dot matrix, 1
Double precision numbers, 6
Double precision numeric variables, 14
DRAW statement, 8
Drawing, see Fitting
Drive A, 10
Drive B, 92
Dummy activities, 233

Editing, 7

Editor, 7
EMU, 324
"END," 266
End of file, see EOF, 266
END statement, 259
ENTER TO CONTINUE, 93
"ENTER" key, 87
  location, 7
  use, 11
EOF mark, 262, 266
Equality constraints, in linear programming, 175-176
Equation(s)
  applying regression to, 72
  differential, see Differential equations
  point-slope for, 141
  simultaneous, 66
  solving problem represented by, 158
  unknown, 66
Erasures, 266
Erlang distributions, 325
  generation, 318, 318f
    frequency histogram, 319f
Error(s), 9
  of estimate, 67
  parity, 4
  squared, sum of, 66
Error message, 4, 5
  in sequential files, 266
  for too many files, 256
Estimate
  "biased," 24
  error of, 67
  standard deviation of, 87
  "unbiased," 24
Execution times, for 8087, 6, 6t
Exercises, answers to, 327-343
EXIT PROGRAM option, 281
Expansion slots, 2
Exponent overflow, 14
Exponential distribution, 325
  generation, 316-317f
    frequency histogram, 317f
EXPONENTIAL GENERATOR subroutine, 324
Exponential smoothing, forecasting with, 193, 221
  examples and solutions, 212-221, 221-223
  program, 197-210
  seasonal, 195-197
  single, 193-194
  with a trend, 194
Exponentiation, of matrix, 55-56
Expression(s)
  applying regression to, 72
  in exponential smoothing
    coefficients in, 194
    manipulation of, 194

F, 260
Factoring, 137
"Feasible region," in linear programming, 171
Features of IBM Personal Computer, 1-8
Field(s), 268
  assigning of, 273-274
FIELD statement, 268, 269, 270
FIFO, 293, 297, 319
  data processing in, 288
File(s), 257
  closing of, 259
  opening of, 260, 261
  temporary, 266
File allocation, table and directory, 256, 257
File cabinet, 285
File index method, 275
File name(s), 256, 257
  directory for, 256
  rules, 257-258
File reference number, 269
File specifications, reduction of, 259
First in-first out, see FIFO
First order differential equation, 158
Fitting (drawing), of straight line, 65
Floppy disk(s), 5, 255
Forecast horizon, 195
Forecast lead time, 195, 213
Forecasting, 193
  with exponential smoothing, see Exponential smoothing
  linear regression and, 88
FORTRAN, 9, 10, 325
Frequency histogram, 26, 42f
  exponential distribution, 317f
  normal distribution, 315f
  plotting subroutines for, 39
  Poisson distribution, 316f
  uniform distribution, 313-314f
Frequency plots, 41
  main program, 42f
  subroutines, 30-34f, 40
FRNT array, 294
Function
  integral of, 147
  root of, 137
    see also Polynomials, roots of
Function key, 7

Games, 305
Gauss-Jordan method, for solving linear equations, 119, 123-126, 134
  subroutines, 126-129f
Gauss-Seidel method, for solving linear equations, 119, 129-132
  subroutines, 132-134
Gaussian distribution, see Normal distribution
Gaussian elimination, 56, 57, 60, 123
GET statement, 267, 270
Graphics
  capabilities, 1
  in linear programming solution, 171
Graphics adapter card, 1, 2
Graphics resolution, 1-2
Graphics statement, 8

H variable, 13
Hardware, 8

High resolution graphics, 2
Histogram, *see* Frequency histogram
Holt's method, for forecasting, 196

I, 40
I/O, *see* Input and output
IBM BASIC, 7
IBM Personal Computer
　IBM BASIC interpreter, 9-17
　significant features, 1-8
　types of applications, 8-9
IDAT, 287
Identity matrices, 59, 60
IFL, 287
Independent variable, 114
　entering name of, 89
　entering number of, 89
　equations with, 71-72
　in linear regression program, 87
　removal of, 87
　in stepwise regression program, 106
　values required for, 90, 102, 107
"Infeasible" solution, in linear programming, 171
Information, recording of, on diskette, 255-258
Information transfer rate, 9
INITIALIZE DIRECTORY option, 280
Input, 259
INPUT mode, 259
　omitting of, 267
Input and output (I/O), 5
INPUT TO QUEUE, 291
INPUT subroutine, 295
Insertion(s), to linked lists, 297
Instruction(s), accepting of, 11
Instruction set, 5
Integer data, 298
Integer numeric variable, 14
Integer plot, 39
Integer value, positive, 315
Integer variable, 40, 269
Integral(s)
　closed-form expression, 147
　definite, 147
　indefinite, 147
Integrating function, 147
　see also Numerical integration
Intel 8088 microprocessor, 5
Intel 8087 numeric data processor, 5-6
Interest, variables of, 170
Interpreter, 9
　see also BASIC interpreter
INUM, 39, 40
Inverse of a matrix, finding of, 59-62

J, 40
Jacobi method, 130
Job
　removal from file, 266, 281
　storage of information, 272
Job inventory file, creation of, 280
Joint observations plotting subroutines, 34-36f,
40, 41, 43-44f

K, 86
Key(s), 7
Keyboard, 1, 5
　entering data from, 86, 209, 211, 233
　features, 6-7
Keyed access method, 272, 273f
Keyed index method, 281
KILL, 266

L$ variable, 268
Last in-first out order, *see* LIFO
Lead time, forecast, 195, 213
LEF array, 247, 248
Left-justify, of strings, 269
Legendre-Gauss quadrature, for numeric integration, 147, 151-152, 154
　main program, 153-154f
　subroutines, 152-153f
Letter quality printer, 1
LIFO, 286, 293, 297
LIMITS ON PREDICTIONS, 112, 113, 115
Line, slope of, 141
Line feed sequence, 260
Linear algebra, for linear programming solution, 169, 175-177
Linear equations, *see* Simultaneous linear equations
Linear functions, of interest of variables, 170
Linear programming (LP), 169-170, 191
　basic formulation, 170-171
　graphic solution, 171-173
　program example, 173-175
　solution methods, 175
　　simplex procedures, 177-191
　　via linear algebra, 175-177
Linear regression, 116-118, 193
　assumptions in, 114
　for curve fitting, 65-71f, 116
　　multiple regression, 71-72
　　program options, 85-88
　　stepwise regression, 72-85f, 88-116
LINK array, 292, 294
Linked list(s)
　linked queues, 293-297f, 303
　multiple, 293, 293f
　use, 292-293
Linked queues, 293-294, 303
　subroutines, 294-297f
List(s), 292
　linked, *see* Linked list
　sorting of, *see* Quicksort
LIST, 7
LIST DATA AT CRT option, 234
LIST DATA option, 281
　in stepwise regression program, 190
LIST DATA AT PRINTER option, 234
LIST DIRECTORY CONTENTS option, 281
List option, 263
LOAD command, 7, 11
LOG, removal of, 105

Logarithmic transformation, 72
  natural, 102
Logical record number (REC%), 271
Lower triangular matrix subroutine, 61f
Lower triangular subroutines, in solving simultaneous linear equations, 126
LPRINT statement, 324
LQDAT array, 293
LSET statement, 269

Machine language, translation to, 9-10
Mainframe computers, 8
  interfacing to, 8
Mathematical computations, 5
Mathematical-logical model, developing of, 218
Mathematical operations, precision and, 14-16
Mathematical programming, *see* Linear programming
Matrix, matrices, 47
  definitions of, 47-48
  inverse of, finding of, 123
  storage of, 47
Matrix dimensions, subroutines, 48f
Matrix form, in simultaneous linear equations, 119-120
Matrix inverse method, for solving simultaneous linear equation, 119, 120-121, 123, 134
  program, 122f
  subroutines, 121, 122f
Matrix multiplication, 121
Matrix operations, 48
  addition and subtraction, 50-52
  determinant, 56-59
  exponentiation, 55-56
  inverse, 59-62, 121
    see also Matrix inverse method
      multiplication, 53-55
    by scaler, 52-53
  transpose, 48-50
Maximum queue number (MXQ), 324
Maximum size (SIZ), of array, 287
"Maximization" problem, in linear programming, 171
Mean, 24
  calculation of, 20-21, 22, 23, 44
  deviation from, 24
Mean arrival rate (EMU), 324
Mean service rate (NMU), 324
Median, calculation of, 21-22, 23, 44
Median value, 248
Medium resolution graphics, 2
Memory, 2, 9
  addressing of, 5
  amount of, in quickest procedure, 248
  bytes of, 1
  extra, 2
Memory storage devices, 255
  types, 255
MERGE command, use, 13
Microcomputer(s), 1
  limitations, 8-9
Microprocessor, 8088, 2, 5

Microprocessor chip, 1
  8088, 1
Microsoft, 9
Microsoft BASIC, 7
Minicomputer(s), 8
Minimization objective, in linear programming solution, 175
MKD$ function, 268, 269
MKI$ function, 268, 269
MKS$ function, 268, 269
Mode
  calculation, 44
  use, for central tendency measurement, 22-23
MODIFY JOB DATA option, 281
MODIFY option, 263, 266
Monochrome, 1
Monochrome display, 1
Monochrome monitor, 1
MS-DOS, 1, 7
Multiple regression, 71-72, 87
  program, 88, 106, 108-109, 114
  stepwise, 72
MULTIPLE REGRESSION option, 93, 94, 96, 99, 104, 113
Multiplication, 5
  of matrix, 53-55, 121
  by scaler, 52
MXQ, 324

N data points, 21
N-period average, 193-194
N$ variable, 14
Name (s), variable, 14
Natural log transformation, 102
Network data, in CPM program
  inputting of, 233
  storing of, 235
NEW JOB option, 280
Newton-Raphson Method, for finding root of polynomial, 144
  general concept, 138-139
  operational requirements, 139-142
  program, 142-144
Next event simulation, 318-324, 325
NMU, 324
NOM variable, 269
Nonbasic variables, conversion to basic variables, 178
Non-integer plot, 39
Nonlinear expressions
  regression applied to, 72
  transformation, 87
Normal distribution, use of random number generator for, 314-315f
  frequency histogram, 315f
NORMAL GENERATOR subroutine, 324
Number(s)
  conversion of numeric values to, 271
  for linear regression program, 86
  plotting of, 40
  sorting of
    bubble sort, 244-245f
    quicksort procedure, 245, 246

Index

Numeric data processor, 5-6
   8087, 6, 9, 62
   socket for, 5, 6f
Numeric function, *see also* RND
Numeric keypad, 7
Numeric values
   changing to string values, 268
   conversion to number, 271
   representation of, 261
Numeric variables, 14
   types, 14
Numerical integration
   basics of, 147, 154-155
      numerical methods
         Legendre-Gauss quadrature, 151-154f
         trapezoidal rule, 148-151f

Object program, 10
Objective function, in linear programming, 171, 172f, 187, 190, 191
   value of, 189
Odd parity, 2, 4
OK prompt, 11
ON ERROR GOTO statement, 266
ON ERROR statement, 266
OPEN statement, 259, 260, 267, 268, 270
Opening a file, 259, 261
Operating system, 1
Optimal point, in linear programming, 172
"Optimal solution," in linear programming, 171, 172, 173f, 177, 191
Option(s), in sequential file program, 263
Optional diskette, 13
Orders, in stepwise regression program, 92, 93, 97
   data elements for, 90
Ordinary differential equation, 158
   first-order, 159
ORGRND subroutine, 308, 309, 325
   use, 324-325
      for number generation, 311-312f
Output, *see* Input and output
OUTPUT mode, 259
   omitting of, 268

"(|)," 13
Packed binary format, 267
PAINT statement, 8
Parity, 2
   odd, 2, 4
   use, 4
Parity bit, 2
Parity error, 4
Partial differential equations, 157
PERFORM CPM CALCULATIONS option, 234
PERFORM FORECAST COMPUTATION option, 212
PERFORM REGRESSION COMPUTATIONS option, 93
Performance data, collecting of, 19
Physical system(s)
   analyzing of, 318

differential equations in, 157
"Pivot" element, in quicksort procedure, 245, 246-247, 248, 249
Pivot number, in quicksort program, 245
Plot(s), plotting, *see* Data plots, Plotting subroutines
Plotting subroutines
   data analysis and, 26-30f, 41, 44-46
      frequency, 30-34f
   format, 40
   frequency histogram, 41, 42f
   frequency plots, 42, 42f
   for joint observations, 34-36f, 43-44f
   requirements, 39-40
   scaling subroutines, 40-41f
   sorting, 37-39f
Point(s), in linear programming solution, 177
Point-slope, of equation, 141
Pointer
   in linked lists, 292, 293, 297, 298
   in queue structure, 289
   record, 269
Poisson distributions, 325
   generation, 315-316
      frequency histogram, 316f
Polynomials, roots of, 137-138, 144, 145
   Newton-Raphson method, 138-144
Population statistics, 19
Population variance, true, 24
Positive integer value, 315
Power supply, 2
Precision, 14
   loss of, 14-17
Precision numbers, double, 6
Predecessor, in bidirectional linked lists program, 297, 298
Predicted values, confidence limits of, 109, 110, 112
Predictions, limits on, 112, 113, 115
PRESS ENTER TO CONTINUE, 89, 115
Price(s), 8
PRINT CHR$ statement, 41
PRINT# statement, use, 260
PRINT USING statement
   in linear programming solution, 190
   for plot subroutines, 40
PRINT# USING statement, 260
Printer, 1, 87
   in exponential smoothing forecasting program, 212
   listing on, 7
Problem, formulation, 172
   via linear programming, 169
Process control computer, 8
Product rejection example, 70-71f
Program
   documenting of, 14
   terminating of, 281
Program files, 257
Programming language statements, translation to machine language, 9
Project
   minimum duration, 226

planning and scheduling, with CPM, *see* Critical path method
Project network, drawing of, 225f, 226
PrtSc key, 7
Pseudo-random number, 305
  generation, 307-308
Pseudo-random number generator, 313
PUT statement, 267, 269, 271

QDAT$, 298
QSIZ, 291
Quadrature formulas, *see* Legendre-Gauss quadrature
Queue(s), 288-290, 303
  linked, 293-297f
  simulation, 319-324f
  subroutines, 290-291f, 324
QUEUE DELETE routine, 291
Queue number, 295
QUEUE PRINT subroutine, 291
Quicksort procedure, 245, 252
  program, 249-252f
  refinement of, 248-249
QUIT option, 97, 263
Quotation marks, 261

R option, 11, 260, 268
Radioactive substance, decay, 157-158
RAM, 2, 5, 255
  devices, problems in, 255
  memory, 257
Random access files, 258, 267, 282
  management program, 272-282
  reading from, 270-272
  writing to, 267-270
Random access memory, *see* RAM
Random access mode, specifying of, 268
Random buffer, 267, 268, 269
Random number(s)
  generation, 305-306, 309, 309f, 324-325
  testing random number series, 309-319
  pseudo, 305, 307
Random number generator, 312-313
  developing of, 307-309f
  reseeding of, 308
  for specific distribution
    Erlang distribution, 317
    exponential distribution, 316-317f
    normal distribution, 314-315f
    Poisson differential, 315-316f
    uniform distribution, 313-314f
Random number seed, 305, 308
Random number series, testing of, 309-312
RANDOMIZE statement, 305
Randomly ordered list, sorting of, *see* Quicksort
Randomness, 312-313
Range, 23, 44
  calculation of, 23-24
"RCST," (reduced cost), in linear programming solution, 189
Read Only Memory, *see* ROM
Reading

  in linked list program, 295
  from random access file, 270-272
  of sequential files, 259-262
Real data, 298
Real number, storing of, 14
Real-world systems, analyzing of, 318
REAR array, 294
REC % variable, 269, 271
Record, 267
  length, 274
  locating of, 272
Record pointer, setting of, 269
Recording, of information, on diskette, 255-258
Reduced cost ("Rcst"), in linear programming solution, 189
Reduction procedures, 19
  see also Data reduction
Reference(s), 10
Reference number, in sequential files, 259, 261
Regression
  in forecasting, 193
  linear, *see* Linear regression
Regression analysis option, 68
Regression equation, 111
  developing of, 114-116
Regression line
  deviation about, 65, 66f
  form, 68
REGRESSION subroutine, 68
Remote job entry terminal (RJE), 8
REMOVE subroutine, 295
Resources, scarce, allocation of, 170
Restrictions, in linear programming, 170
RIT array, 247, 248
RND function, 309, 324
  use, 305-306
  in number generation, 311
RNDR variable, 314
Robot, controller for, 8
ROM, 2, 5, 8, 255
Roots, of polynomials, *see* Polynomials, roots of
Row operations, application to matrix, 57, 60
RSET statement, 269
RUFM variable, 313
RUN command, use, 11
Runge-Kutta method, for solving differential equations, 159-160, 161
  application of, 161-166

SALES data, in stepwise regression programs, 92, 93-94, 97
Sample statistics, 19
SAVE, 7
Scaler
  determinant of matrix, 56
  for matrix addition, 50
  matrix multiplication by, 52, 53, 60
  reciprocal of, 59
Scaling subroutine, 40-41f
Scientific analysis, 1
SD, 324
Search process
  improvement of, 272

in random access memory program, 281
Seasonal exponential smoothing, 195-197, 221
Seasonal pattern, 195
Second-order differential equation, 158
Secondary storage devices, 255, 257
Sequential access files, 258, 272, 282
　buffer, 267
　buffer area, 258, 258f, 259
　management program, 262-267
　writing and reading of, 259-262
Shift key, 7
　location, 7
Simplex procedure, in linear programming
　　solution, 177-178, 191
　use
　　program coverage, 191
　　program input, 185-188
　　program output, 188-191
　　using program, 178-185
Simulation
　next event, 318-324, 325
　use, 312-313
Simulation models, 305
Simulation run length (STOPSIM), 324
Simultaneous linear equations, solving of, 134, 135
　Gauss-Jordan method, 123-129
　Gauss-Seidel method, 129-134
　general form, 119-120
　matrix inverse method, 120-123
Single exponential smoothing, for forecasting, 193-194
Single-matrix manipulation, 47
Single precision variable, 269
　conversion to string variable, 271
　numeric, 14
SIZ, *see* Maximum size
Slack, for activity, in CPM program, 227, 228
"Slack" variable, in linear programming
　　solution, 175-176, 186
SLAM, 325
Slope
　point, 141
　near zero, 139
　zero, 139, 140
Smoothing constant(s), 194, 214
　optimizing of, 212
　optimum, determining of, 213
Software, 1, 8
　word processing, 7
SORT program, 13
Sorting, 243, 252-253
　bubble sort, 243-245f
　plot subroutine, 37-39f
　quicksort, 245-252
Source program, 10
Square(s), sum of, 67
Square matrices, inverses for, 59-62
Stack(s), 303
　use, 293
Stack data structure, 286, 286f
　subroutines, 286-288f
　use, 286

STACK POP subroutine, 287
STACK PUSH subroutine, 287
Standard deviation (SD), 26, 44, 324
　definition of, 27
　values for, 314
Statistical measure, categories, 20
Statistical studies, 305
Statistics, 19
　calculation of, 19
　types, 19
　see also Central tendency, Dispersion
STD DEVIATION OF ESTIMATE, 109, 110
STEPWISE file, loading of, 88
STEPWISE MULTIPLE REGRESSION, 99
Stepwise regression, 72-85f, 87, 116-118
　program, 88-116
STOPSIM, 324
Storage area, temporary, 258
Storage devices, secondary, 255, 257
Storage space
　disk, 9
　limited, 292
STORE DATA ON DISK option, 92, 234
Straight line
　equation of, 72
　fitting of, 65
String(s)
　alphabetical order, 298
　delimiting of, 261
　left-justifying of, 269
String data, movement into random buffer, 269
String values, changing of numeric values to, 268
String variable, 14
　conversion to single precision variable, 271
STUDY ANOTHER MODEL option, 97
Subintervals, variable, in trapezoidal rule for
　　numeric integration, 149, 151f
Submatrix, in Gauss-Seidel method, 132
Subroutine(s)
　plotting, *see* Plotting subroutines
　in sequential file program, 263
Subset(s), in quickset procedure, 246, 247, 248, 249
Subtraction, of matrix, 50, 51
Successor, in bidirectional linked list program, 297
Sum
　matrix, 50
　of squares, 67
Summarization techniques, 19
System, reducing of, 169
System board, 2
SYSTEM command, 11
System diagnostics, 4, 5
System unit, 1
　design, 2, 4f, 5f
Systems Network Architecture (SNA), 8

T-statistics, 93, 115
T superscript, 48
Table, 272
　file allocation, 256

"Tableau," in linear programming solution, 196, 187, 188f, 189, 190, 190f
Tangent line, slope, 139, 140
Tape, sequential access file, 258
Taylor-series expansion, 160
Temporary file, 266
Terminal, 8
Text editing, 7
Text editing keys, 7
Time series data, 218
   entering of, 211
Time series data management options, 211
Time-series forecasting, techniques, 193
Time Series Forecasting Program, 212
Time system, 324
Too many files message, 256
Tracks, concentric, 255
TRANSFORMATION DATA option, in stepwise regression program, 102
Transpose of a matrix, 48, 50
   subroutine, 49f
Trapezoidal rule, for numeric integration, 148-149, 154
   integration of EXP(x) over 0-1, 150, 150f
   main program, 150f, 150, 151f
   subroutines, 149, 149f
Trend(s), exponential smoothing with, 195, 214, 221
Triangular matrix
   lower, 61f
   upper, 56-57, 58f, 59f
Triangular subroutines, 126
Type-ahead buffer, 7

"Unbiased" estimate, 24
Uncertainty, in simulation models, 312-313
Uniform distribution, use of random number generator for, 313-313f
   frequency histograms, 313f
Unit(s), of IBM Personal Computer, 1
Upper triangular matrix, 56-57
   creation, 57, 58f, 59f
Upper triangular routines, in solving simultaneous linear equations, 126

Value(s)
   actual versus predicted, 97
   for dependent variable, 102, 107
   for independent variable, 102, 107
   predicted, 109, 110
   predicting of, 113
VALUE variable, 271
Variable(s)
   adding of, in stepwise regression program, 72, 73

   dependent, *see* Dependent variable
   independent, *see* Independent variable
   in linear programming solution, 188-189
      artificial, 191
   passing of, 13
VARIABLE IDENTIFYING DATA, name of, 103
Variable names, 14
   length, 14
   storing of, 40
Variable types, 14
Variance, 24, 44
   calculation of, 24-26
   defining of, 24
Vector(s), 47, 62
   operations, 62-63
   right-hand side, in solving simultaneous equations, 123
Video display adapter card, 5

W variable, 13
Waiting line (queue), simulation, 319-324f
Weight(s), in exponential smoothing, 194
WEIGHT DATA option, 97
Weighting, of data, 100
   in linear regression program, 97, 98, 99
Winter's method, for forecasting, 196, 197, 221
Word processing, 1
   software, 7
   work station, 8
Work space, 1
WRITE# statement, 260, 261
Writing
   to random access file, 267-270
   to sequential files, 259-262

X-axis, intersecting of, 139
X variable, 13, 40, 106
X2 variable, values required for, 107
X3 variable, values required for, 107
XM, 40
XT, 40
XT version, system board, 2
XV$, 40
XVAR, 40
XVAR array, 40
XVAR variable, 39, 40

Y variable, 13, 37, 40, 65, 106
   values required for, 107
YV variable, 39

Z variable, 13
Zero slope, 139, 140

Documentation for the Program Diskette

About the Diskette

This diskette contains the subroutines and stand-alone programs which are presented and discussed in *BASIC Engineering and Scientific Programs for the IBM PC*. There is some specific information about the diskette which should be useful to you.

This diskette is almost full; only four sectors (2048 bytes) are not used. It contains all of the subroutines and stand-alone programs in the book. There are 38 major subroutines and programs and a host of supporting subroutines. You must write the main programs to utilize the subroutines. Main programs are presented in the book which you can use.

The subroutines have unique line numbers. This allows you to have more than one subroutine in memory at one time. To use a subroutine(s) with a main program you have entered, you can MERGE it. An example of this command is

```
MERGE "B:MATOP"
```

Make sure your main program does not have any line numbers that correspond to line numbers of subroutines you are MERGEing. Only the subroutines can be MERGEd; they have been saved on the diskette in ASCII format (SAVEd with the A option). In order to save space on the diskette (since it is almost full), the stand-alone programs were not saved in ASCII format (programs saved in the ASCII format require more space). One of the first things you will want to do is prepare a backup copy of the programs on another diskette to prevent their inadvertent loss.

Subroutine and Program Index

Below is a summary of the contents of each file on the diskette. The format used to describe each file is:

FILENAME, designation of program type (program or subroutine), reference to chapter in text where a detailed explanation can be formed, and a short description of what the program does.

CONTENTS, program, No chapter reference

This program is a handy reference to the programs and subroutines of this book. It provides a brief explanation of the subroutines and programs, including chapter location and line numbers. Using this program you may access information describing a specific subroutine.

ANPLOT, subroutines, CHAPTER 2

Subroutines are included here for data analysis and plotting. Sample statistics (mean, median, mode, range, standard deviation, variance) are calculated in one subroutine. Other subroutines generate frequency histograms and joint plots. Sorting and scaling subroutines are also included.

MATOP, subroutines, CHAPTER 3

This series of subroutines can be used for matrix operations. Matrix addition, multiplication, exponentiation, determinant, and inverse are included. Many auxiliary subroutines are used to assist in these operations.

SIMREG, program, CHAPTER 4

This program will estimate a line $Y=A+BX$ using simple linear regression. X is the independent variable and Y is the dependent variable.

STEPWISE, program, CHAPTER 4

This program will perform multiple or stepwise linear regression. Some data transforms are provided. Data management routines are also included. A forecast can be made using the estimated regression equation.

SIMLQ, subroutines, CHAPTER 5

This presents a series of subroutines to solve simultaneous linear equations. Two analytic techniques, the matrix inverse and Gauss-Jordan methods, are presented as well as a numerical technique, Gauss-Seidel.

NEWRAP, subroutine, CHAPTER 6

This subroutine presents the Newton-Raphson procedure for locating the real roots of a polynomial.

NUMINT, subroutines, CHAPTER 7

This contains subroutines to perform numerical integration. Two different techniques are presented, integration by the trapezoidal rule and Legendre-Gauss quadrature.

DIFEQ, subroutines, CHAPTER 8

Two subroutines are presented for the numerical solution to ordinary first-order differential equations. The Runge-Kutta method is used. One subroutine uses a fixed step size while another uses a variable step size.

SIMPLEX, program, CHAPTER 9

This program uses the simplex algorithm to solve linear programming problems. An editing subroutine is included to assist in data input.

EXSMOOTH, program, CHAPTER 10

This program will perform time series forecasting using Winter's method for exponential smoothing. The time series can contain trend and seasonal factors. Data management subroutines are provided.

CPM, program, CHAPTER 11

This program uses the critical path method (CPM) to determine the estimated project completion time and the associated critical path. Data management subroutines are also provided.

BUBSORT, subroutine, CHAPTER 12

This subroutine will sort a series of numbers into increasing or decreasing order using the bubble sort algorithm.

QICKSORT, subroutine, CHAPTER 12

This subroutine will sort a series of numbers into increasing or decreasing order using the quicksort algorithm. In general, this algorithm is faster than the bubble sort algorithm. One exception is when the series of numbers is initially "almost" in the proper sequence.

SEQMAN, program, CHAPTER 13

This program illustrates the use of the sequential access method to manage data. A menu is provided that lets the user interactively add, delete, modify, and list the data using a sequential access disk file.

RANMAN, program, CHAPTER 13

This program illustrates the use of the random access method to manage data. A menu is provided that lets the user interactively add, delete, modify, and list data using a random access disk file.

STACK, subroutines, CHAPTER 14

These subroutines can be used to implement a stack data structure. Subroutines are provided that will initialize a stack and perform additions and deletions to a stack.

QUEUE, subroutines, CHAPTER 14

These subroutines can be used to implement a queue data structure. Subroutines are provided that will initialize a queue and perform additions and deletions to a queue.

LLQUEUE, subroutines, CHAPTER 14

These subroutines illustrate the use of a linked list. Each list is assumed to be a queue data structure. Each data element is assumed to be an integer. Subroutines are provided for initialization, addition and deletion using linked lists (queues).

BIDILIST, subroutines, CHAPTER 14

These subroutines illustrate the use of a bidirectional linked list. Each list consists of a set of alphanumeric strings which are maintained in alphabetical order. Subroutines are provided that will initialize, and perform additions and deletions to, bidirectional linked lists.

ORGRND, subroutine, CHAPTER 15

This subroutine will generate uniformly distributed random numbers in the range of $0 \leq X \leq 1.0$, using the congruential method. The user can specify the initial random number seed.

CHISQ, subroutine, CHAPTER 15

This subroutine can be used to perform a chi-square goodness of fit test using a series of numbers in the range of $0 \leq X \leq 1.0$. This test can be used to determine if the hypothesis can be accepted that a series of numbers is not uniformly distributed.

DISTGEN, subroutines, CHAPTER 15

A set of subroutines is provided that will generate random numbers from any of the following distributions: uniform, normal, Poisson, exponential, and Erlang. Parameters necessary to specify a particular distribution, such as mean and variance, must be provided.

QUEUESIM, program, CHAPTER 15

This program illustrates next event simulation. A single-server queueing system is simulated. The time between arrivals to the system is exponentially distributed and the service times are normally distributed.